Fundamentals of Soil Physics

DANIEL HILLEL

DEPARTMENT OF PLANT AND SOIL SCIENCES
UNIVERSITY OF MASSACHUSETTS
AMHERST, MASSACHUSETTS

Academic Press
San Diego New York Boston
London Sydney Tokyo Toronto

Find Us on the Web! http://www.apnet.com

COPYRIGHT © 1980, BY ACADEMIC PRESS, INC.
ALL RIGHTS RESERVED.
NO PART OF THIS PUBLICATION MAY BE REPRODUCED OR
TRANSMITTED IN ANY FORM OR BY ANY MEANS, ELECTRONIC
OR MECHANICAL, INCLUDING PHOTOCOPY, RECORDING, OR ANY
INFORMATION STORAGE AND RETRIEVAL SYSTEM, WITHOUT
PERMISSION IN WRITING FROM THE PUBLISHER.

Academic Press, Inc.
A Division of Harcourt Brace & Company
525 B Street, Suite 1900, San Diego, California 92101-4495

United Kingdom Edition published by
ACADEMIC PRESS, INC. (LONDON) LTD.
24/28 Oval Road, London NW1 7DX

Library of Congress Cataloging in Publication Data

Hillel, Daniel.
    Fundamentals of soil physics.

　　Bibliography:
　　Includes indexes.
　　1. Soil physics.　I. Title.
S592.3.H54　　　631.4'3　　　　80-16688
ISBN 0-12-348560-6

Transferred to digital printing 2004

*Dedicated to Rachel,
who has made my labor of seven years,
as that of Patriarch Jacob,
seem but a few days.*

Contents

PREFACE xiii
ACKNOWLEDGMENTS xvii

Part I: BASIC RELATIONSHIPS

1. The Task of Soil Physics 3

2. General Physical Characteristics of Soils

    A. Introduction 6
    B. Soil Physics 7
    C. Chemical and Physical Aspects of Soil Productivity 7
    D. Soil as a Disperse Three-Phase System 8
    E. Volume and Mass Relationships of Soil Constituents 9
    F. The Soil Profile 14
        Sample Problems 17

3. Properties of Water in Relation to Porous Media

    A. Introduction 21
    B. Molecular Structure 22
    C. Hydrogen Bonding 24
    D. States of Water 25
    E. Ionization and pH 27
    F. Solvent Properties of Water 29
    G. Osmotic Pressure 31
    H. Solubility of Gases 34
    I. Adsorption of Water on Solid Surfaces 35
    J. Vapor Pressure 37

K.	Surface Tension	40
L.	Curvature of Water Surfaces and Hydrostatic Pressure	42
M.	Contact Angle of Water on Solid Surfaces	43
N.	Capillarity	45
O.	Density and Compressibility	47
P.	Viscosity	48
	Sample Problems	50

Part II: THE SOLID PHASE

4. Texture, Particle Size Distribution, and Specific Surface

A.	Introduction	55
B.	Soil Texture	56
C.	Textural Fractions (Separates)	56
D.	Soil Classes	59
E.	Particle Size Distribution	60
F.	Mechanical Analysis	62
G.	Specific Surface	65
	Sample Problems	68

5. Nature and Behavior of Clay

A.	Introduction	71
B.	Structure of Clay	71
C.	The Principal Clay Minerals	74
D.	Humus: The Organic Constituent of Soil Colloids	76
E.	The Electrostatic Double Layer	77
F.	Ion Exchange	81
G.	Hydration and Swelling	84
H.	Flocculation and Dispersion	87
	Sample Problems	90

6. Soil Structure and Aggregation

A.	Introduction	93
B.	Types of Soil Structure	94
C.	Structure of Granular Soils	95
D.	Structure of Aggregated Soils	97
E.	Additional Factors Affecting Aggregation	101
F.	Characterization of Soil Structure	103
G.	Aggregate Stability	108
H.	Soil Crusting	112
I.	Soil Conditioners	113
J.	Hydrophobization of Soil Aggregates	115
	Sample Problems	117

Part III: THE LIQUID PHASE

7. Soil Water: Content and Potential

A.	Introduction	123
B.	The Soil-Water Content (Wetness)	124
C.	Measurement of Soil Wetness	125
D.	Energy State of Soil Water	134
E.	Total Soil-Water Potential	136
F.	Thermodynamic Basis of the Potential Concept	138
G.	Gravitational Potential	140
H.	Pressure Potential	141
I.	Osmotic Potential	144
J.	Revised Terminology	144
K.	Quantitative Expression of Soil-Water Potential	145
L.	Soil-Moisture Characteristic Curve	148
M.	Hysteresis	152
N.	Measurement of Soil-Moisture Potential	155
	Sample Problems	162

8. Flow of Water in Saturated Soil

A.	Laminar Flow in Narrow Tubes	166
B.	Darcy's Law	168
C.	Gravitational, Pressure, and Total Hydraulic Heads	172
D.	Flow in a Vertical Column	174
E.	Flow in a Composite Column	176
F.	Flux, Flow Velocity, and Tortuosity	177
G.	Hydraulic Conductivity, Permeability, and Fluidity	178
H.	Limitations of Darcy's Law	181
I.	Relation of Conductivity and Permeability to Pore Geometry	183
J.	Homogeneity and Isotropy	185
K.	Measurement of Hydraulic Conductivity of Saturated Soils	187
L.	Equations of Saturated Flow	188
	Sample Problems	190

9. Flow of Water in Unsaturated Soil

A.	Introduction	195
B.	Comparison of Flow in Unsaturated versus Saturated Soil	195
C.	Relation of Conductivity to Suction and Wetness	198
D.	General Equation of Unsaturated Flow	202
E.	Hydraulic Diffusivity	204
F.	The Boltzmann Transformation	207
G.	Theoretical Calculation of the Hydraulic Conductivity Function	209
H.	Measurement of Unsaturated Hydraulic Conductivity and Diffusivity in the Laboratory	212
I.	Measurement of Unsaturated Hydraulic Conductivity of Soil Profiles in Situ	213

		J.	Vapor Movement	221
			Sample Problems	223

10. Movement of Solutes and Soil Salinity

	A.	Introduction	233
	B.	Convective Transport of Solutes	234
	C.	Diffusion of Solutes	236
	D.	Hydrodynamic Dispersion	238
	E.	Miscible Displacement and Breakthrough Curves	240
	F.	Combined Transport of Solutes	242
	G.	Effects of Solutes on Water Movement	245
	H.	Soil Salinity and Alkalinity	250
	I.	Salt Balance of the Soil Profile	251
	J.	Leaching of Excess Salts	254
		Sample Problems	257

Part IV: THE GASEOUS PHASE

11. Soil Air and Aeration

	A.	Introduction	265
	B.	Volume Fraction of Soil Air	266
	C.	Composition of Soil Air	268
	D.	Convective Flow of Soil Air	269
	E.	Diffusion of Soil Air	272
	F.	Soil Respiration and Aeration Requirements	277
	G.	Measurement of Soil Aeration	280
		Sample Problems	283

Part V: COMPOSITE PROPERTIES AND BEHAVIOR

12. Soil Temperature and Heat Flow

	A.	Introduction	287
	B.	Modes of Energy Transfer	288
	C.	Energy Balance for a Bare Soil	290
	D.	Conduction of Heat in Soil	291
	E.	Volumetric Heat Capacity of Soils	293
	F.	Thermal Conductivity of Soils	295
	G.	Simultaneous Transport of Heat and Moisture	300
	H.	Thermal Regime of Soil Profiles	303
	I.	Modification of the Soil Thermal Regime	309
		Sample Problems	313

13. Stress–Strain Relations and Soil Strength

A.	Introduction	318
B.	The Concept of Strain and Stress	319
C.	Elasticity and Plasticity	322
D.	Rheology of Liquids	325
E.	Rheological Models	327
F.	Stress Distribution in Soil	328
G.	The Mohr Circle of Stresses	331
H.	Stress–Strain Relations and Failure of Soil Bodies	334
I.	The Concept of Soil Strength	337
J.	Measurement of Soil Strength	338
K.	Soil Consistency	347
	Sample Problems	352

14. Soil Compaction and Consolidation

A.	Introduction	355
B.	Two Opposing Views of Soil Compaction: Engineering and Agronomic	356
C.	Soil Compactibility in Relation to Wetness	357
D.	Occurrence of Soil Compaction in Agricultural Fields	360
E.	Pressures Caused by Machinery	361
F.	Soil Compaction under Machinery-Induced Stresses	367
G.	Occurrence and Consequences of Soil Compaction	371
H.	Control of Soil Compaction	375
I.	Soil Consolidation	376
	Sample Problems	382

Bibliography 387

INDEX 407

> "And furthermore, my son, be admonished
> Of the making of many books there is no end
> and much study is a weariness of the flesh."
>
> Ecclesiastes XII: 12

Preface

Tradition has it that wise King Solomon, using Ecclesiastes as *nom de plume*, reached his sorrowful conclusion (cited above) only in old age; otherwise, we might have been deprived of the enchanting "Song of Songs" of his youth and of the worldly "Proverbs" of his middle age.

This book is not, in any case, in total defiance of the Wise Old Man's admonition, for it is not an entirely new book. Rather, it is an outgrowth of a previous treatise, written a decade ago, entitled "Soil and Water: Physical Principles and Processes." Though that book was well enough received at the time, the passage of the years has inevitably made it necessary to either revise and update the same book, or to supplant it with a fresh approach in the form of a new book which might incorporate still-pertinent aspects of its predecessor without necessarily being limited to the older book's format or point of view.

After some deliberation, I have decided to follow the second course. In so doing, I have also endeavored to enlarge the scope of the book so as to encompass a number of topics that were omitted or only scantily treated in the original book, such as the properties of clay, soil structure, soil aeration, soil heat, soil rheology and mechanics, and solute movement. Consequently, the present book is an attempt at a comprehensive, albeit elementary, exposition of the foundations of soil physics as a whole, rather than a restricted treatment of soil–water relations alone. A companion volume, entitled "Applications of Soil Physics," deals specifically with the field-water cycle and associated phenomena.

In writing this book, I have attempted to answer the need for an upper-level undergraduate textbook in soil physics for students of the agricultural as well as of the environmental and engineering sciences. Toward this end, I have made a conscious effort to avoid unnecessary technical or mathematical jargon and unfamiliar notation, and to explain each development explicitly without assuming anything more than general undergraduate knowledge of the basic concepts of calculus, physics, chemistry, and biology. This book is thus meant to be as autonomous and self-sustaining as possible. When necessary, however, the reader is referred to outside sources for supplementary study, particularly where the topic relates to a field of science outside the scope of our own necessarily limited coverage. Sample problems are presented at the end of each chapter and are worked out explicitly (in what some might consider *excruciating* detail) in an effort to help students transmute the vague abstractions of unfamiliar theory into actual working knowledge.

Some students of agriculture and biology are deterred by the very sight of mathematical equations, and tend to skip over them (as if they were merely irrelevant clusters of meaningless symbols . . .) and to read only the narrative text. This is a mistake. Equations should be deciphered and digested, for they provide essential information on the quantitative relationships among factors and variables, and they do so with precision and logic. Mathematics, as Josiah Willard Gibbs once defined it so aptly, is a *language*. In fact, it is the indispensible language of science. Ordinary language is too cumbersome and imprecise to replace the elegance and economy of mathematics. Still, the fear of mathematics is so pervasive that it alone can prevent highly intelligent people from understanding science. We have therefore attempted in this text to meet our friends half-way, by refraining from excessive or avoidable mathematical niceties and by going to some length to explain the mathematics we could not avoid.

Some students may even feel uneasy about the amount of physics involved in a study of soil physics. They, too, have nothing to fear. The concepts are elementary and ought to be understood by all naturalists. These concepts include the conservation of mass, energy, and momentum; velocity and acceleration; force and force fields; pressure and viscosity; as well as potential and kinetic energy. They also include a few basic physicochemical and thermodynamic concepts—all essential to the understanding not merely of soil physics but of all natural systems. At this point the bewildered student might be tempted to ask: Is that all? Well, that is *practically* all . . . .

A textbook on so vital a subject as soil physics ought by right to capture and convey the special fascination and excitement of the soil physicist's quest for knowledge and understanding of his (or her) complex system, and hence should give some pleasure in the reading. It is my hope that this book might be read, not merely consulted, and that the reader might discover in it a few insights as well as facts.

Preface

While any book written by an individual author inevitably reflects his particular point of view, it is in the nature of the ongoing process of scientific exchange that one's own ideas cannot easily be distinguished from those of numerous others. Some of the concepts elucidated herein have had their roots in my formal studies in various universities, mainly in my native U.S.A. Others were conceived during the course of my rather extensive travels, which included assignments and sojourns as observer, consultant, researcher, and teacher in such diverse places as Japan, India, Southeast Asia, Australia, Europe, Africa, and the Americas. An important nursery of this book is the State of Israel, where I witnessed and took part in the development of intensive land and water management methods which have enabled that country, despite its arid climate, to multiply its agricultural production severalfold within a single generation.

Being still in midcareer, I cannot yet assume the mantle of the Wise Old Man and encapsulate my experience as an ultimate truth in the form of a terse maxim of my own. Suffice it to say that I have already discovered the truth of the ancient Talmudic adage: "Much have I learned from my teachers, and yet more from my colleagues, but most of all from my students."

Acknowledgments

Thanks are due to the following colleagues who read and commented on various parts of the book during its formative stages: Professors John Baker, Allen Barker, Bernard Berger, Haim Gunner, and Mack Drake, as well as my graduate student David Leland, of the University of Massachusetts; and to Professors David Elrick of Guelph University and Peter Wierenga of New Mexico State University. I hereby absolve them of any responsibility for the book's undoubtedly numerous shortcomings, for which I alone am to blame. I am grateful to Lisa Cohn for her careful typing of the manuscript. Finally, an acknowledgment is also due to the draftsman who prepared the illustrations. Indeed he seemed to have worked harder on doing these than did the author on writing the text. However, being something of an amateur and still unsure of the quality of the results, he chooses to remain anonymous.

Part I:

BASIC RELATIONSHIPS

> To see a world in a grain of sand
> and heaven in a wild flower
> Hold Infinity in the palm of your hand
> and eternity in an hour.
> William Blake
> *Innocence and Experience*
> 1789–1794

1 *The Task of Soil Physics*

The soil beneath our feet is the basic substrate of all terrestrial life. The intricate and fertile mix composing the soil, with its special life-giving attributes, is a most intriguing field of study. The soil serves not only as a medium for plant growth and for microbiological activity per se but also as a sink and recycling factory for numerous waste products which might otherwise accumulate to poison our environment. Moreover, the soil supports our buildings and provides material for the construction of earthen structures such as dams and roadbeds.

The attempt to understand what constitutes the soil and how it operates within the overall biosphere, which is the essential task of soil science, derives both from the fundamental curiosity of man, which is his main creative impulse, and from urgent necessity. Soil and water are, after all, the two fundamental resources of our agriculture, as well as of our natural environment. The increasing pressure of population has made these resources scarce or has led to their abuse in many parts of the world. Indeed, the necessity to manage these resources efficiently on a sustained basis is one of the most vital tasks of our age.

That knowledge of the soil is imperative to ensure the future of civilization has been proven repeatedly in the past, at times disastrously. In many regions we find shocking examples of once-thriving agricultural fields reduced to desolation by man-induced erosion or salinization resulting from injudicious management of the soil–water system. Add to that the shortsighted depletion of unreplenished water resources as well as the dumping of poisonous wastes—and indeed we see a consistent pattern of mismanagement. In view of the population–environment–food crisis facing the world, we can ill afford to continue squandering and abusing such precious resources.

The soil itself is of the utmost complexity. It consists of numerous solid components (mineral and organic) irregularly fragmented and variously associated and arranged in an intricate geometric pattern that is almost indefinably complicated. Some of the solid material consists of crystalline particles, while some consists of amorphous gels which may coat the crystals and modify their behavior. The adhering amorphous material may be iron oxide or a complex of organic compounds which attaches itself to soil particles and binds them together. The solid phase further interacts with the fluids, water and air, which permeate soil pores. The whole system is hardly ever in a state of equilibrium, as it alternately wets and dries, swells and shrinks, disperses and flocculates, compacts and cracks, exchanges ions, precipitates and redissolves salts, and occasionally freezes and thaws.

To serve as a favorable medium for plant growth, the soil must store and supply water and nutrients and be free of excessive concentrations of toxic factors. The soil–water–plant system is further complicated by the facts that plant roots must respire constantly and that most terrestrial plants cannot transfer oxygen from their aerial parts to their roots at a rate sufficient to provide for root respiration. Hence the soil itself must be well aerated, by the continuous exchange of oxygen and carbon dioxide between the air-filled pores and the external atmosphere. An excessively wet soil will stifle roots just as surely as an excessively dry soil will desiccate them.

These are but a few of the issues confronting the relatively new science of soil physics, a field of study which has really come into its own only in the last generation. Definable as the study of the state and transport of all forms of matter and energy in the soil, soil physics is an inherently difficult subject, a fact which may account for its rather late development.

Because of the soil's complexity, we find it practically impossible to define completely its exact physical state at any time. In dealing with any particular problem, therefore, we are generally obliged to take the easy way, which is to simplify our system by concentrating upon the factors which appear to have the greatest and most direct bearing upon the problem at hand, while, at least for the moment, disregarding as extraneous complications the factors which may seem to be of secondary importance.

In many cases, the theories and equations employed in soil physics describe not the soil itself, but an ideal and well-defined model which we construct to simulate the soil. Thus, for example, at different times and for different purposes, the soil may be compared to a collection of small spheres, or to a bundle of capillary tubes, or to a collection of parallel colloidal platelets, or to an homogeneous mechanical continuum. The value of each of these models depends upon the degree of approximation or of realism with which the model portrays the pertinent phenomena in each case. However, even at best such models cannot provide anything but a partial explanation of soil

1. The Task of Soil Physics

behavior. The complications which we may choose to disregard do not in fact disappear. Having once defined the most important (or "primary") effects, we find that to refine our model we must now consider the next to the most important ("secondary") effects, and so on ad infinitum. Our developing knowledge of the soil, as of other complex systems, is achieved by successive approximations.

Our present-day knowledge of the soil physical system is still rather fragmentary. Hence, we continue to search and re-search for answers to the numerous newly arising questions. The business, and fun, and occasional agony of science is the continuing endeavor to achieve a coordinated understanding and explanation of observable phenomena without ever resting on yesterday's conclusions. Consequently a valid book on soil physics should reflect the complexity of the system even while attempting to present a coordinated and logical description of what is admittedly only a partial knowledge of it.

As G. Ferrero (1895) wrote almost a century ago:

> Therefore theory, which gives facts their value and significance, is often very useful, even if it is partially false, because it throws light on phenomena which no one has observed, it forces an examination, from many angles, of facts which no one has hitherto studied, and provides the impulse for more extensive and more productive researches . . . .
>
> It is a moral duty for the man of science to expose himself to the risk of committing error, and to submit to criticism in order that science may continue to progress . . . . Those who are endowed with a mind serious and impersonal enough not to believe that everything they write is the expression of absolute and eternal truth will approve of this theory, which puts the aims of science well above the miserable vanity and paltry *amour propre* of the scientist.

> To the wise man, the whole world's a soil.
> Ben Jonson
> 1573–1637

2 General Physical Characteristics of Soils

A. Introduction

The term *soil* refers to the weathered and fragmented outer layer of the earth's terrestrial surface. It is formed initially through disintegration and decomposition of rocks by physical and chemical processes, and is influenced by the activity and accumulated residues of numerous species of microscopic and macroscopic plants and animals. The physical weathering processes which bring about the disintegration of rocks into small fragments include expansion and contraction caused by alternating heating and cooling, stresses resulting from freezing and thawing of water and the penetration of roots, and scouring or grinding by abrasive particles carried by moving ice or water and by wind. The chemical processes tending to decompose the original minerals in the parent rocks include hydration, oxidation and reduction, solution and dissociation, immobilization by precipitation or removal of components by volatilization or leaching, and various physicochemical exchange reactions. The loose products of these weathering processes are often transported by running water, glaciers, or wind, and deposited elsewhere.

Soil formation processes continue beyond the initial weathering of rocks and minerals. In the course of soil development, the original character of the material is further modified by the formation of secondary minerals (e.g., clay minerals) and the growth of organisms which contribute organic matter and bring about a series of ongoing physicochemical and biochemical reactions in addition to those experienced by the original mineral material.

The process of soil development culminates in the formation of a characteristic *soil profile*, to be described later in this chapter.

B. Soil Physics

Throughout this book we shall be considering the soil from the viewpoint of soil physics, which can be described as the branch of soil science dealing with the physical properties of the soil, as well as with the measurement, prediction, and control of the physical processes taking place in and through the soil. As physics deals with the forms and interrelations of matter and energy, so soil physics deals with the state and movement of matter and with the fluxes and transformations of energy in the soil.

On the one hand, the fundamental study of soil physics aims at achieving a basic understanding of the mechanisms governing the behavior of the soil and its role in the biosphere, including such interrelated processes as the terrestrial energy exchange and the cycles of water and transportable materials in the field. On the other hand, the practice of soil physics aims at the proper management of the soil by means of irrigation, drainage, soil and water conservation, tillage, aeration and the regulation of soil heat, as well as the use of soil material for engineering purposes. Soil physics is thus seen to be both a basic and an applied science with a very wide range of interests, many of which are shared by other branches of soil science and by other interrelated sciences including terrestrial ecology, hydrology, microclimatology, geology, sedimentology, botany, and agronomy. Soil physics is likewise closely related to the engineering profession of soil mechanics, which deals with the soil mainly as a building and support material.

C. Chemical and Physical Aspects of Soil Productivity

A soil which contains adequate amounts of the various substances required for plant nutrition, in available forms, and which is not excessively acidic or alkaline and is free of toxic agents, can be considered to possess *chemical fertility*. However, the overall suitability of a soil as a medium for plant growth depends not only upon the presence and quantity of chemical nutrients, and on the absence of toxicity, but also upon the state and mobility of water and air and upon the mechanical attributes of the soil and its thermal regime. The soil must be loose and sufficiently soft and friable to permit germination and root development without mechanical obstruction. The pores of the soil should be of the volume and size distribution that will allow sufficient entry, movement, and retention of both water and air to meet plant needs. The thermal regime of the root zone should remain within the range which is optimal for plant growth. In short, in addition to chemical fertility, the soil should possess a *physical fertility*, since both attributes are equally essential to overall soil *productivity*. The problem for soil physicists is to define, quantify, and optimize the physical factors of soil productivity, in relation to the interactive chemical and biological factors.

D. Soil as a Disperse Three-Phase System

Natural systems can consist of one or more substances and of one or more phases. A system comprised of a single substance is also monophasic if its physical properties are uniform throughout. An example of such a system is a body of water consisting entirely of uniform ice. Such a system is called homogeneous. A system comprised of a single chemical compound can also be heterogeneous if that substance exhibits different properties in different regions of the system. A region inside a system which is internally uniform physically is called a phase. A mixture of ice and water, for instance, is chemically uniform but physically heterogeneous, as it includes two phases. The three ordinary phases in nature are the solid, liquid, and gaseous phases.

A system containing several substances can also be monophasic. For example, a solution of salt and water is a homogeneous liquid. A system of several substances can obviously also be heterogeneous. In a heterogeneous system the properties differ not only between one phase and another, but also among the internal parts of each phase and the boundary between the phase and its neighboring phase or phases. Interfaces between phases exhibit specific phenomena resulting from the interaction of the phases. The importance of these phenomena, which include adsorption, surface tension, and friction, depends on the magnitude of the interfacial area per unit volume of the system. Systems in which at least one of the phases is subdivided into numerous minute particles, which together exhibit a very large interfacial area per unit volume, are called *disperse systems*. Colloidal sols, gels, emulsions, and aerosols are examples of disperse systems.

The soil is a heterogeneous, polyphasic, particulate, disperse, and porous system, in which the interfacial area per unit volume can be very large. The disperse nature of the soil and its consequent interfacial activity give rise to such phenomena as adsorption of water and chemicals, ion exchange, adhesion, swelling and shrinking, dispersion and flocculation, and capillarity.

The three phases of ordinary nature are represented in the soil as follows: the solid phase constitutes the *soil matrix*; the liquid phase consists of soil

E. Volume and Mass Relationships of Soil Constituents

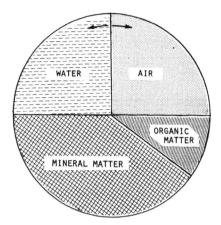

Fig. 2.1. Schematic composition (by volume) of a medium-textured soil at a condition considered optimal for plant growth. Note that the solid matter constitutes 50% and the pore space 50% of the soil volume, with the latter divided equally between water and air. The arrows indicate that these components can vary widely, and in particular that water and air are negatively related so that an increase in one is associated with a decrease of the other.

water, which always contains dissolved substances so that it should properly be called the *soil solution*; and the gaseous phase is the *soil atmosphere*. The solid matrix of the soil includes particles which vary in chemical and mineralogical composition as well as in size, shape, and orientation. It also contains amorphous substances, particularly organic matter which is attached to the mineral grains and often binds them together to form aggregates. The organization of the solid components of the soil determines the geometric characteristics of the pore spaces in which water and air are transmitted and retained. Finally, soil water and air vary in composition, both in time and in space.

The relative proportions of the three phases in the soil vary continuously, and depend upon such variables as weather, vegetation, and management. To give the reader some general idea of these proportions, we offer the rather simplistic scheme of Fig. 2.1, which represents the volume composition of a medium-textured soil at a condition considered to be approximately optimal for plant growth.

E. Volume and Mass Relationships of Soil Constituents

Let us now consider the volume and mass relationships among the three phases, and define some basic parameters which have been found useful in characterizing the physical condition of a soil.

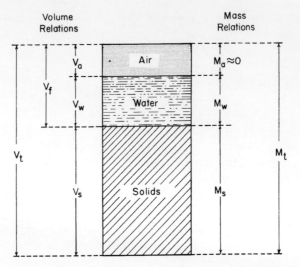

Fig. 2.2. Schematic diagram of the soil as a three-phase system.

Figure 2.2 is a schematic representation of a hypothetical soil showing the volumes and masses of the three phases in a representative sample. The masses of the phases are indicated on the right-hand side: the mass of air M_a, which is negligible compared to the masses of solids and water; the mass of water M_w; the mass of solids M_s; and the total mass M_t. These masses can also be represented by their weights (the product of the mass and the gravitational acceleration). The volumes of the same components are indicated on the left-hand side of the diagram: volume of air V_a, volume of water V_w, volume of pores $V_f = V_a + V_w$, volume of solids V_s, and the total volume of the representative soil body V_t.

On the basis of this diagram, we can now define terms which are generally used to express the quantitative interrelations of the three primary soil constituents.

1. DENSITY OF SOLIDS (MEAN PARTICLE DENSITY) ρ_s

$$\rho_s = M_s/V_s \qquad (2.1)$$

In most mineral soils, the mean density of the particles is about 2.6–2.7 gm/cm^3, and is thus close to the density of quartz, which is often prevalent in sandy soils. Aluminosilicate clay minerals have a similar density. The presence of iron oxides, and of various heavy minerals, increases the average value of ρ_s, whereas the presence of organic matter lowers it. Sometimes the density is expressed in terms of the *specific gravity*, being the ratio of the density of the material to that of water at 4°C and at atmospheric pressure.

E. Volume and Mass Relationships of Soil Constituents

In the metric system, since the density of water at standard temperature is assigned the value of unity, the specific gravity is numerically (though not dimensionally) equal to the density.

2. Dry Bulk Density ρ_b

$$\rho_b = M_s/V_t = M_s/(V_s + V_a + V_w) \qquad (2.2)$$

The dry bulk density expresses the ratio of the mass of dried soil to its total volume (solids and pores together). Obviously, ρ_b is always smaller than ρ_s, and if the pores constitute half the volume, ρ_b is half of ρ_s, namely 1.3–1.35 gm/cm^3. In sandy soils, ρ_b can be as high as 1.6, whereas in aggregated loams and in clay soils, it can be as low as 1.1 gm/cm^3. The bulk density is affected by the structure of the soil, i.e., its looseness or degree of compaction, as well as by its swelling and shrinkage characteristics, which are dependent upon clay content and wetness. Even in extremely compacted soil, however, the bulk density remains appreciably lower than the particle density, since the particles can never interlock perfectly and the soil remains a porous body, never completely impervious.

3. Total (Wet) Bulk Density ρ_t

$$\rho_t = M_t/V_t = (M_s + M_w)/(V_s + V_a + V_w) \qquad (2.3)$$

This is an expression of the total mass of a moist soil per unit volume. The wet bulk density depends even more strongly than the dry bulk density upon the wetness or moisture content of the soil.

4. Dry Specific Volume v_b

$$v_b = V_t/M_s = 1/\rho_b \qquad (2.4)$$

The volume of a unit mass of dry soil (cubic centimeter per gram) serves as another index of the degree of looseness or compaction of the soil.

5. Porosity f

$$f = V_f/V_t = (V_a + V_w)/(V_s + V_a + V_w) \qquad (2.5)$$

The porosity is an index of the relative pore volume in the soil. Its value generally lies in the range 0.3–0.6 (30–60%). Coarse-textured soils tend to be less porous than fine-textured soils, though the mean size of individual pores is greater in the former than in the latter. In clayey soils, the porosity is highly variable as the soil alternately swells, shrinks, aggregates, disperses, compacts, and cracks. As generally defined, the term porosity refers to the volume fraction of pores, but this value should be equal, on the average, to the areal porosity (the fraction of pores in a representative cross-sectional area) as well as to the average lineal porosity (being the fractional length of pores along a straight line passing through the soil in any direction). The total porosity, in any case, reveals nothing about the *pore size distribution*, which is itself an important property to be discussed in a later section.

6. Void Ratio e

$$e = (V_a + V_w)/V_s = V_f/(V_t - V_f) \tag{2.6}$$

The void ratio is also an index of the fractional volume of soil pores, but it relates that volume to the volume of solids rather than to the total volume of soil. The advantage of this index over the previous one (f) is that a change in pore volume changes the numerator alone, whereas a change of pore volume in terms of the porosity will change both the numerator and denominator of the defining equation. Void ratio is the generally preferred index in soil engineering and mechanics, whereas porosity is the more frequently used index in agricultural soil physics. Generally, e varies between 0.3 and 2.0.

7. Soil Wetness

The wetness, or relative water content, of the soil can be expressed in various ways: relative to the mass of solids, relative to the total mass, relative to the volume of solids, relative to the total volume, and relative to the volume of pores. The various indexes are defined as follows (the most commonly used are the first two).

a. *Mass Wetness w*

$$w = M_w/M_s \tag{2.7}$$

This is the mass of water relative to the mass of dry soil particles, often referred to as the *gravimetric water content*. The term *dry soil* is generally defined as a soil dried to equilibrium in an oven at 105°C, though clay will often retain appreciable quantities of water at that state of dryness. Mass wetness is sometimes expressed as a decimal fraction but more often as a percentage. Soil dried in "ordinary" air will generally contain several per cent more water than oven-dry soil, a phenomenon due to vapor adsorption and often referred to as soil *hygroscopicity*. In a mineral soil that is saturated, w can range between 25 and 60% depending on the bulk density. The saturation water content is generally higher in clayey than in sandy soils. In the case of organic soils, such as peat or muck, the saturation water content on the mass basis may exceed 100%.

E. Volume and Mass Relationships of Soil Constituents

b. *Volume Wetness* θ

$$\theta = V_w/V_t = V_w/(V_s + V_f) \tag{2.8}$$

The volume wetness (often termed volumetric water content or volume fraction of soil water) is generally computed as a percentage of the total volume of the soil rather than on the basis of the volume of particles alone. In sandy soils, the value of θ at saturation is on the order of 40–50%; in medium-textured soils, it is approximately 50%; and in clayey soils, it can approach 60%. In the latter, the relative volume of water at saturation can exceed the porosity of the dry soil, since clayey soils swell upon wetting. The use of θ rather than of w to express water content is often more convenient because it is more directly adaptable to the computation of fluxes and water quantities added to soil by irrigation or rain and to quantities subtracted from the soil by evapotranspiration or drainage. Also, θ represents the depth ratio of soil water, i.e., the depth of water per unit depth of soil.

c. *Water Volume Ratio* v_w

$$v_w = V_w/V_s \tag{2.9}$$

For swelling soils, in which porosity, and hence total volume, change markedly with wetness, it may be preferable to refer the volume of water present to the volume of particles rather than to total volume.

d. *Degree of Saturation* s

$$s = V_w/V_f = V_w/(V_a + V_w) \tag{2.10}$$

This index expresses the volume of water present in the soil relative to the volume of pores. The index s ranges from zero in dry soil to unity (or 100%) in a completely saturated soil. However, complete saturation is seldom attained, since some air is nearly always present and may become trapped in a very wet soil.

8. AIR-FILLED POROSITY (FRACTIONAL AIR CONTENT) f_a

$$f_a = V_a/V_t = V_a/(V_s + V_a + V_w) \tag{2.11}$$

This is a measure of the relative air content of the soil, and as such is an important criterion of soil aeration. The index is related negatively to the degree of saturation s (i.e., $f_a = f - s$).

9. Additional Interrelations

From the basic definitions given, it is possible to derive the relation of the various parameters to one another. The following are some of the most useful interrelations.

(1) Relation between porosity and void ratio:

$$e = f/(1 - f) \tag{2.12}$$

$$f = e/(1 + e) \tag{2.13}$$

(2) Relation between volume wetness and degree of saturation:

$$\theta = sf \tag{2.14}$$

$$s = \theta/f \tag{2.15}$$

(3) Relation between porosity and bulk density:

$$f = (\rho_s - \rho_b)/\rho_s = 1 - \rho_s/\rho_b \tag{2.16}$$

$$\rho_b = (1 - f)\rho_s \tag{2.17}$$

(4) Relation between mass wetness and volume wetness:

$$\theta = w\rho_b/\rho_w \tag{2.18}$$

$$w = \theta\rho_w/\rho_b \tag{2.19}$$

Here ρ_w is the density of water (M_w/V_w), approximately equal to 1 gm/cm^3. Since the bulk density ρ_b is generally greater than water density ρ_w, it follows that volume wetness exceeds mass wetness (the more so in compact soils of higher bulk density).

(5) Relation between volume wetness, fractional air content, and degree of saturation:

$$f_a = f - \theta = f(1 - s) \tag{2.20}$$

$$\theta = f - f_a \tag{2.21}$$

A number of these relationships are derived or proven at the end of this chapter, and the derivation or proof of the others is left as a useful exercise for students. Of the various parameters defined, the most commonly used in characterizing soil physical properties are the porosity f, bulk density ρ_b, volume wetness θ, and mass wetness w.

F. The Soil Profile

Having defined the soil's components and their proportions, let us now consider a composite soil body as it appears in nature. The most obvious,

F. The Soil Profile

and very important, part of the soil is its surface zone. An examination of that zone will reveal much about processes taking place through the surface, but will not necessarily reveal the character of the soil as a whole. To get at the latter, we must examine the soil in depth, and we can do this, for instance, by digging a trench and sectioning the soil from the surface downward. The vertical cross section of the soil is called the *soil profile*.

The soil profile is seldom uniform in depth, and typically consists of a succession of more-or-less distinct layers, or strata. Such layers may result from the pattern of deposition, or sedimentation, as can be observed in wind-deposited (aeolian) soils and particularly in water-deposited (alluvial) soils. If, however, the layers form in place by internal soil-forming (pedogenic) processes, they are called *horizons*. The top layer, or *A horizon*, is the zone of major biological activity and is therefore generally enriched with organic matter and often darker in color than the underlying soil. Next comes the *B horizon*, where some of the materials migrating from the A horizon (such as clay or carbonates) tend to accumulate. Under the B horizon lies the *C horizon*, which is the soil's parent material. In the case of a residual soil formed in place from the bedrock, the C horizon consists of the weathered and fragmented rock material. In other cases, the C horizon may consist of alluvial, aeolian, or glacial sediments.

The A, B, C sequence of horizons is clearly recognizable in some cases, as for example in a typical zonal soil such as a *podzol*. In other cases, no clearly developed B horizon may be discernible, and the soil is then characterized by an A, C profile. In still other cases, as in the case of very recent alluvium, hardly any profile differentiation is apparent. The character of the profile depends primarily on the climate, and secondarily on the parent material, the vegetation, the topography, and time.

The typical development of a soil and its profile, called *pedogenesis*, can be summarized as follows: The process begins with the physical disintegration or "weathering" of the exposed rock formation, which thus forms the soil's parent material. Gradual accumulation of organic residues near the surface brings about the development of a discernible A horizon, which may acquire a granular structure stabilized to a greater or lesser degree by organic matter cementation. (This process is retarded in desert regions.) Continued chemical weathering (e.g., hydration, oxidation, and reduction), dissolution, and reprecipitation may bring about the formation of clay. Some of the clay thus formed tends to migrate, along with other transportable materials (such as soluble salts) downward from the A horizon and to accumulate in an intermediate zone (namely, the B horizon) between the A horizon and the deeper parent material of the so-called C horizon. Important aspects of soil formation and profile development are the twin processes of *eluviation* and *illuviation* (washing out and washing in, respectively) wherein clay and other substances emigrate from the overlying *eluvial* A horizon and accumulate in the underlying *illuvial* B horizon, which therefore differs from the A horizon in composition and structure. Throughout these processes, the profile as a whole deepens as the upper part of the C horizon is gradually transformed, until eventually a quasi-stable condition is approached in which the counter processes of soil formation and of soil erosion are more or less in balance. In arid regions, salts such as calcium sulfate and calcium carbonate, dissolved from the upper part of the soil, may precipitate at some depth to form a cemented "pan." Numerous variations of these processes are possible, depending on local conditions. The characteristic depth of the soil, for instance, varies from location to location. Valley soils are typically deeper than mountain soils, and the depth of the latter depends on slope steepness. In some cases, the depth of the soil is a moot question, as the soil blends into its parent material without any distinct boundary. However, the biological activity zone seldom extends below 2–3 m, and in some cases is shallower than 1 m.

A hypothetical soil profile is presented in Fig. 2.3. This is not a typical soil, for in the myriad of greatly differing soil types recognized by pedologists it is well nigh impossible to define a single typical soil. Our illustration is only meant to suggest the sort of differences in appearance and structure likely to be encountered in a soil profile between different depth strata. Pedologists classify soils by their genesis and recognizable characteristics (see Fig. 2.4). However, pedological profile characterization is still somewhat qualitative and not sufficiently based on exact measurements of pertinent

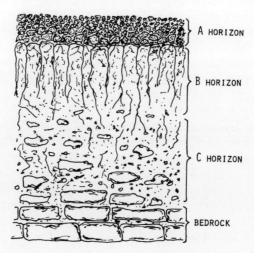

Fig. 2.3. Schematic representation of an hypothetical soil profile, with its underlying parent rock. The A horizon is shown with an aggregated crumblike structure, and the B horizon with columnar structure.

F. The Soil Profile

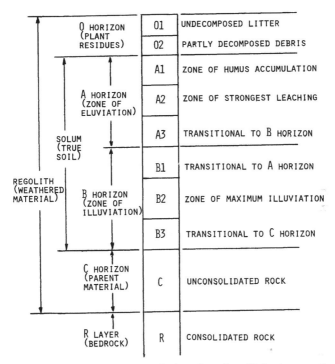

Fig. 2.4. Descriptive terminology for soil profile horizons.

physical properties such as hydraulic, mechanical, and thermal characteristics.

Sample Problems

1. Prove the following relation between porosity, particle density, and bulk density:

$$f \stackrel{?}{=} (\rho_s - \rho_b)/\rho_s = 1 - \rho_b/\rho_s$$

Substituting the respective definitions of f, ρ_s, and ρ_b, we can rewrite the equation as

$$V_f/V_t \stackrel{?}{=} 1 - (M_s/V_t)/(M_s/V_s)$$

Simplifying the right-hand side, we obtain

$$V_f/V_t \stackrel{?}{=} 1 - (V_s/V_t) = (V_t - V_s)/V_t$$

But since $V_t - V_s = V_f$, we have

$$V_f/V_t = V_f/V_t$$

Q.E.D.

2. Prove the following relation between volume wetness, mass wetness, bulk density, and water density ($\rho_w = M_w/V_w$):

$$\theta \stackrel{?}{=} w\rho_b/\rho_w$$

Again, we start by substituting the respective definitions of θ, w, ρ_b, and ρ_w:

$$V_w/V_t \stackrel{?}{=} [(M_w/M_s)(M_s/V_t)]/(M_w/V_w)$$

Rearranging the right-hand side,

$$\frac{V_w}{V_t} = \frac{V_w}{M_w}\frac{M_w}{M_s}\frac{M_s}{V_t} = \frac{V_w}{V_t}$$

Q.E.D.

3. A sample of moist soil having a wet mass of 1000 gm and a volume of 640 cm³ was dried in the oven and found to have a dry mass of 800 gm. Assuming the typical value of particle density for a mineral soil, calculate the bulk density ρ_b, porosity f, void ratio e, mass wetness w, volume wetness θ, water volume ratio v_w, degree of saturation s, and air-filled porosity f_a.

Bulk density:

$$\rho_b = \frac{M_s}{V} = \frac{800 \text{ gm}}{640 \text{ cm}^3} = 1.25 \text{ gm/cm}^3$$

Porosity:

$$f = 1 - \frac{\rho_b}{\rho_s} = 1 - \frac{1.25 \text{ gm/cm}^3}{2.65 \text{ gm/cm}^3} = 1 - 0.472 = 0.528$$

Alternatively,

$$f = V_f/V_t = (V_t - V_s)/V_t$$

Since

$$V_s = \frac{M_s}{\rho_s} = \frac{800 \text{ gm}}{2.65 \text{ gm/cm}^3} = 301.9 \text{ cm}^3$$

Hence

$$f = \frac{640 \text{ cm}^3 - 301.9 \text{ cm}^3}{640 \text{ cm}^3} = 0.528 = 52.8\%$$

F. The Soil Profile

Void ratio:
$$e = \frac{V_f}{V_s} = \frac{V_t - V_s}{V_s} = \frac{640 \text{ cm}^3 - 301.9 \text{ cm}^3}{301.9 \text{ cm}^3} = 1.12$$

Mass wetness:
$$w = \frac{M_w}{M_s} = \frac{M_t - M_s}{M_s} = \frac{1000 \text{ gm} - 800 \text{ gm}}{800 \text{ gm}} = 0.25 = 25\%$$

Volume wetness:
$$\theta = \frac{V_w}{V_t} = \frac{200 \text{ cm}^3}{640 \text{ cm}^3} = 0.3125 = 31.25\%$$

(*Note*: $V_w = M_w/\rho_w$, wherein ρ_w, the density of water, equals approximately 1 gm/cm^3.) Alternatively,

$$\theta = w \frac{\rho_b}{\rho_w} = 0.25 \frac{1.25 \text{ gm/cm}^3}{1 \text{ gm/cm}^3} = 0.3125 = 31.25\%$$

Water volume ratio:
$$v_w = \frac{V_w}{V_s} = \frac{200 \text{ cm}^3}{301.9 \text{ cm}^3} = 0.662$$

Degree of saturation:
$$s = \frac{V_w}{V_t - V_s} = \frac{200 \text{ cm}^3}{640 \text{ cm}^3 - 301.9 \text{ cm}^3} = 0.592 = 59.2\%$$

Air-filled porosity:
$$f = \frac{V_a}{V_t} = \frac{600 \text{ cm}^3 - 200 \text{ cm}^3 - 301.9 \text{ cm}^3}{640 \text{ cm}^3} = 0.216 = 21.6\%$$

4. How many centimeters (equivalent depth) of water are contained in a soil profile 1 m deep if the mass wetness of the upper 40 cm is 15% and that of the lower 60 cm is 25%? The bulk density is 1.2 gm/cm^3 in the upper layer and 1.4 in the deeper layer. How much water does the soil contain in cubic meters per hectare of land?

Recall that $\theta = w(\rho_b/\rho_w)$ (where $\rho_w = 1$).
Volume wetness in upper layer: $\theta_1 = 0.15 \times 1.2 = 0.18$.
Equivalent depth of water in upper 40 cm = $0.18 \times 40 = 7.2$ cm.
Volume wetness in lower layer: $\theta_2 = 0.25 \times 1.4 = 0.35$.

Equivalent depth of water in lower 60 cm = 0.35 × 60 = 21.0 cm.

Total equivalent depth of water in 100 cm profile = 7.2 + 21.0 = 28.2 cm.

Area of hectare = 10,000 m^2; volume of soil (1 m deep) per hectare = 10,000 m^3.

Volume of water contained in 1 m deep soil per hectare = 10,000 × 0.282 = 2820 m^3.

وَجَعَلْنَا مِنَ الْمَاءِ كُلَّ شَيْءٍ حَيٍّ

And with water we have made all living things.
The Koran

3 *Properties of Water in Relation to Porous Media*

A. Introduction

Water is the most prevalent substance on earth, covering more than two-thirds of its surface in oceans, seas, and lakes. The continental areas are themselves frequently charged with and shaped by water. In vapor form, water is always present in the atmosphere. Finally, water is the principal constituent of all living organisms, plants and animals alike. Water and its ionization products, hydrogen and hydroxyl ions, are important factors determining the structure and biological properties of proteins and other cell and tissue components. The basic processes of life are intrinsically dependent upon water's unique attributes. Far from being a bland, inert liquid, water is, in fact, a highly reactive substance and an exceedingly effective solvent and transporter of numerous substances.

Notwithstanding its ubiquity, water remains something of an enigma, possessing unusual and anomalous attributes still not entirely understood. Perhaps the first anomaly is that water, despite its low molecular weight, is a liquid and not a gas at normal temperatures, (Its sister compound, H_2S, has its boiling point at $-60.7°C$.) Compared with other common liquids, furthermore, water has unusually high melting and boiling points, heats of fusion and vaporization, specific heat, dielectric constant, viscosity, and surface tension. (In particular, the specific heat capacity of liquid water is equal to 1 cal/deg gm at 15°C, whereas that of ice is about 0.5, of iron 0.106, of mercury 0.033, of air 0.17, and of dry soil about 0.2 cal/deg gm.) The heat of vaporization (which is a measure of the amount of kinetic energy required to overcome the attractive forces between adjacent molecules in a liquid so that individual molecules can escape from the denser liquid phase and enter the more rarefied gaseous state) is much higher in water (540 cal/gm at the boiling point) than in methanol (263), ethanol (204), acetone (125), benzene (94), or chloroform (59 cal/gm), all of which have higher molecular weights than water. These properties suggest that there must be some particularly strong force of attraction between molecules in liquid water to impart to this of all liquids such a high internal cohesion. The secret, as we shall see, is in the molecular structure and intermolecular bonding of the peculiar substance called water.

B. Molecular Structure

One cubic centimeter of liquid water contains about 3.4×10^{22} (34,000 billion billion) molecules, the diameter of which is about 3 Å (3×10^{-8} cm). The chemical formula of water is H_2O, which signifies that each molecule consists of two atoms of hydrogen and one atom of oxygen. There are three isotopes of hydrogen (^1H, ^2H, ^3H), as well as three isotopes of oxygen (^{16}O, ^{17}O, ^{18}O), which can form 18 different combinations. However, all isotopes but ^1H and ^{16}O are quite rare.

The hydrogen atom consists of a positively charged proton and a negatively charged electron. The oxygen atom consists of a nucleus having the positive charge of eight protons, surrounded by eight electrons, of which six are in the outer shell. Since the outer electron shell of the hydrogen lacks one electron and that of the oxygen lacks two electrons, one atom of oxygen can combine with two atoms of hydrogen in an electron-sharing molecule.

The strong intermolecular forces in liquid water are caused by the electrical polarity of the water molecule, which in turn is the consequence of the specific arrangement of electrons in its oxygen and hydrogen atoms (Fig. 3.1). The oxygen atom shares a pair of electrons with each of the two hydrogen atoms, through overlap of the 1s orbitals of the hydrogen atoms with two hybridized sp^3 orbitals of the oxygen atom. Each of these electron pairs has about one-third ionic and two-thirds covalent character. From spectroscopic and x-ray analyses, the precise bond angles and lengths have been determined. The H—O—H bond in water is not linear but bent

B. Molecular Structure

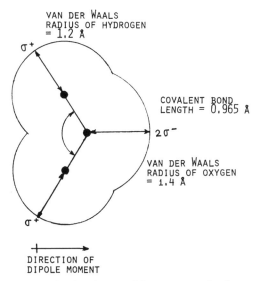

Fig. 3.1. Model of a water molecule. The curved lines represent borders at which van der Waals attractions are counterbalanced by repulsive forces.

at an angle of 104.5°, which represents a slight deviation from a perfectly tetrahedral arrangement of the four possible sp³ orbitals of the oxygen atom, which would have an angle of 109.5°. One explanation for this deviation is that the unpaired electrons of oxygen tend to repel the paired electrons. The average H–O interatomic distance is 0.965 Å.

This arrangement of electrons in the water molecule gives it electrical asymmetry. The more electronegative oxygen atom tends to attract the single electrons of the hydrogen atoms, leaving the hydrogen nuclei bare. As a result, each of the two hydrogen atoms has a local partial positive charge (designated σ^+). The oxygen atom, in turn, has a local partial negative charge (designated σ^-), located in the zone of the unshared orbitals. Thus, although the water molecule has no net charge, it is an electrical dipole. The degree of separation of positive and negative charges in dipolar molecules is given by the *dipole moment*, a measure of the tendency of a molecule to orient itself in an electrical field. From the dipole moment of water, it has been calculated that each hydrogen atom has a partial positive charge of about $+0.33 \times 10^{-10}$ esu (electrostatic unit) and the oxygen atom a partial negative charge of about -0.66×10^{-10} esu.

It is the polarity of water molecules which makes them mutually attractive. It is also the reason why water is so good a solvent and essentially why it adsorbs readily upon solid surfaces and hydrates ions and colloids.

C. Hydrogen Bonding

Every hydrogen proton, while it is attached primarily to a particular molecule, is also attracted to the oxygen of the neighboring molecule, with which it forms a secondary link known as a *hydrogen bond*. Although this intermolecular link resulting from dipole attraction is not as strong as the primary attachment of the hydrogen to the oxygen of its own molecule, water can be regarded as a polymer of hydrogen-bonded molecules. This structure is most characteristically complete in ice crystals, in which each molecule is linked to four neighbors by means of four hydrogen bonds, thus forming a hexagonal lattice that is a rather open structure (Fig. 3.2). When the ice melts, this rigid structure collapses partially, so that additional molecules can enter into the intermolecular spaces and each molecule thus can have more than four near neighbors. For this reason, liquid water can be more dense than ice at the same temperature, and thus lakes and ponds develop a surface ice sheet in winter rather than freeze solid from bottom to top as they would if ice were denser than liquid water.

An important property of hydrogen bonds is that they are much weaker than covalent bonds. The H bonds in liquid water are estimated to have a bond energy of only about 4.5 kcal/mole, compared with 110 kcal/mole for the H—O electron-pair bonds in water. (*Note:* Bond energy is the energy required to break a bond.) Another important property of hydrogen bonds

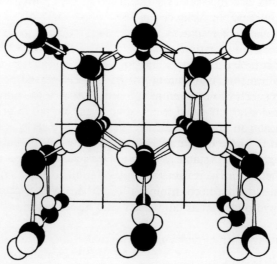

Fig. 3.2. Schematic structure of an ice crystal. The oxygen atoms are shown in black and the hydrogen atoms in white. The pegs linking adjacent molecules represent hydrogen bonds.

D. States of Water

is that they have a high degree of directionality, which is conferred by the characteristic arrangement of the bonding orbitals of the hydrogen and oxygen atoms (Fig. 3.1). Hydrogen bonds also possess a specific bond length, which differs from one type of H bond to another, depending on the structural geometry and the electron distribution in the molecules involved. In ice, each water molecule is hydrogen-bonded: the length of the hydrogen bond is 1.77 Å. Hydrogen bonds, therefore, form and remain stable only under specific geometrical conditions.

Water is of course not the only substance capable of hydrogen bonding. Ammonia, hydrogen fluoride, and alcohols also exhibit hydrogen bonds (i.e., exposed hydrogen nuclei in these compounds are attracted by concentrations of negative charges on adjacent molecules). Molecules of many compounds, in fact, are so structured that one part has an excess of positive and the other part an excess of negative charges. All such molecules are called *polar*, since they act somewhat like minute magnets and are characterized by a *dipole moment* (a measure of the tendency of a polar molecule to be affected by an electrical or magnetic field). The uniqueness of water lies in the ease with which its molecules form extended, three-dimensional hydrogen-bonded aggregates.

D. States of Water

In the vapor or gaseous state, water molecules are largely independent of one another and occur mostly as *monomers* signified as $(H_2O)_1$. Occasionally, colliding molecules may fuse to form *dimers*, $(H_2O)_2$, or even *trimers*, $(H_2O)_3$, but such combinations are rare.

In the solid state, at the other extreme, a rather rigidly structured lattice forms with a tetrahedral configuration (Fig. 3.2) which can be schematically represented as sheets of puckered hexagonal rings (Fig. 3.3). In actual fact, as many as nine alternative structures, or ice forms, can occur when water freezes, depending on prevailing temperature and pressure conditions which can be charted in phase diagrams. Figure 3.3 pertains to ice I, the familiar form which occurs and is stable at ordinary atmospheric pressure.

The orderly structure of ice does not totally disappear in the liquid state[1], as the molecules do not become entirely independent of one another. The polarity and hydrogen bonds continue to bond water molecules together,

---

[1] It has been estimated from the heat of fusion of ice that only about 15% of the hydrogen bonds are broken when ice is melted to liquid water at 0°C. Strong attractions between water molecules still exist in liquid water even at 100°C, as is indicated by the high heat of vaporization. Hydrogen bonding between water molecules may not be completely overcome, in fact, until the water vapor is heated to nearly 600°C.

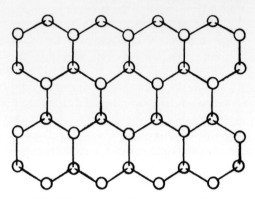

Fig. 3.3. The crystalline structure of ice.

though the structural forms which develop in the liquid state are much more flexible and transitory than is the case in the rigidly structured solid state.

Although a number of different models have been proposed in an attempt to account for the thermodynamic, spectroscopic, and transport properties of liquid water, no single model explains all of its known properties. Considerable evidence supports the idea that hydrogen bonds in liquid water form an extensive three-dimensional network, the detailed features of which are probably short-lived. According to the "flickering cluster" model, for instance, the molecules of liquid water associate and dissociate repeatedly in transitory or flickering polymer groups, designated $(H_2O)_n$, having a quasi-crystalline internal structure. These microcrystals, as it were, form and melt so rapidly and randomly that, on a macroscopic scale, water appears to behave as a homogeneous liquid. The typical cluster is conjectured to have an average n value of about 40 (at 20°C and atmospheric pressure) and a short life of 10^{-10}–10^{-11} sec, and to be continuously exchanging molecules with the surrounding unstructured phase. This model is illustrated hypothetically in Fig. 3.4. The average cluster size n probably decreases with increasing temperature or pressure, a factor which may account for such phenomena as the maximum in the temperature dependence of the density and the minimum in the pressure dependence of the relative viscosity. The value of n probably also depends on the type and concentration of solute present.

In transition from the solid to the liquid, and from the liquid to the gaseous state, hydrogen bonds must be disrupted (while in condensation and freezing, they must be reestablished). Hence it requires relatively high temperatures and energy values to achieve these transitions. To thaw 1 gm of ice, 80 cal must be supplied; and conversely, the same energy (the latent heat of fusion) is released in freezing.

At the boiling point (100°C at atmospheric pressure), water passes from

E. Ionization and pH

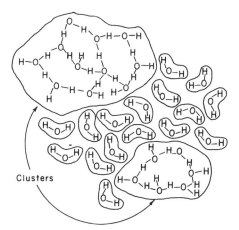

Fig. 3.4. Schematic illustration of the "flickering clusters" of polymeric associations and monomeric molecules in liquid water.

the liquid to the gaseous state and in so doing it absorbs 540 cal/gm. This amount of heat, known as the latent heat of vaporization, destroys the intermolecular structure and separates the molecules. Water can be vaporized at temperatures below 100°C, but such vaporization requires greater heat. At 25°C, for instance, the latent heat is 580 cal/gm. *Sublimation* is the direct transition from the solid state to vapor, and the heat required to effect it is equal to the sum of the latent heats of fusion and of vaporization.

E. Ionization and pH

Because of its small mass and the tightness with which its single electron is bound to the oxygen atom, the nucleus of the hydrogen atom in the water molecule exhibits a finite tendency to dissociate from the oxygen with which it is covalently associated and to "jump" to the adjacent water molecule to which it is hydrogen bonded. Such an event produces two ions: the hydronium[2] ion (H_3O^+) and the hydroxyl ion (OH^-).

The reaction described is reversible, and should be written as $2H_2O \rightleftharpoons (H_3O)^+ + OH^-$. However, by convention it is written simply as

$$H_2O \rightleftharpoons H^+ + OH^-$$

and one speaks of hydrogen ions rather than of hydronium ions.

[2] Recent evidence suggests that the ionized hydrogen nucleus (a proton) in aqueous solutions is even more heavily hydrated and should be represented as bonded to four water molecules: $H_9O_4^+$ or $H(H_2O)_4^+$.

Although the self-ionization of water is small, its consequences are extremely important. Since the ionization of water is reversible, it tends to an equilibrium state in which the rate of dissociation of water molecules into ions equals the rate of ionic reassociation to form molecules once again. For such a system in equilibrium (in this case, when the concentration of each of the species H_2O, H^+, and OH^- remains constant), the law of mass action applies; i.e., the ratio of concentrations of the products and the reactants must be constant. Using brackets to denote concentration, we can write this in the following way:

$$K_{eq} = [H^+][OH^-]/[H_2O] \qquad (3.1)$$

Since the number of water molecules undergoing dissociation at any given time is very small relative to the total number of water molecules present, $[H_2O]$ can be considered constant. Assuming this concentration to be 55.5 mole/liter (the number of grams per liter divided by the gram molecular weight: 1000/18 = 55.5 mole/liter), we can simplify the equilibrium constant expression as follows:

$$55.5 \times K_{eq} = [H^+][OH^-] \quad \text{or} \quad K_w = [H^+][OH^-] \qquad (3.2)$$

in which K_w is a composite constant called the *ion product* of water. In actual fact, the concentrations of H^+ and OH^- ions in pure water at 25°C have been found to be 1×10^{-7} mole/liter, an extremely small value when compared to the overall concentrations of (largely undissociated) water, namely, 55.5 mole/liter. Thus, the value of K_w at 25°C is 1.0×10^{-14}. If the hydroxyl ion concentration $[OH^-]$ is changed, the hydrogen ion concentration $[H^+]$ changes automatically to maintain the constancy of the product, and vice versa. An excess concentration of hydrogen ions over the corresponding concentration of hydroxyl ions imparts to the aqueous medium the property of *acidity*, whereas a predominance of hydroxyl ions produces the opposite property of *alkalinity* or *basicity*. A condition in which the concentrations of H^+ and OH^- are equal is called *neutrality*.

The ion product of water, K_w, is the basis for the pH scale, a means of designating the concentration of H^+ (and thus of OH^- as well) in any aqueous solution in that concentration range between $1.0M$ (mole per liter) H^+ and $1.0M$ OH^-. The term pH is defined as

$$\text{pH} = \log_{10} 1/[H^+] = -\log_{10}[H^+] \qquad (3.3)$$

As already stated, in a precisely neutral solution at 25°C,

$$[H^+] = [OH^-] = 1.0 \times 10^{-7} M$$

The pH of such a solution is

$$\text{pH} = \log_{10}[1/(1 \times 10^{-7})] = 7.0$$

The value of 7.0 for the pH of a neutral solution is thus not an arbitrarily chosen number, but one which derives from the absolute value of the ion product of water at 25°C. It is especially important to note that the pH scale is logarithmic, not arithmetic. To say that two solutions differ in pH by one unit means that one solution has ten times the hydrogen ion concentration of the other. A pH of 6, for instance, implies a hydrogen ion concentration of 10^{-6} and a hydroxyl ion concentration of 10^{-8}. A pH of 5 indicates $[H^+] = 10^{-5}$ (i.e., an acidity ten times greater than the preceding case) and $[OH^-] = 10^{-9}$.

The pH of soil water can vary greatly, from 2 or less in acid soils to 9 or more in sodic ("alkali") soils. Highly leached, humid region soils tend to be acidic (especially if the parent material is derived from such acid rocks as granite). So do organic soils. On the other hand, desert-region soils tend to have an alkaline reaction. Soil pH is strongly related to *sodicity*, which affects soil structure, and thus pH is also related to physical, as well as to chemical, soil conditions.

F. Solvent Properties of Water

Water dissolves or disperses many substances because of its polar nature. It is a much better solvent than most common liquids and has indeed been called the universal solvent. All chemical substances have a finite solubility in water, ranging from compounds like ethanol which, as cocktail mixers know, is miscible with water in all proportions, to the so-called insoluble substances whose saturated solutions may contain less than one molecule or ion per liter. Many crystalline salts and other ionic compounds readily dissolve in water but are nearly insoluble in nonpolar liquids such as chloroform or benzene. Since the crystal lattice of salts, such as sodium chloride, is held together by very strong electrostatic attractions between alternating positive and negative ions, considerable energy is required to pull these ions away from each other. However, water dissolves sodium chloride because the strong electrostatic attraction between water dipoles and the Na^+ and Cl^- ions, forming stable hydrated Na^+ and Cl^- ions, exceeds the tendency of these ions to attract each other. In the case of Na^+, this *hydration* is represented by the equation

$$Na^+ + (n + 4)H_2O \rightarrow Na(H_2O)_4(H_2O)_n^+$$

as illustrated in Fig. 3.5. In addition to hydration, there is also the *hydrolysis* of metal species, a reaction in which the metal ion displaces one of the protons (hydrogen) of water to form *basic* substances (i.e., hydroxides).

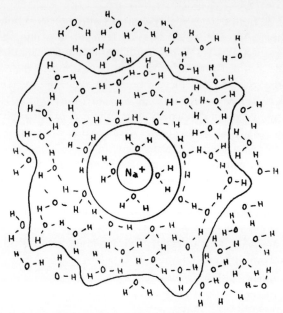

Fig. 3.5. A model of the hydration "atmosphere" of sodium ion: an inner shell of more or less rigidly structured water surrounded by a cluster of looser but still structure-enhanced water, the whole floating in a sea of "free" water.

Ion solvation is also aided by the tendency of the solvent to oppose the electrostatic attraction between ions of opposite charges. This is characterized by the *dielectric constant D*, which is defined by the relationship

$$F = e_1 e_2 / Dr^2 \tag{3.4}$$

where F is the attractive force between two ions of opposite charge, e_1 and e_2 are charges on the ions, and r is the distance between them. Water has an extremely high dielectric constant, as can be seen in Table 3.1. (For instance, the attractive force between Na^+ and Cl^- ions at a given distance in water is less than one-third that in ethanol and only one-fortieth that in benzene.) This fact greatly facilitates hydration of ions and dissolution of the crystal lattice of salts in water.

The presence of an ionic solute such as NaCl causes a distinct change in

Table 3.1

DIELECTRIC CONSTANTS OF SOME LIQUIDS (20°C)

Water	80	Acetone	21.4
Methanol	33	Benzene	2.3
Ethanol	24	Hexane	1.9

the structure of liquid water since each Na^+ and Cl^- ion is surrounded by a shell of water dipoles. These hydrated ions have a geometry somewhat different from the clusters of hydrogen-bonded water molecules: they are more highly ordered and regular in structure. Generally speaking, the greater the charge density (i.e., the ratio of charge to surface area) of an ion, the more heavily hydrated it will be, and as a rule negative ions (anions) are less heavily hydrated than positive ions (cations) because of the greater difficulty of crowding water molecules about the anions with the positive charges of the water protons oriented inward; that is to say, it is easier to orient the water with the protons outward and the negative end inward toward the positively charged cation at the center of the hydration cluster, or "envelope." A typical Coulombic (electrostatic) hydration envelope may consist of an inner sphere of strongly bound molecules (called primary hydration), which is a zone of enhanced water structure. This inner zone is surrounded by an outer shell of disrupted water structure (Fig. 3.5). So strongly held are the innermost water molecules that they may stick to the solute even when it is crystallized out of solution—the so-called *water of hydration*. The water molecules thus bound lose energy, and the heat released is known as the heat of solution.

The effect of a solute on the solvent is manifest in another set of properties, namely, the *colligative properties* of solutions, which are dependent on the number of solute particles per unit volume of solvent. Solutes produce such characteristic effects in the solvent as depression of the freezing point, elevation of the boiling point, and depression of the vapor pressure. They also endow a solution with the property of osmotic pressure. One gram molecular weight of an ideal nondissociating nonassociating solute dissolved in 1,000 grams of water at a pressure of 760 mm of mercury depresses the freezing point by 1.86°C and elevates the boiling point by 0.543°C. Such a solution also yields an osmotic pressure of 22.4 atm in an appropriate apparatus. Since aqueous solutions usually deviate considerably from ideal behavior, these relationships are quantitative only at infinite dilution, i.e., on extrapolation to zero concentration of solute.

G. Osmotic Pressure

Owing to their constant thermal motion, molecules of any given species in a heterogeneous solution tend to migrate from a zone where their concentration is higher to where it is lower, in a spontaneous tendency toward a state of equal concentration and composition throughout.[3] This migration of molec-

---

[3] The reader is invited to ponder the fact that thermal motion of each molecule is random, yet the overall tendency of a large number of randomly moving and colliding molecules is to produce a net directional flux of each species in the solution from a high toward a low concentration region.

ular species in response to spatial differences in concentration is called *diffusion*. If a physical barrier is interposed between the two regions, across the path of diffusion, and if that barrier is permeable to molecules of the solvent but not to those of the solute, the former will diffuse through the barrier in a process called osmosis,[4] tending as in the case of unhindered diffusion toward a state of uniformity of composition even across the barrier. Barriers permeable to one substance in a solution but not to another are called *selective* or *semipermeable membranes*. Membranes surrounding cells in living organisms, for example, exhibit selective permeability to water while restricting the diffusion of solutes between the cells' interior and their exterior environment. Water molecules can and, in fact, do cross the membrane in both directions, but the net flow is from the more dilute solution to the more concentrated.

Figure 3.6a is a schematic representation of a pure solvent separated from a solution by a semipermeable membrane. Solvent will pass through the membrane and enter the solution compartment, driving the solution level up the left-hand tube until the hydrostatic pressure of the column of dilute solution on the left is sufficient to counter the diffusion pressure of the solvent molecules drawn into the solution through the membrane. The hydrostatic pressure at equilibrium, when solvent molecules are crossing the membranes in both directions at equal rates, is the *osmotic pressure* of the solution. To measure the osmotic pressure of a solution, it is not necessary to wait until the flow stops and equilibrium is established. Theoretically, a counter pressure can be applied with a piston on the solution side, and when the counter pressure thus applied (Fig. 3.6b) is just sufficient to prevent the osmosis of water into the solution, it represents the osmotic pressure of the solution.[5]

The expression osmotic pressure of a solution can be misleading. What a solution exhibits relative to the pure solvent (say, water) is not an excess pressure but, on the contrary, a "suction" such that will draw water from a reservoir of pure water brought into contact with the solution across a semipermeable boundary. Hence we ought perhaps to speak of an osmotic suction as the characteristic property of a solution rather than of osmotic pressure, which, however, is the conventional term. The actual *process* of osmosis will obviously take place only in the presence of a semipermeable membrane separating a solution from its pure solvent, or from another solution of different concentration. In principle, however, the *property* of the solution which induces the process of osmosis exists whether or not a

[4] From the Greek ωσμοσ, meaning push.

[5] If a pressure greater than the osmotic pressure is applied on the solution side, some solvent will be forced out of the solution compartment through the membrane. This is the basis of the *reverse osmosis* method of water purification.

G. Osmotic Pressure

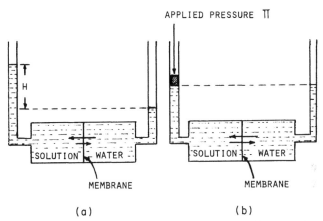

Fig. 3.6. Osmosis and osmotic pressure. (a) Osmosis: the flow of water molecules through the membrane into the solution is at first greater than the reverse flow from the solution into the water compartment. The hydrostatic pressure due to the column of expanded solution increases the rate of water flow from the solution to the water compartment until at equilibrium, the opposite flows are equal. (b) The osmotic pressure of the solution is equal to the hydrostatic pressure Π which must be applied to the solution to equalize the rate of flow to and from the solution and produce a net flow of zero.

membrane happens to be present, as it derives fundamentally from the decrease of potential energy (the so-called free energy, to be discussed in Chapter 7) of water in solution relative to that of pure water. The other manifestations of this property are a decrease in vapor pressure, a rise of the boiling point, and a depression of the freezing point (popularly known as the "antifreeze" effect).

The term semipermeable membrane, first coined by van't Hoff in 1886 and since hallowed by tradition, is itself something of a misnomer. Membrane selectivity toward different species in a solution is not an absolute but a relative property. As such, it can be characterized by means of a parameter known as the selectivity or reflection coefficient, which varies from zero in the case of nonselectivity to unity in the (hypothetical) case of perfect selectivity. The nearest thing to a perfectly selective membrane is probably an air space separating two aqueous solutions of a completely nonvolatile solute. Such an air barrier allows free passage of solvent (water) molecules from liquid to vapor to liquid (from the more dilute to the more concentrated solution) while restricting the solute. However, most known porous membranes, including the various biological membranes, are "leaky," in the sense that they transmit molecules of the solute, as well as of the solvent, to a greater or lesser degree. The mechanism of selectivity is often related to pore size, as is evident from the often observed fact that solutes of small molecular weight tend to pass through membranes more readily than do solutes of large molecular weight (e.g., polymers). However, in many cases membranes are not merely molecular sieves. Often the molecules of solute are unable to penetrate the complex network of the membrane's pores owing to the preferential adsorption of solvent molecules on the membrane's inner surfaces, which makes the pores effectively smaller in diameter. In some cases the solute is subject to "negative adsorption" or repulsion by the membrane. In still other cases the semipermeable membrane may consist of a large number of fine capillaries not wetted by the liquid solution but through which molecules of the solvent can pass in the vapor phase. Osmosis can thus occur by distillation through the membrane from the region of higher to lower vapor pressure (i.e., from the more dilute to the more concentrated solution).

In dilute solutions, the osmotic pressure is proportional to the concentration of the solution and to its temperature according to the following equation:

$$\Pi = MRT \tag{3.5}$$

where Π is the osmotic pressure in atmospheres, M the total molar concentration of solute particles (whether molecules or dissociated ions), T the temperature in degrees Kelvin, and R the gas constant (0.08205 liter atm/deg mole).[6]

H. Solubility of Gases

The concentration of gases in water generally increases with pressure and decreases with temperature. According to Henry's law, the mass concentration of the dissolved gas c_m is proportional to the partial pressure of the gas p_i:

$$c_m = s_c p_i / p_0 \tag{3.6}$$

where s_c is the solubility coefficient of the gas in water and p_0 is the total pressure of the atmosphere. The volume concentration is similarly proportional:

$$C_v = s_v p_i / p_0 \tag{3.7}$$

[6] The osmotic pressure increase with temperature is associated with the corresponding increase of the molecular diffusivity (self-diffusion coefficient) of water, D_w. According to the Einstein–Stokes equation,

$$D_w = kT/6\pi r\eta$$

where $k = R/N$, the Boltzmann constant (1.38×10^{-16} erg/°K); r is the rotation radius of the molecule (~ 1.5 Å), and η is the viscosity.

I. Adsorption of Water on Solid Surfaces

Table 3.2

SOLUBILITY COEFFICIENTS OF GASES IN WATER

Temperature (°C)	Nitrogen (N_2)	Oxygen (O_2)	Carbon dioxide (CO_2)	Air (without CO_2)
0	0.0235	0.0489	1.713	0.0292
10	0.0186	0.0380	1.194	0.0228
20	0.0154	0.0310	0.878	0.0187
30	0.0134	0.0261	0.665	0.0156
40	0.0118	0.0231	0.530	—

where s_v is the solubility expressed in terms of volume ratios (i.e., C_v is the volume of dissolved gas relative to the volume of water). The values of s_c, s_v are determined experimentally, and if the gas does not react chemically with the liquid, they should remain constant over a range of pressures, especially at low partial pressures of the dissolved gases. Solubility is, however, strongly influenced by temperature.

Table 3.2 gives the values of s_v for several atmospheric gases at various temperatures.

Note that carbon dioxide constitutes a special case, its solubility being about 50 times greater than that of nitrogen and about 25 times that of oxygen. The high solubility of carbon dioxide in water is the result of its chemical reaction with water. The entry of carbon dioxide into solution is accompanied by hydrolysis, the consequence of which is the formation of carbonic acid with hydrogen and bicarbonate ions:

$$CO_2 + H_2O \rightleftharpoons H^+ + HCO_3^-$$

The extent of this reaction is determined by the partial pressure of CO_2 gas in the ambient atmosphere, and, in turn, determines the pH of the solution and its capability of dissolving such minerals as calcium carbonate.

The solubilities of various gases (particularly oxygen) in varying conditions strongly influence such vital soil processes as oxidation and reduction, and respiration by roots and microorganisms.

I. Adsorption of Water on Solid Surfaces

Adsorption is an interfacial phenomenon resulting from the differential forces of attraction or repulsion occurring among molecules or ions of different phases at their exposed contact surfaces. As a result of both cohesive and adhesive forces coming into play, the contact zone may exhibit a concentration or a density of material different from that inside the phases themselves. According to the different phases which may come in contact, various types of adsorption can occur, such as the adsorption of gases upon solids, of gases upon liquid surfaces, or of liquids upon solids. A distinction should perhaps be made between *adsorption*, being a surface attachment or repulsion, and its complementary term *absorption*, which refers to cases in which one phase penetrates or permeates another. In actual fact it is often impossible to separate the phenomenon of adsorption from that of absorption, particularly in the case of highly porous systems, and hence the noncommittal term *sorption* is frequently employed. Quite another distinction has been attempted between *physical adsorption*, involving mainly van der Waals forces and characterized by low energies of adsorption (about 5 kcal/mole or less), and *chemical adsorption*, involving the formation of stronger and more permanent bonds of a chemical nature (i.e., analogous to valence bonds) with energies of adsorption of the order of 20–100 kcal/mole. However, this distinction is often arbitrary, as both categories can occur simultaneously, and some types of bonding can be classified as either physical or chemical.

The interfacial forces of attraction or repulsion may themselves be of different types, including electrostatic or ionic (Coulombic) forces, intermolecular forces such as van der Waals and London forces, and short-range repulsive (Born) forces. The adsorption of water upon solid surfaces is generally of an electrostatic nature. The polar water molecules attach to the charged faces of the solids. The adsorption of water is the mechanism causing the strong retention of water by clay soils at high suctions.

The interaction of the charges of the solid with the polar water molecules may impart to the adsorbed water a distinct and rigid structure in which the water dipoles assume an orientation dictated by the charge sites on the solids. Some investigators believe that the adsorbed layer or "phase" has a quasicrystalline, icelike structure and can assume a thickness of 10–20 Å (i.e., from three to seven molecular layers) from the adsorbing surface. This adsorbed water layer may have mechanical properties of strength and viscosity which differ from those of ordinary liquid water at the same temperature. Other investigators, however, doubt that such a distinct phase exists or that it extends beyond the first or second molecular layer.

The adsorption of water upon clay surfaces is an exothermic process, resulting in the liberation of an amount of heat known as the heat of wetting. Anderson (1926) found a linear relationship between heat of wetting and exchange capacity. Janert (1934) traced the relationship between the heat of wetting and the nature of the exchange cations. The distinction between

J. Vapor Pressure

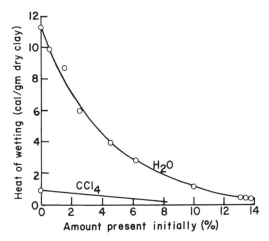

Fig. 3.7. Heat of wetting in relation to initial liquid content. (After Janert, 1934.)

polar and nonpolar adsorption is illustrated in Fig. 3.7, in which water and carbon tetrachloride are compared for a "brick clay."

A further discussion of adsorption theory and of adsorption isotherms will be found in Chapter 4, in connection with measuring the specific surface of porous media.

J. Vapor Pressure

According to the kinetic theory, molecules in a liquid are in constant motion, which is an expression of their thermal energy. These molecules collide frequently, and occasionally one or another of them absorbs sufficient momentum to leap out of the liquid and into the atmosphere above it. Such a molecule, by virtue of its kinetic energy, thus changes from the liquid to the gaseous phase. This kinetic energy is then lost in overcoming the potential energy of intermolecular attraction while escaping from the liquid. At the same time, some of the molecules in the gaseous phase may strike the surface of the liquid and be absorbed in it.

The relative rates of these two directions of movement depend upon the concentration of vapor in the atmosphere relative to its concentration at a state of equilibrium (i.e., when the movement in both directions is equal). An atmosphere that is at equilibrium with a body of pure water which is at atmospheric pressure is considered to be saturated with water vapor, and the partial pressure of the vapor in such an atmosphere is called the saturation (or equilibrium) vapor pressure. The vapor pressure at equilibrium with any body of water depends upon the physical condition of the water (pressure and temperature) and its chemical condition (solutes) but does not depend upon the absolute or relative quantity of liquid or gas in the system.

The saturation vapor pressure increases with increasing temperature. As the kinetic energy of the molecules in the liquid increases, the evaporation rate increases and a higher concentration of vapor in the atmosphere is required for the rate of return to the liquid to match the rate of escape from it. A liquid arrives at its boiling point when the vapor pressure becomes equal to the atmospheric pressure. If the temperature range is not too wide, the dependence of saturation vapor pressure on temperature is expressible by the equation[7] (see Table 3.3)

$$\ln p_0 = a - b/T \tag{3.8}$$

where $\ln p_0$ is the logarithm to the base e of the saturation vapor pressure p_0, T is the absolute temperature, and a and b are constants.

As mentioned earlier, the vapor pressure depends also upon the pressure of the liquid water. At equilibrium with drops of water which have a hydrostatic pressure greater than atmospheric, the vapor pressure will be greater than in a state of equilibrium with free water, which has a flat interface with the atmosphere. On the other hand, in equilibrium with adsorbed or capillary water under an hydrostatic pressure smaller than atmospheric, the vapor pressure will be smaller than that in equilibrium with free water. The curvature of drops is considered to be positive, as these drops are convex toward the atmosphere, whereas the curvature of capillary water menisci is considered negative, as they are concave toward the atmosphere.[8]

Water present in the soil invariably contains solutes, mainly electrolytic salts, in highly variable concentrations. Thus, soil water should properly be called the *soil solution*. The composition and concentration of the soil solution affect soil behavior. While in humid regions the soil solution may have a concentration of but a few parts per million, in arid regions the con-

---

[7] This is a simplified version of the Clausius–Clapeyron equation $dp/dT = \Delta H_v/T(\bar{v}_v - \bar{v}_l)$, where ΔH_v is the latent heat of vaporization, and $\bar{v}_v$, $\bar{v}_l$ are the specific volumes of the vapor and liquid, respectively. If the vapor behaves as an ideal gas, $\bar{v}_v = RT/p$, and hence $\ln p = \Delta H_v/RT + \text{const}$.

[8] For water in capillaries, in which the air–water interface is concave, the Kelvin equation applies:

$$-(\mu_1 - \mu_1^\circ) = RT \ln(p_1^\circ/p_1) = 2\gamma \bar{v}_1 \cos \alpha / r_c$$

in which $\mu_1 - \mu_1^\circ$ is the change in potential of the water due to the curvature of the air–water interface, γ the surface tension of water, α the contact angle, $\bar{v}_1$ the partial molar volume of water, and r_c the radius of the capillary.

The concept of water potential will be elucidated more fully in Chapter 7.

J. Vapor Pressure

Table 3.3

PHYSICAL PROPERTIES OF WATER VAPOR

Temperature (°C)	Saturation vapor pressure (torr)		Vapor density in saturated air (gm/m^3)		Diffusion coefficient (cm^2/sec)
	Over liquid	Over ice	Over liquid	Over ice	
−10	2.15	1.95	2.36	2.14	0.211
−5	3.16	3.01	3.41	3.25	—
0	4.58	4.58	4.85	4.85	0.226
5	6.53	—	6.80	—	—
10	9.20	—	9.40	—	0.241
15	12.78	—	12.85	—	—
20	17.52	—	17.30	—	0.257
25	23.75	—	23.05	—	—
30	31.82	—	30.38	—	0.273
35	42.20	—	39.63	—	—
40	55.30	—	51.1	—	0.289
45	71.90	—	65.6	—	—
50	92.50	—	83.2	—	—

centration may become as high as several percent. The ions commonly present are H^+, Ca^{2+}, Mg^{2+}, Na^+, K^+, NH_4^+, OH^-, Cl^-, HCO_3^-, NO_3^-, SO_4^{2-}, and CO_3^{2-}. Since the vapor pressure of electrolytic solutions is lower than that of pure water,[9] soil water also has a lower vapor pressure, even when the soil is saturated. In an unsaturated soil, the capillary and adsorptive effects further lower the potential and the vapor pressure, as will be shown in the next chapter.

Vapor pressure can be expressed in units of dynes per square centimeter, or bars, or millimeters of mercury, or in other convenient pressure units. The *vapor content* of the atmosphere can also be expressed in units of *relative humidity* (the ratio of the existing vapor pressure to the saturation vapor pressure at the same temperature), *vapor density* (the mass of water vapor per unit volume of the air), *specific humidity* of the air (the mass of water vapor per unit mass of the air), *saturation* (or vapor pressure) *deficit* (the difference between the existing vapor pressure and the saturation vapor

---

[9] The equation is

$$\bar{v}_1 \Pi_0 = RT \ln(p_1^\circ/p_1) = \mu_1 - \mu_1^\circ$$

where Π_0 is the osmotic pressure of a nonvolatile solute, μ_1° and p_1° are the chemical potential and vapor pressure of the liquid in its standard (pure) state, and μ_1 and p_1 are the same for the solution.

pressure at the same temperature), and *dew-point temperature* (the temperature at which the existing vapor pressure becomes equal to the saturation vapor pressure, i.e., the temperature at which a cooling body of air with a certain vapor content will begin to condense dew).

K. Surface Tension

Surface tension is a phenomenon occurring typically, but not exclusively, at the interface of a liquid and a gas. The liquid behaves as if it were covered by an elastic membrane in a constant state of tension which tends to cause the surface to contract. To be sure, no such membrane exists, yet the analogy is a useful one if not taken too literally. If we draw an arbitrary line of length L on a liquid surface, there will be a force F pulling the surface to the right of the line and an equal force pulling the surface leftwards. The ratio F/L is the surface tension and its dimensions are those of force per unit length (dynes per centimeter, or grams per second-squared). The same phenomenon can also be described in terms of energy. Increasing the surface area of a liquid requires the investment of energy, which remains stored in the enlarged surface, just as energy can be stored in a stretched spring, and which can perform work if the enlarged surface is allowed to contract again. Energy per unit area has the same dimensions as force per unit length (ergs per square centimeter or grams per second-squared).

An explanation for occurrence of surface tension is given in Fig. 3.8. Molecule A inside the liquid is attracted in all directions by equal cohesive forces, while molecule B at the surface of the liquid is attracted into the denser liquid phase by a force greater than the force attracting it into the much less dense gaseous phase. This unbalanced force draws the surface

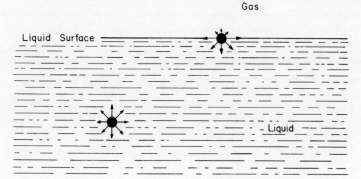

Fig. 3.8. Cohesive forces acting on a molecule inside the liquid and at its surface.

K. Surface Tension

molecules inward into the liquid and results in the tendency for the surface to contract.

As we shall see later, surface tension is associated with the phenomenon of capillarity. When the interface of the liquid and the gas is not planar but curved (concave or convex), a pressure difference between the two phases is indicated, since the surface-tension forces have a resultant normal to the surface, which, in equilibrium, must be counteracted by a pressure difference across the interface. If we stretch a rubber membrane as a boundary between two air cells of different pressure, this membrane will bulge into the side having the lower pressure. Similarly, a liquid with an interface which is convex toward the atmosphere is under a pressure greater than atmospheric; a liquid with an interface concave toward the gaseous phase is at a pressure smaller than atmospheric; and a liquid with a flat interface is at the same pressure as the atmosphere.[10]

Different liquids differ in their surface tension, as illustrated in the following list:

Water, 72.7 dyn/cm (at 20°C);
Ethyl ether, 17 dyn/cm;
Ethyl alcohol, 22 dyn/cm;
Benzene, 29 dyn/cm;
Mercury, 430 dyn/cm.

Surface tension also depends upon temperature, generally decreasing almost linearly as the temperature rises. Thermal expansion tends to decrease the density of the liquid, and therefore to reduce the cohesive forces at the surface as well as inside the liquid phase. The decrease of surface tension is accompanied by an increase in vapor pressure.

Soluble substances can influence surface tension in either direction. If the affinity of the solute molecules or ions to water molecules is greater than the affinity of the water molecules to one another, then the solute tends to be drawn into the solution and to cause an increase in the surface tension (e.g., electrolytes).[11] If, on the other hand, the cohesive attraction between water molecules is greater than their attraction to the solute molecules, then the latter tend to be relegated or concentrated more toward the surface, reducing its tension (e.g., many organic solutes, particularly detergents).

---

[10] An important difference between a rubber membrane and a liquid surface is that the former increases its tension as it is stretched and reduces its tension as it is allowed to contract, while the liquid surface retains a constant surface tension regardless of curvature.

[11] For example, a 1.0% NaCl concentration increases the surface tension by 0.17 dyn/cm at 20°C.

L. Curvature of Water Surfaces and Hydrostatic Pressure

In order to illustrate the relationship between surface curvature and pressure (see Table 3.4), we shall carry out a hypothetical experiment, as illustrated in Fig. 3.9. This figure shows a bubble of gas A blown into a liquid B through a capillary C. If we neglect the influence of gravitation and the special conditions occurring at the edge of the tube, we can expect the bubble to be spherical (a shape that is obtained because it affords the smallest surface area for a given volume), with a radius R.

If we now add a small amount of gas by lowering the piston D under a pressure greater than that of the liquid by a magnitude ΔP, the radius of the bubble will increase to $R + dR$. This will in turn increase the surface area of the bubble by $4\pi(R + dR)^2 - 4\pi R^2 = 8\pi R \, dR$ (neglecting the second-order differential terms). Increasing the surface area of the bubble required the investment of work against the surface tension γ, and the amount of this work is $\gamma 8\pi R \, dR$. Simultaneously, we have increased the volume of the bubble by $\frac{4}{3}\pi(R + dR)^3 - \frac{4}{3}\pi R^3 = 4\pi R^2 \, dR$. This increase in volume against the incremental pressure involved work in the amount $\Delta P 4\pi R^2 \, dR$. The two expressions for the quantity of work performed must be equal; i.e., $\gamma 8\pi R \, dR = \Delta P 4\pi R^2 \, dR$. Therefore,

$$\Delta P = 2\gamma/R \tag{3.9}$$

Table 3.4

RELATION OF PRESSURE (OR TENSION) OF WATER UNDER CURVED SURFACES TO VAPOR PRESSURE[a]

Radius of curvature (cm)	Hydrostatic pressure (bar)	Height of capillary rise (cm)	Relative vapor pressure at 15°C
10^{-6}	1.5×10^2	-1.5×10^5	1.114
10^{-5}	1.5×10	-1.5×10^4	1.011
10^{-4}	1.5	-1.5×10^3	1.001
10^{-3}	1.5×10^{-1}	-1.5×10^2	1.0001
—	0	0	1.0000
-10^{-1}	-1.5×10^{-3}	1.5	1.0000
-10^{-2}	-1.5×10^{-2}	1.5×10	1.0000
-10^{-3}	-1.5×10^{-1}	1.5×10^2	1.0000
-10^{-4}	-1.5	1.5×10^3	0.9989
-10^{-5}	-1.5×10	1.5×10^4	0.9890
-10^{-6}	-1.5×10^2	1.5×10^5	0.8954
-10^{-7}	-1.5×10^3	1.5×10^6	0.3305
-10^{-8}	-1.5×10^4	1.5×10^7	0.000016

[a] Negative values of capillary rise represent capillary depression (e.g., mercury in glass capillaries).

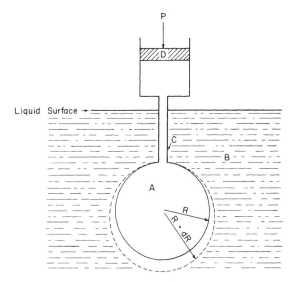

Fig. 3.9. A trial illustrating the relation between surface tension, radius of curvature, and bubble pressure.

This important equation shows that the difference ΔP between the pressure of the bubble and the pressure of the water surrounding it is directly proportional to the surface tension and inversely proportional to the radius of the bubble. Thus, the smaller the bubble is, the greater is its pressure.[12]

If the bubble is not spherical, then instead of Eq. (3.9) we will obtain

$$\Delta P = \gamma(1/R_1 + 1/R_2) \tag{3.10}$$

where R_1 and R_2 are the principal radii of curvature for a given point on the interface. This equation reduces to the previous one whenever $R_1 = R_2$.

M. Contact Angle of Water on Solid Surfaces

If we place a drop of liquid upon a dry solid surface, the liquid will usually displace the gas which covered the surface of the solid and spread over that

---

[12] The reader is invited to ponder the fact that, in the experiment described in Fig. 3.9, blowing additional air into the bubble by lowering the piston results in a decrease, not an increase, of internal pressure of the bubble. And what should happen if a larger bubble were to meet a smaller one? Social justice would demand that the larger one contribute some of its gas to the smaller one until the two become equal. Alas, nature is heartless. In fact, the smaller bubble, having greater pressure, is bound to empty into the larger one. Thus, the rich get richer and the poor poorer.... The reader can verify or refute this prediction by close observation of his next glass of soda (provided it contains nothing else that might becloud his vision).

surface to a certain extent. Where its spreading will cease and the edge of the drop will come to rest, its interface with the gas will form a typical angle with its interface with the solid. This angle, termed contact angle, is illustrated in Fig. 3.10.

Viewed two dimensionally on a cross-sectional plane, the three phases meet at a point A and form three angles with the sum of 360°. If we assume the angle in the solid to be 180°, and if we designate the angle in the liquid as x, the angle in the gaseous phase will be $180° - x$.

We can perhaps simplify the matter by stating that, if the adhesive forces between the solid and liquid are greater than the cohesive forces inside the liquid itself and greater than the forces of attraction between the gas and solid, then the solid–liquid contact angle will tend to be acute and the liquid will wet the solid. A contact angle of zero would mean the complete flattening of the drop and perfect wetting of the solid surface by the liquid. It would be as though the solid surface had an absolute preference for the liquid over the gas. A contact angle of 180° (if it were possible) would mean a complete nonwetting or rejection of the liquid by the gas-covered solid; i.e., the drop would retain its spherical shape without spreading over the surface at all (assuming no gravity effect). Surfaces on which water exhibits an obtuse contact angle are called *water repellent*, or *hydrophobic* (Greek: water hating).

In order for a drop resting on a solid surface to be in equilibrium with that surface and with a gas phase, the vector sum of the three forces arising from the three types of surface tension present must be zero. On the solid surface drawn in Fig. 3.11 the sum of the forces pulling leftward at the edge of the drop must equal the sum of the forces pulling to the right:

$$\gamma_{gs} = \gamma_{sl} + \gamma_{lg} \cos \alpha$$

and therefore

$$\cos \alpha = (\gamma_{gs} - \gamma_{sl})/\gamma_{lg} \tag{3.11}$$

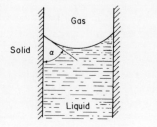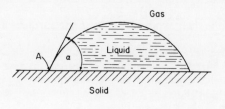

Fig. 3.10. The contact angle of a meniscus in a capillary tube and of a drop resing upon a plane solid surface.

N. Capillarity

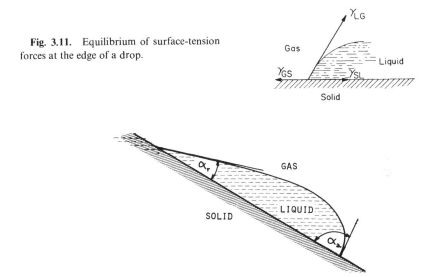

Fig. 3.11. Equilibrium of surface-tension forces at the edge of a drop.

Fig. 3.12. Hypothetical representation of a drop resting on an inclined surface. The contact angle α_a at the advancing edge of the drop is shown to be greater than the corresponding angle α_r at the receding edge.

where γ_{sl} is surface tension between solid and the liquid, γ_{gs} is surface tension between gas and solid, and γ_{lg} is the surface tension between liquid and gas. Each of these surface tensions tends to decrease its own interface. Reducing the interfacial tensions γ_{lg} and γ_{sl} (as with the aid of a detergent) can increase $\cos \alpha$ and decrease the contact angle α, thus promoting the wetting of the solid surface by the liquid.

The contact angle of a given liquid on a given solid is generally characteristic of their interaction under given physical conditions. This angle, however, can be different in the case of a liquid that is advancing upon the solids (the wetting or advancing angle) than for a liquid that is receding upon the solid surface (the retreating or receding angle). The wetting angle of pure water upon clean and smooth inorganic surfaces is generally zero, but where the surface is rough or coated with adsorbed surfactants of a hydrophobic nature, the contact angle, and especially the wetting angle, can be considerably greater than zero. This is illustrated in Fig. 3.12.

N. Capillarity

A capillary tube dipped in a body of free water will form a meniscus as the result of the contact angle of water with the walls of the tube. The curvature

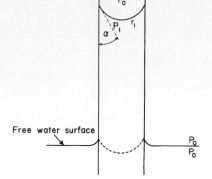

Fig. 3.13. Capillary rise.

of this meniscus will be greater (i.e., the radius of curvature smaller) the narrower the tube. The occurrence of curvature causes a pressure difference to develop across the liquid–gas interface. A liquid with an acute contact angle (e.g., water on glass) will form a meniscus concave toward the air, and therefore the liquid pressure under the meniscus P_1 will be smaller than the atmospheric pressure P_0 (Fig. 3.13). For this reason, the water inside the tube, and the meniscus, will be driven up the tube from its initial location (shown as a dashed curve in Fig. 3.13) by the greater pressure of the free water[13] outside the tube at the same level, until the initial pressure difference between the water inside the tube and the water under the flat surface outside the tube is entirely countered by the hydrostatic pressure of the water column in the capillary tube.

In a cylindrical capillary tube, the meniscus assumes a spherical shape. When the contact angle of the liquid on the walls of the tube is zero, the meniscus is a hemisphere (and in two-dimensional drawing can be represented as a semicircle) with its radius of curvature equal to the radius of the capillary tube. If, on the other hand, the liquid contacts the tube at an angle greater than zero but smaller than 90°, then the diameter of the tube $(2r)$ is the length of a chord cutting a section of a circle with an angle of π-2α, as shown in Fig. 3.14. Thus,

$$R = r/\cos \alpha \tag{3.12}$$

where R is the radius of curvature of the meniscus, r the radius of the capillary, and α the contact angle.

---

[13] By free water, we refer to water at atmospheric pressure, under a horizontal air–water interface. This is in contrast with water that is constrained by capillarity or adsorption, and is at an equivalent pressure smaller than atmospheric (i.e., suction or tension).

O. Density and Compressibility

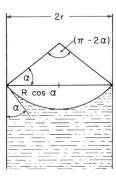

Fig. 3.14. The geometric relation of the radius of curvature R to the radius of the capillary r and the contact angle α.

The pressure difference ΔP between the capillary water (under the meniscus) and the atmosphere, therefore, is

$$\Delta P = (2\gamma \cos \alpha)/r \tag{3.13}$$

Recalling that hydrostatic pressure is proportional to the depth d below the free water surface (i.e., $P = \rho g d$) we can infer that hydrostatic tension (negative pressure) in a capillary tube is proportional to the height h above the free water surface. Hence the height of capillary rise is

$$h_c = (2\gamma \cos \alpha)/g(\rho_l - \rho_g)r \tag{3.14}$$

where ρ_g is the density of the gas (which is generally neglected), ρ_l the density of the liquid, g the acceleration of gravity, r the capillary radius, α the contact angle, and γ the surface tension between the liquid and the air.

When the liquid surface is concave, the center of curvature lies outside the liquid and the curvature, by convention, is regarded as negative. Thus, for a concave meniscus such as that of water in a clean glass capillary, P is negative with reference to the atmosphere, indicating a capillary pressure deficit, or subpressure, called tension. For a convex meniscus (such as that of mercury in glass, or of water in an oily or otherwise water-repellent tube), P is positive and capillary depression, rather than capillary rise, will result.

O. Density and Compressibility

The open packing of water molecules in ice and liquid water accounts for their relatively low densities. If the molecules were close packed, the hypothetical density of water would be nearly 2 gm/cm³, i.e., approximately double what it is in reality. Unlike most substances, water exhibits a point of maximum density (at 4°C) below which the substance expands due to the formation of the hexagonal lattice structure, and above which the expansion is due to the increasing thermal motion of the molecules. The coefficient of thermal expansion of water is rather low, and in the normal temperature range of, say, 4–50°C, the density decreases only slightly from 1.000 to 0.988 gm/cm³. This change is generally considered negligible.

The compressibility of water C_w can be defined as the relative change in density with change in pressure:

$$C_w = \frac{1}{\rho_w} \frac{\partial \rho_w}{\partial P} \tag{3.15}$$

At 20°C and at atmospheric pressure, the compressibility of pure water is about 4.6×10^{-11} cm²/dyn. In soil–water relationships, water can usually be taken to be incompressible. This approximation should not be carried too far, however. The compression of water cannot be ignored, for instance, in the case of confined aquifers, which may be subject to a pressure of, say, 100 atm or more.

P. Viscosity

When a fluid is moved in shear (that is to say, when adjacent layers of fluid are made to slide over each other), the force required is proportional to the velocity of shear. The proportionality factor is called the *viscosity*.

As such, it is the property of the fluid to resist the rate of shearing and can be visualized as an internal friction. The coefficient of viscosity η is defined as the force per unit area necessary to maintain a velocity difference of 1 cm/sec between two parallel layers of fluid which are 1 cm apart.

The viscosity equation is

$$\tau = F_s/A = \eta \, du/dx \tag{3.16}$$

where τ is the shearing stress, consisting of a force F_s acting on an area A; η [dimensions: mass/(length × time)] is the coefficient of viscosity; and du/dx is the velocity gradient perpendicular to the stressed area A.

The ratio of the viscosity to the density of the fluid is called the kinematic viscosity, designated v. It expresses the shearing-rate resistance of a fluid mass independently of the density. Thus, while the viscosity of water exceeds that of air by a factor of about 50 (at room temperature), its kinematic viscosity is actually lower.

Fluids of lower viscosity flow more readily and are said to be of greater fluidity (which is the reciprocal of viscosity). As shown in Table 3.5, the viscosity of water decreases by about 3% per 1°C rise in temperature, and thus decreases by half as the temperature increases from 5 to 35°C. The viscosity is also affected by type and concentration of solutes.

O. Density and Compressibility

Table 3.5

PHYSICAL PROPERTIES OF LIQUID WATER

Temperature (°C)	Density (gm/cm^3)	Specific heat (cal/gm deg)	Latent heat (vaporization) (cal/gm)	Surface tension (gm/sec^2)	Thermal conductivity (cal/cm sec deg) × 10^{-3}	Viscosity (gm/cm sec) × 10^{-2}	Kinematic viscosity (cm^2/sec)
−10	0.99794	1.02	603.0	—	—	—	—
−5	0.99918	1.01	600.0	76.4	—	—	—
0	0.99987	1.007	597.3	75.6	1.34	1.787	0.0179
4	1.00000	1.005	595.1	75.0	1.36	1.567	0.0157
5	0.99999	1.004	594.5	74.8	1.37	1.519	0.0152
10	0.99973	1.001	591.7	74.2	1.40	1.307	0.0131
15	0.99913	1.000	588.9	73.4	1.42	1.139	0.0114
20	0.99823	0.999	586.0	72.7	1.44	1.002	0.01007
25	0.99708	0.998	583.2	71.9	1.46	0.890	0.00897
30	0.99568	0.998	580.4	71.1	1.48	0.798	0.00804
35	0.99406	0.998	577.6	70.3	1.50	0.719	0.00733
40	0.99225	0.998	574.7	69.5	1.51	0.653	0.00661
45	0.99024	0.998	571.9	68.7	1.53	0.596	0.00609
50	0.98807	0.999	569.0	67.9	1.54	0.547	0.00556

An additional discussion of viscosity is included in Chapter 8 in connection with the flow of water in capillary tubes.

Sample Problems

1. Calculate the number of H^+ and OH^- ions present in 100 ml of a solution with a pH value of 5.0 (25°C.)

$$pH = \log_{10}(1/[H^+]) = -\log_{10}[H^+]$$

In this case

$$\log_{10}[H^+] = -5, \quad [H^+] = 10^{-pH}, \quad [H^+] = 10^{-5} \text{ mole/liter}$$

To calculate OH^- ion concentration

$$K_w = [H^+][OH^-] = 10^{-14} \text{ mole/liter},$$
$$[OH^-] = 10^{-14}/10^{-5} = 10^{-9} \text{ mole/liter}$$

The number of moles per 100 ml is

$$[H^+] = 10^{-5} \times 10^{-1} \text{ liter} = 10^{-6} \text{ mole},$$
$$[OH^-] = 10^{-9} \times 10^{-1} \text{ liter} = 10^{-10} \text{ mole}$$

and the number of ions

Avogadro's number $= 6.023 \times 10^{23}$
 $=$ number of H^+ or OH^- ions per mole of ions
$[H^+] = 10^{-6} \times 6.023 \times 10^{23} = 6.023 \times 10^{17}$ ions in 100 ml of solution
$[OH^-] = 10^{-10} \times 6.023 \times 10^{-23} = 6.023 \times 10^{13}$ ions in 100 ml of solution

2. At 25°C and at the partial pressure of CO_2 in normal air, the concentration of CO_2 dissolved in water at equilibrium with the atmosphere is $1.0 \times 10^{-5} M$. What is the pH of CO_2-saturated water (e.g., rainfall)?

All CO_2 dissolved in the water is considered to form carbonic acid according to the equation

$$CO_2(\text{gas}) + H_2O \rightleftharpoons CO_2(\text{aqueous}) \rightarrow H_2CO_3$$

Carbonic acid dissociates as follows:

$$H_2CO_3 \rightleftharpoons H^+ + HCO_3^-$$

Assuming the value of $K = 4.2 \times 10^{-7}$ for the appropriate ion product K,

$$4.2 \times 10^{-7} = [H^+][HCO_3^-]/[H_2CO_3]$$

Assuming, further, that $[H_2CO_3] = 1.0 \times 10^{-5} M$ (actually, only about 0.25% of the dissolved CO_2 forms H_2CO_3 but the value of K used here takes

O. Density and Compressibility

this into account and allows the calculation to be made on the assumption of complete conversion to H_2CO_3), and neglecting the self-ionization of water, we have

$$[H^+] = [HCO_3^-], \quad 4.2 \times 10^{-7} = [H^+]^2/1.0 \times 10^{-5},$$
$$[H^+]^2 = 4.2 \times 10^{-12}, \quad [H^+] = 2.0 \times 10^{-6}$$
$$pH = -\log(2.0 \times 10^{-6}) = -(0.3 - 6.0) = 5.7$$

This example illustrates one reason why humid region soils tend to be acidic. Another reason is the acidity of organic matter and of some types of rocks. Incidentally, the pH of rainfall in industrial areas can be as low as 3.5, owing to industrial emissions of nitrogen and sulfur oxides.

3. Calculate the osmotic pressure of a $0.01 M$ solution of sodium chloride at 20°C, assuming complete dissociation into Na^+ and Cl^- ions.

$$\Pi = MRT = (2 \times 0.01 \text{ mole/liter})(0.08205 \text{ liter atm/mole °K})(293°K)$$
$$= 0.48 \text{ atm}$$

4. A liter of water at 25°C dissolves 0.0283 liter of oxygen when the pressure of oxygen in equilibrium with the solution is 1 atm. From this we can find the proportionality constant s in Henry's law with $P_{O_2} = 1$ atm, as follows.

Restating Henry's law,

$$c = s(P_{O_2}/P_{total})$$

(where c is the amount of gas dissolved at equilibrium with a partial pressure P_{O_2}/P_{total}) we can write

$$s = c/P_{O_2}/P_{total} = 0.0283/1 = 0.0283 \text{ liter/liter atm}$$

With the prevailing pressure of oxygen in normal dry air equal to 159 mm Hg,

$$c = s(P_{O_2}/P_{total}) = (0.0283)(159/760) = 0.00592 \text{ liter/liter } H_2O$$
$$= 5.92 \text{ ml/liter}$$

5. Calculate the hydrostatic pressure of water in raindrops of the following diameters: (a) 5, (b) 1, (c) and 0.2 mm. Assume a temperature of 4°C. What are the corresponding pressures at 25°C? The equation is $\Delta P = 2\gamma/r$.

At 4°C the surface tension γ is equal to 75.0 dyn/cm (Table 3.5).

(a) $\Delta P = 2 \times 75 \text{ dyn/cm}/(0.25 \text{ cm}) = 600 \text{ dyn/sm}^2 = 0.6 \text{ mbar}$.
(b) $\Delta P = 2 \times 75 \text{ dyn/cm}/(0.05 \text{ cm}) = 3000 \text{ dyn/cm}^2 = 3.0 \text{ mbar}$.
(c) $\Delta P = 2 \times 75 \text{ dyn/cm}/(0.01 \text{ cm}) = 1500 \text{ dyn/cm}^2 = 15.0 \text{ mbar}$.

At 25°C the surface tension of pure water is 71.9 dyn/cm.

(a) $\Delta P = 2 \times 71.9$ dyn/cm/(0.25 cm) = 575.2 dyn/cm² = 0.575 mbar.
(b) $\Delta P = 2 \times 71.9$ dyn/cm/(0.05 cm) = 2876 dyn/cm² = 2.876 mbar.
(c) $\Delta P = 2 \times 71.9$ dyn/cm/(0.01 cm) = 14380 dyn/cm² = 14.38 mbar.

We note that the hydrostatic pressure in drops is greatly influenced by drop size but not much influenced by temperature.

6. Calculate the equilibrium height of capillary rise of liquids in glass cylindrical capillary tubes of the following diameters: (a) 2 mm; (b) 0.5 mm; (c) 1 mm. Disregard density of atmosphere.

Water at 20°C: Surface tension = 72.7 dyn/cm, contact angle = 0°, density = 1 gm/cm³.

(a) $r = 0.1$ cm:

$$\Delta h = (2\gamma \cos \alpha)/(\rho_w g r)$$
$$= (2 \times 72.7 \text{ dyn/cm} \times 1)/(1 \text{ gm/cm}^3 \times 980.7 \text{ cm/sec}^2 \times 0.1 \text{ cm})$$
$$= 1.48 \text{ cm}$$

(b) $r = 0.025$ cm:

$$\Delta h = (2 \times 72.7 \times 1)/(1 \times 980.7 \times 0.025) = 5.92 \text{ cm}$$

(c) $r = 0.005$ cm:

$$\Delta h = (2 \times 72.7 \times 1)/(1 \times 980.7 \times 0.005) = 29.6 \text{ cm}$$

Mercury at 20°C: Surface tension = 430 dyn/cm, contact angle = 180°, density = 13.6 gm/cm³.

(a) $r = 0.1$ cm:

$$\Delta h = [2 \times 430 \text{ dyn/cm} \times (-1)]/(13.6 \text{ gm/cm}^3 \times 980.7 \text{ cm/sec}^2 \times 0.1 \text{ cm})$$
$$= -0.64 \text{ cm} \quad \text{(capillary depression)}$$

(b) $r = 0.025$ cm:

$$\Delta h = [2 \times 430 \times (-1)]/(13.6 \times 980.7 \times 0.025) = -2.58 \text{ cm}$$

(c) $r = 0.005$ cm:

$$\Delta h = [2 \times 430 \times (-1)]/(13.6 \times 980.7 \times 0.005) = -12.9 \text{ cm}$$

Part II:

THE SOLID PHASE

> God in the beginning formed matter in solid, massy, hard, impenetrable, movable particles, of such sizes and figures, and in such proportion to space, as most conducted to the end for which he formed them.
>
> Isaac Newton
> 1642–1727

4 Texture, Particle Size Distribution, and Specific Surface

A. Introduction

Having introduced the concept of the soil as a three-phase system, let us now take a closer look at the solid phase, which is, to begin with, the permanent component of the soil and the one which gives substance to the whole. Conceivably, one could have soil without air, or without water, and in a vacuum without both (as is the case with the "soil" found on the moon), but it would be difficult to imagine a soil in any circumstances without the solid phase. The material of which the soil solid phase is composed includes discrete mineral particles of various sizes, as well as amorphous compounds, with the latter generally attached to, and sometimes coating, the particles. As the content of the amorphous material, such as hydrated iron oxides and humus, is generally (though not invariably) small, we can in most cases represent the solid phase as consisting by and large of distinct particles, the largest among which are visible to the naked eye and the smallest of which are colloidal and can only be observed by means of an electron microscope.

In general, it is possible to separate soil particles into groups and to characterize the soil in terms of the relative proportions of its particle-size groups, which may differ from one another in mineral composition as well as in particle size. It is these attributes of the soil solid phase, particle size and mineral composition, which largely determine the behavior of the soil: its interactions with fluids and solutes, as well as its compressibility, strength, and thermal regime.

To characterize the soil material physically and quantitatively, we must define its pertinent and measurable properties. A distinction should be made in this context between static and dynamic soil properties. *Static properties* are intrinsic to the material itself and are unaffected by any external variables. *Dynamic properties*, on the other hand, are manifested in the response of the body to externally imposed effects, such as mechanical stresses tending to cause deformation and failure or the entry of water. In this chapter we shall concentrate our attention upon static properties of the soil solid phase, such as texture, particle size distribution, and specific surface, insofar as these are permanent and immutable attributes of the soil material capable of being measured objectively and having a bearing upon soil behavior.

B. Soil Texture

The term *soil texture* refers to the size range of particles in the soil, i.e., whether the particles of which a particular soil is composed are mainly large, small, or of some intermediate size or range of sizes. As such, the term carries both qualitative and quantitative connotations. Qualitatively, it represents the "feel" of the soil material, whether coarse and gritty or fine and smooth. An experienced soil classifier can tell, by kneading or rubbing the moistened soil with his fingers, whether it is coarse textured or fine textured and can also assess in a semiquantitative way to which of the several intermediate textural "classes" the particular soil might belong. In a more rigorously quantitative sense, however, the term soil texture denotes the measured distribution of particle sizes or the proportions of the various size ranges of particles which occur in a given soil. As such, soil texture is a permanent, natural attribute of the soil and the one most often used to characterize its physical makeup.

C. Textural Fractions (Separates)

The traditional method of characterizing particle sizes in soils is to divide the array of possible particle sizes into three conveniently separable size ranges known as *textural fractions* or *separates*, namely, *sand*, *silt*, and *clay*. The actual procedure of separating out these fractions and of measuring their proportions is called *mechanical analysis*, for which standard techniques have been devised. The results of this analysis yield the *mechanical composition* of the soil, a term which is often used interchangeably with soil texture.

Unfortunately, there is as yet no universally accepted scheme for classification of particle sizes, and the various criteria used by different workers are rather arbitrary. For instance, the classification standardized in America by the U.S. Department of Agriculture differs from that of the International Soil Science Society (ISSS), as well as from those promulgated by the American Society for Testing Materials (ASTM), the Massachusetts Institute of Technology (MIT), and various national institutes abroad. The classification followed by soil engineers often differs from that of agricultural soil scientists. The same terms are used to designate differing size ranges, an inconsistency which can result in considerable confusion.

A number of the often used particle size classification schemes are compared in Fig. 4.1.

In the first place, the problem is what should be the upper limit of particle size which can properly be included in the definition of soil material. Some soils contain large rocks which obviously do not behave like soil although, if numerous, might affect the behavior of the soil in bulk. It is more or less conventional to define *soil material* as particles smaller than 2 mm in diameter. Larger particles are generally referred to as *gravel*, and still larger rock fragments, several centimeters in diameter, are variously called *stones*, *cobbles*, or if very large, *boulders*. Where gravel and stones occupy enough of the soil's volume to influence soil physical processes significantly, their volume fraction and size range should be reported along with the specification of the finer soil material.

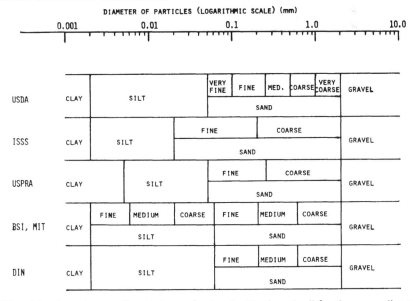

Fig. 4.1. Several conventional schemes for the classification of soil fractions according to particle diameter ranges: U.S. Department of Agriculture (USDA); International Soil Science Society (ISSS); U.S. Public Roads Administration (USPRA); German Standards (DIN); British Standards Institute (BSI); Massachusetts Institute of Technology (MIT).

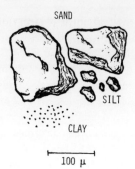

Fig. 4.2. A visual representation of the comparative sizes and shapes of sand, silt, and clay particles.

The largest group of particles generally recognized as soil material is *sand*, which is defined as particles ranging in diameter from 2000 μm (2 mm) down to 50 μm (USDA classification) or to 20 μm (ISSS classification). The sand fraction is often further subdivided into subfractions such as coarse, medium, and fine sand. Sand grains usually consist of quartz, but may also be fragments of feldspar, mica, and occasionally heavy minerals such as zircon, tourmaline, and hornblende, though the latter are rather rare. In most cases, sand grains have more or less uniform dimensions and can be represented as spherical, though they are not necessarily smooth and may in fact have quite jagged surfaces (Fig. 4.2), which, together with their hardness[1], account for their abrasiveness.

The next fraction is *silt*, which consists of particles intermediate in size between sand and *clay*, which, in turn, is the smallest sized fraction. Mineralogically and physically, silt particles generally resemble sand particles, but since they are smaller and have a greater surface area per unit mass and are often coated with strongly adherent clay, they may exhibit, to a limited degree, some of the physicochemical attributes of clay.

The clay fraction, with particles ranging from 2 μm downwards, is the colloidal fraction. Clay particles are characteristically platelike or needlelike in shape and generally belong to a group of minerals called the *aluminosilicates*. These are *secondary minerals*, formed in the soil itself in the course of its evolution from the *primary minerals* contained in the original rock. In some cases, however, the clay fraction may include considerable concentrations of fine particles which do not belong to the aluminosilicate clay mineral category, e.g., iron oxide or calcium carbonate.

Because of its far greater surface area per unit mass and its resulting physicochemical activity, clay is the decisive fraction which has the most influence on soil behavior. Clay particles adsorb water and hydrate, thereby

---

[1] Quartz particles have a hardness of 7 on the so-called *Mohs scale of hardness*, and will readily abrade steel (of hardness 5.5), as can be commonly observed with tillage implements.

D. Soil Classes

causing the soil to swell upon wetting and then shrink upon drying. Clay particles typically carry a negative charge and when hydrated form an *electrostatic double layer* with *exchangeable ions* in the surrounding solution. Another expression of surface activity is the heat which evolves when a dry clay is wetted, called the *heat of wetting*. A body of clay will typically exhibit plastic behavior and become sticky when moist and then cake up and crack to form cemented hard fragments when desiccated. (So important is the clay fraction, in fact, that we have devoted our entire next chapter to it.) The relatively inert sand and silt fractions can be called the "soil skeleton," while the clay, by analogy, can be thought of as the "flesh" of the soil. Together, all three fractions of the solid phase, as they are combined in various configurations, constitute the *matrix* of the soil.

The expressions light soil and heavy soil are used in common parlance to describe the physical behavior of sandy versus clayey soils. Since a sandy soil tends to be loose, well drained, well aerated, and easy to cultivate, it is called light. A clayey soil, on the other hand, tends to absorb and retain much more water and to become plastic and sticky when wet, as well as tight and cohesive when dry, and is thus difficult to cultivate. It is therefore called heavy. These can be misleading expressions, however, since in actual fact it is the coarse-textured soils which are generally more dense (i.e., have a lower porosity) than the fine-textured soils and thus are heavier, rather than lighter, in weight (at least in the dry state).

D. Soil Classes

The overall textural designation of a soil, called the *textural class*, is conventionally determined on the basis of the mass ratios of the three fractions. Soils with different proportions of sand, silt, and clay are assigned to different classes, as shown in the triangular diagram of Fig. 4.3. To illustrate the use of the textural triangle, let us assume that a soil is composed of 50% sand, 20% silt, and 30% clay. Note that the lower left apex of the triangle represents 100% sand and the right side of the triangle represents 0% sand. Now find the point of 50% sand on the bottom edge of the triangle and follow the diagonally leftward line rising from that point and parallel to the zero line for sand. Next, identify the 20% line for silt, which is parallel to the zero line for silt, namely, the left edge of the triangle. Where the two lines intersect each other, as well as the 30% line for clay, is the point we are seeking. In this particular example, it happens to fall within the realm of "sandy clay loam."

Note that a class of soils called *loam* occupies a rather central location in the textural triangle. It refers to a soil which contains a "balanced" mixture of coarse and fine particles, so that its properties are intermediate among

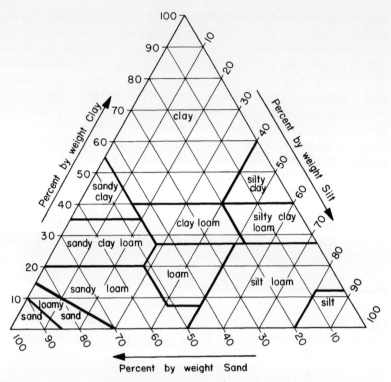

Fig. 4.3. Textural triangle, showing the percentages of clay (below 0.002 mm), silt (0.002–0.05 mm), and sand (0.05–2.0 mm) in the basic soil textural classes.

those of a sand, a silt, and a clay. As such, loam is often considered to be the optimal soil for plant growth and for agricultural production, as its capacity to retain water and nutrients is better than that of sand while its drainage, aeration, and tillage properties are more favorable than those of clay. This is an oversimplification, however, as under different environmental conditions and for different plant species a sand or a clay may be more suitable than a loam.

E. Particle Size Distribution

Any attempt to divide into distinct preconceived fractions what is usually in nature a continuous array of particle sizes is arbitrary to begin with, and the further classification of soils into discrete textural classes is doubly so. Although this approach is widely followed and evidently useful, it seems

E. Particle Size Distribution

better to measure and display the complete distribution of particle sizes. This method of representing soil texture is used mostly by soil engineers.

Figure 4.4 presents typical *particle-size distribution curves*. The ordinate of the graph indicates the percentage of soil particles with diameters smaller than the diameter denoted in the abscissa, which is drawn on a logarithmic scale to encompass several orders of magnitude of particle diameters while allowing sufficient space for the representation of the fine particles. Note that this graph gives an integral, or cumulative, representation. In practice, the particle-size distribution curve is constructed by connecting a series of n points, each expressing the cumulative fraction of particles finer than each of the n diameters measured $(F_1, F_2, \ldots, F_i, \ldots, F_n)$. Thus,

$$F_i = (M_s - \sum_1^i M_i)/M_s \qquad (4.1)$$

in which M_s is the total mass of the soil sample analyzed and $\sum M_i$ is the cumulative mass of particles finer than the ith diameter measured.

The information obtainable from this representation of particle-size distribution includes the diameter of the largest grains in the assemblage, and the grading pattern, i.e., whether the soil is composed of distinct groups of particles each of uniform size or whether it consists of a more or less continuous array of particle sizes. Soils which contain a preponderance of particles of one or several distinct sizes, indicating a steplike distribution curve, are called *poorly graded*. Soils with a flattened and smooth distribution curve (without apparent discontinuities) are called *well graded*.

This aspect of the particle size distribution can be expressed in terms of the

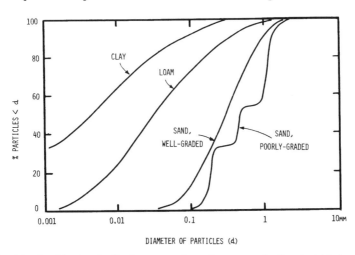

Fig. 4.4. Particle-size distribution curves for various types of soil material (schematic).

so-called *uniformity index* I_u, defined as the ratio of the diameter d_{60} which includes 60% of the particles to the smaller diameter d_{10} which includes 10% of the particles (as shown in Fig. 4.4). This index, also called the *uniformity coefficient*, is used mostly with coarse-grained soils. For a soil material consisting entirely of equal-sized particles, if such were to exist, I_u would be unity. Some sand deposits may have uniformity indexes smaller than 10. Some well-graded soils, on the other hand, have I_u values greater than 1000.

The particle size distribution curve can also be differentiated graphically to yield a frequency distribution curve for grain sizes, with a peak indicating the most prevalent grain size. Attempts have been made to correlate this index, as well as the harmonic mean diameter of grains, with various soil properties such as permeability.

F. Mechanical Analysis

Mechanical analysis is the procedure for determining the particle-size distribution of a soil sample. The first step in this procedure is to disperse the soil sample in an aqueous suspension. The primary soil particles, often naturally aggregated, must be separated and made discrete by removal of cementing agents (such as organic matter, calcium carbonate, or iron oxides) and by deflocculating the clay. Removal of organic matter is usually achieved by oxidation with hydrogen peroxide, and calcium carbonate can be dissolved by addition of hydrochloric acid. Deflocculation is carried out by means of a chemical *dispersing agent* (such as sodium metaphosphate) and by mechanical agitation (shaking, stirring, or ultrasonic vibration). The function of the dispersing agent is to replace the cations adsorbed to the clay, particularly divalent or trivalent cations, with sodium, which has the effect of increasing the hydration of the clay micelles, thus causing them to repel each other rather than coalesce, as they do in the flocculated state. Failure to disperse the soil completely will result in flocs of clay or aggregates settling as if they were silt-sized or sand-sized primary particles, thus biasing the results of mechanical analysis to indicate an apparent content of clay lower than the real value.

Actual separation of particles into size groups can be carried out by passing the suspension through graded sieves, down to a particle diameter of approximately 0.05 mm. To separate and classify still finer particles, the method of sedimentation is usually used, based on measuring the relative settling velocity of particles of various sizes from the aqueous suspension.

According to *Stokes' law*, the terminal velocity of a spherical particle settling under the influence of gravity in a fluid of a given density and viscosity is proportional to the square of the particle's radius. We shall now

F. Mechanical Analysis

proceed to derive this law, as it governs the method of sedimentation analysis.

A particle falling in a vacuum will encounter no resistance as it is accelerated by gravity and thus its velocity will increase as it falls. A particle falling in a fluid, on the other hand, will encounter a frictional resistance proportional to the product of its radius and velocity and to the viscosity of the fluid.

The resisting force due to friction F_r was shown by Stokes in 1851 to be

$$F_r = 6\pi\eta r u \tag{4.2}$$

where η is the viscosity of the fluid and r and u are the radius and velocity of the particle. Initially, as the particle begins its fall, its velocity increases. Eventually, a point is reached at which the increasing resistance force equals the constant downward force, and the particle then continues to fall without acceleration at a constant velocity known as the terminal velocity u_t.

The downward force due to gravity F_g is

$$F_g = \tfrac{4}{3}\pi r^3(\rho_s - \rho_f)g \tag{4.3}$$

where $\tfrac{4}{3}\pi r^3$ is the volume of the spherical particle, ρ_s is its density, ρ_f is the density of the fluid, and g is the acceleration of gravity.

Setting the two forces equal, we obtain Stokes' law:

$$u_t = \frac{2}{9}\frac{r^2 g}{\eta}(\rho_s - \rho_f) = \frac{d^2 g}{18\eta}(\rho_s - \rho_f) \tag{4.4}$$

where d is the diameter of the particle. Assuming that the terminal velocity is attained almost instantly, we can obtain the time t needed for the particle to fall through a height h:

$$t = 18h\eta/d^2 g(\rho_s - \rho_f) \tag{4.5}$$

Rearranging and solving for the particle diameter gives

$$d = [18h\eta/tg(\rho_s - \rho_f)]^{1/2} \tag{4.6}$$

Since all terms on the right-hand side of the equation, except t, are constants, we can combine them all into a single constant A and write

$$d = A/t^{1/2} \quad \text{or} \quad t = B/d^2 \tag{4.7}$$

where $B = A^2$.

One way of measuring particle-size distribution is to use a pipette to draw samples of known volume from a given depth in the suspension at regular times after sedimentation is begun. An alternative method is to use an *hydrometer* to measure the density of the suspension at a given depth as a function of time. With time this density decreases as the largest particles, and then progressively smaller ones, settle out of the region of the suspension being measured.

To visualize the process, let us focus upon some hypothetical plane at a depth h_i below the surface of the suspension. Using Eq. (4.5) we can calculate the time t_i necessary for all particles with a diameter equal to or greater than d_i to fall below the given plane. A moment's contemplation will convince the reader that the concentration of all smaller particles (with a velocity smaller than h_i/t_i) will have remained constant in the plane under consideration, since the number of these particles reaching the plane from above must be equal to the number falling through it. Therefore, measuring the concentration of particles d_i at this plane at time t_i allows us to calculate the total mass of these particles originally present in the suspension from the known volume of water in the cylinder. With the standardized *Bouyoucos hydrometer*, for instance, a settling time of about 40 sec is needed at 20°C to measure the concentration of clay and silt (all the sand having settled through), and a time of about 8 hr is needed to measure the clay content alone. Correction factors must be applied whenever differences in temperature, or particle density, or initial suspension concentration, arise.

The use of Stokes' law for measurement of particle sizes is dependent upon certain simplifying assumptions which may not be in accord with reality. Among these are the following:

(1) The particles are sufficiently large to be unaffected by the thermal (Brownian) motion of the fluid molecules;

(2) the particles are rigid, spherical, and smooth;

(3) all particles have the same density;

(4) the suspension is sufficiently dilute that particles do not interfere with one another and each settles independently;

(5) the flow of the fluid around the particles is laminar, i.e., no particle exceeds the critical velocity for the onset of turbulence.

In fact, we know that soil particles, while indeed rigid, are neither spherical nor smooth, and some may be platelike. Hence, the diameter calculated from the settlement velocity does not necessarily correspond to the actual dimensions of the particle. Rather, we should speak of an effective or equivalent settling diameter. The results of a mechanical analysis based on sieving may differ from those of a sedimentation analysis of the same particles. Moreover, soil particles are not all of the same density. Most silicates have values of 2.6–2.7 gm/cm^3, whereas iron oxides and other heavy minerals may have density values of 5 gm/cm^3 or even more. For all these reasons, the mechanical analysis of soils yields only approximate results. Its greatest shortcoming, however, is that it does not account for differences in *type* of clay, which can be of decisive importance in determining soil behavior.

G. Specific Surface

The specific surface of a soil material can variously be defined as the total surface area of particles per unit mass (a_m), or per unit volume of particles (a_v), or per unit bulk volume of the soil as a whole (a_b):

$$a_m = A_s/M_s \tag{4.8}$$

$$a_v = A_s/V_s \tag{4.9}$$

$$a_b = A_s/V_t \tag{4.10}$$

where A_s is the total surface area of a mass of particles M_s having a volume V_s and contained in a bulk volume V_t of soil.

Specific surface is commonly expressed in terms of square meters per gram, or per cubic centimeter, of particles. It depends in the first place upon the size of the particles. It also depends upon their shape. Flattened or elongated particles obviously expose greater surface per volume or per mass than do equidimensional (e.g., cubical or spherical) particles. Since clay particles are generally platy, they contribute even more to the overall specific surface area of a soil than is indicated by their small size alone. In addition to their external surfaces, certain types of clay crystals exhibit internal surface areas, such as those which form when the open lattice of montmorillonite expands on imbibing water. Whereas the specific surface of sand is often less than 1 m^2/gm, that of clay can be as high as several hundred square meters per gram. In fact, it is the clay fraction, its content and mineral composition, that largely determines the specific surface of a soil.

The specific surface of a soil material is a fundamental and intrinsic property which has been found to correlate with important phenomena such as cation exchange, retention and release of various chemicals (including nutrients and certain potential pollutants of the environment), swelling, retention of water, and such mechanical properties as plasticity, cohesion, and strength. Hence, it is a highly pertinent property to study, and its measurement can help provide a basis for evaluating and predicting soil behavior. It is probable that the measurement of soil specific surface, though not yet as common as the measurement of soil texture by the traditional methods, may eventually prove to be a more meaningful and pertinent index for characterizing a soil than are the percentages of sand, silt, and clay.

The usual procedure for determining surface area is to measure the amount of gas or liquid needed to form a *monomolecular layer* over the entire surface in a process of adsorption. The standard method is to use an inert gas such as nitrogen. Water vapor and organic liquids (e.g., glycerol and ethylene glycol) are also used.

The adsorption phenomenon was described by de Boer (1953). At low gas pressures, the amount of a gas adsorbed per unit area of adsorbing surface, σ_a, is related to the gas pressure P, the temperature T, and the heat of adsorption Q_a by the equation

$$\sigma_a = k_i P \exp(Q_a/RT) \tag{4.11}$$

where R is the gas constant and k_i is also a constant. Thus, the amount of adsorption increases with pressure, but decreases with temperature.

The *equation of Langmuir* (1918) indicates the relation between the gas pressure P and the volume of gas adsorbed per gram of adsorbent v at constant temperature:

$$P/v = 1/k_2 v_m + P/v_m \tag{4.12}$$

where v_m is the volume of adsorbed gas which forms a complete monomolecular layer over the adsorbent, and can be obtained by plotting P/v versus P. The specific surface of the adsorbent can then be calculated by determining the number of molecules in v_m and multiplying this by the cross-sectional area of these molecules. The Langmuir equation is based on the assumption that only one layer of molecules can be adsorbed, and that the heat of adsorption is uniform during the process.

Brunauer *et al.* (1938) derived what has come to be known as the *BET equation*, based on multilayer adsorption theory:

$$P/v(P_0 - P) = (1/v_m C) + (C - 1)P/v_m C P_0 \tag{4.13}$$

where v is the volume of gas adsorbed at pressure P, v_m is the volume of a single layer of adsorbed molecules over the entire surface of the adsorbent, P_0 is the gas pressure required for monolayer saturation at the temperature of the experiment, and C is a constant for the particular gas, adsorbent, and temperature. The volume v_m can be obtained from the BET theory by plotting $P/v(P_0 - P)$ versus P/P_0. The density of the adsorbed gas is usually assumed to be that of the liquefied or the solidified gas.

Polar adsorbents (such as water) may not obey the BET or Langmuir equations (which are similar at low pressures), since their molecules or ions may tend to cluster at charged sites rather than to spread out evenly over the adsorbent surface. The use of various adsorbents and techniques for the measurement of the specific surface area of soil materials was described by Mortland and Kemper (1965).

An estimation of the specific surface area can also be made by calculation based on the sizes and shapes of the particles. We shall proceed to give some examples of such calculations, for particles of definable geometry.

For a sphere of diameter d, the ratio of surface to volume is

$$a_v = \pi d^2/(\pi d^3/6) = 6/d \tag{4.14}$$

G. Specific Surface

and the ratio of surface to mass is

$$a_m = 6/\rho_s d \tag{4.15}$$

Where the particles have a density ρ_s of about 2.65 gm/cm^3, approximately,

$$a_m \approx 2.3/d \tag{4.16}$$

For a cube of edge L, the ratio of surface to volume is

$$a_v = 6L^2/L^3 = 6/L \tag{4.17}$$

and the ratio of surface to mass is, again,

$$a_m = 6/\rho_s L \tag{4.18}$$

Thus, the expressions for particles of nearly equal dimensions, such as most sand and silt grains, are similar, and knowledge of the particle size distribution can allow us to calculate the approximate specific surface by the summation equation:

$$a_m = (6/\rho_s)\sum c_i(d_i^2/d_i^3) = (6/\rho_s)\sum(c_i/d_i) \tag{4.19}$$

where c_i is the mass fraction of particles of average diameter d_i.

Now let us consider a platy particle. For the sake of argument, we can assume that our plate is square shaped, with sides L and thickness l. The surface-to-volume ratio is

$$a_v = (2L^2 + 4Ll)/L^2l \tag{4.20}$$

and the surface-to-mass ratio

$$a_m = 2(L + 2l)/\rho_s Ll \tag{4.21}$$

If the platelet is very thin, so that its thickness l is negligible compared to principal dimension L, and if $\rho_s = 2.65$ gm/cm^3, then

$$a_m \approx 2/\rho_s l \approx 0.75/l \quad \text{cm}^2/\text{gm} \tag{4.22}$$

Thus, the specific surface area of a clay can be estimated if the thickness of its platelets is known. For example, the thickness of a platelet of fully dispersed montmorillonite is approximately 10 Å, or 10^{-7} cm. Therefore, $a_m = 0.75/10^{-7}$, or 750 m^2/gm, which compares closely with the measured value of about 800 m^2/gm. Often, however, montmorillonite particles are not in a state of ultimate dispersion, and their platelets are several unit-layers thick. The average platelet thickness for illite clay is about 50 Å, and for kaolinite clays it is a few hundred angstroms. The measured specific surface area for kaolinite is about 15 m^2/gm, which is low among the clays. Although kaolinite seems to have a somewhat more active surface in that it is characterized by a higher density of electrostatic charges, by and large the activity of clay is more or less proportional to the specific surface.

Sample Problems

1. Determine the textural class designations for soils with the following distributions of particle sizes:

	<0.0002	0.0002–0.002	0.002–0.01	0.01–0.05	0.05–0.25	0.25–2.0 (mm)
(a)	5%	10%	20%	25%	20%	20%
(b)	6%	9%	30%	30%	15%	10%
(c)	10%	30%	30%	10%	10%	10%
(d)	4%	6%	10%	20%	30%	30%

Using the USDA classification (Fig. 4.1) and the textural triangle (Fig. 4.3), we obtain the following textural classes:

(a) % sand = 40, % silt = 45, % clay = 15. Soil class: loam.
(b) % sand = 25, % silt = 60, % clay = 15. Soil class: silt-loam.
(c) % sand = 20, % silt = 30, % clay = 50. Soil class: clay.
(d) % sand = 60, % silt = 30, % clay = 10. Soil class: sandy loam.

2. Using Stokes' law, calculate the time needed for all sand particles (diameter >50 μm) to settle out of a depth of 20 cm in an aqueous suspension at 30°C. How long would it take for all silt particles to settle out? How long for "coarse" clay (>1 μm)?

We use Eq. (4.5):

$$t = 18\, h\eta/d^2 g(\rho_s - \rho_f)$$

Substituting the appropriate values for depth h (20 cm), viscosity η (0.008 gm/cm sec, obtainable from Table 3.5, p. 49), particle diameter d (50 μm, 2 μm, and 1 μm for the lower limits of sand, silt, and coarse clay, respectively), gravitational acceleration g (981 cm/sec^2), average particle density ρ_s (2.65 gm/cm^3), and water density (1.0 gm/cm^3), we can write:

(a) For all sand to settle out, leaving only silt and clay in suspension:

$$t = \frac{18 \times 20 \times (8 \times 10^{-3})}{(50 \times 10^{-4})^2 \times 981 \times (2.65 - 1.0)} \cong 71 \text{ sec}.$$

(b) For all silt to settle out, leaving only clay in suspension:

$$t = \frac{18 \times 20 \times (8 \times 10^{-3})}{(2 \times 10^{-4})^2 \times 981 \times 1.65} \cong 44500 \text{ sec} = 12.36 \text{ hr}.$$

Sample Problems

(c) For all coarse clay to settle out, leaving only fine clay in suspension:

$$t = \frac{18 \times 20 \times (8 \times 10^{-3})}{(1 \times 10^{-4})^2 \times 981 \times 1.65} \cong 178000 \text{ sec} = 49.44 \text{ hr.}$$

3. Calculate the approximate specific surface of a sand composed of the following array of particle sizes:

Average diameter: 1 mm,   0.5 mm,   0.2 mm,   0.1 mm.
Percentage by mass: 40%,   30%,   20%,   10%.

We use Eq. (4.19):

$$a_m = \frac{6}{2.65} \sum \frac{c_i}{d_i}$$
$$= 2.264 \left(\frac{0.4}{0.1} + \frac{0.3}{0.05} + \frac{0.2}{0.02} + \frac{0.1}{0.01} \right) = 67.92 \text{ cm}^2/\text{gm}$$

Note that the smallest-diameter fraction, constituting only a tenth of the mixture's mass, accounts for a third of its specific surface ($2.264 \times 0.1/0.01 = 22.64 \text{ cm}^2/\text{gm}$).

4. Estimate the approximate specific surface (m^2/gm) of a soil composed of 10% coarse sand (average diameter 0.1 cm), 20% fine sand (average diameter 0.01 cm), 30% silt (average diameter 0.002 cm), 20% kaolinite clay (average platelet thickness 400 Å), 10% illite clay (average thickness 50 Å), and 10% montmorillonite (average thickness 10 Å).

For the sand and silt fractions, we use Eq. (4.19):

$$a_m = \frac{6}{2.65} \left(\frac{0.1}{0.1} + \frac{0.2}{0.01} + \frac{0.3}{0.002} \right) = 387 \text{ cm}^2/\text{gm} = 0.0387 \text{ m}^2/\text{gm}.$$

For the clay fraction, we use Eq. (4.22) in summation form to include the partial specific surface values for kaolinite, illite, and montmorillonite, respectively:

$$a_m = 0.2 \times 0.75/(400 \times 10^{-8}) + 0.1 \times 0.75/(50 \times 10^{-8})$$
$$+ 0.1 \times 0.75/(10 \times 10^{-8})$$
$$= 3.78 \text{ m}^2/\text{gm (kaol.)} + 15.09 \text{ m}^2/\text{gm (ill.)} + 75.45 \text{ m}^2/\text{gm (mont.)}$$
$$= 94.32 \text{ m}^2/\text{gm}$$

Total for the soil $= 0.0387 + 94.32 \cong 94.36 \text{ m}^2/\text{gm}$.

Note that the clay fraction, which constitutes 40% of the soil mass, accounts for 99.96% of the specific surface (i.e., 94.32/94.36). The montmorillonite constituent alone (10% of the mass) accounts for nearly 80% of the soil's specific surface.

> We know more about the movement of celestial bodies
> than about the soil underfoot.
> Leonardo da Vinci
> 1452–1519

5 Nature and Behavior of Clay

A. Introduction

The term *clay* carries several connotations, which are not necessarily mutually consistent. In daily language, it suggests a soil that tends to retain water and to become soft and moldable when wet. In a more exact sense, in the context of soil texture, it designates a range of particle sizes (namely, particles smaller than 2 μm), or a soil material in which this particle-size range predominates. Finally, in the mineralogical sense, it refers to a large group of minerals, many of which occur in the clay fraction of the soil. That fraction differs from the sand and silt fractions not only in particle-size range but also in mineralogical composition. Sand and silt consist mainly of weathering-resistant *primary minerals*, i.e., minerals present in the original rock from which the soil was formed, whereas clay includes *secondary minerals* formed in the soil by decomposition of the primary minerals and their recomposition into new ones.

The numerous *clay minerals* exhibit great variation in prevalence and properties, and in the way they affect soil behavior. Rarely do any of these minerals occur in homogeneous deposits, and in the soil they generally occur in mixtures, the specific composition of which depends in each case upon the nature of the soil forming processes.

To understand why and how the clay fraction serves as the active constituent of soils, we must consider the structure and function of clay minerals.

B. Structure of Clay

The forerunners of modern soil science assumed at first that clay consists of particles which are essentially similar to those of sand and silt, differing from them only in size. Later, they began to notice that when clays were dried from aqueous suspensions they tended to form thin flakes, and also that moist clay could be skimmed and polished to form a smooth and shiny surface. These observations suggested that clay particles may be flat and capable of being oriented in different ways, but it was not until the advent of x-ray diffraction, and later of electron microscopy, that the crystalline nature of clay minerals was proven and their structures were described.

The most prevalent minerals in the clay fraction of temperate region soils are the *silicate* clays, whereas in tropical regions hydrated oxides of iron and aluminum may be more prevalent. The typical *aluminosilicate clay minerals* appear as laminated microcrystals, composed primarily of two basic structural units, namely, a tetrahedron of four oxygen atoms surrounding a central cation, usually Si^{4+}, and an octahedron of six oxygen atoms or hydroxyls surrounding a larger cation of lesser valency, usually Al^{3+} or Mg^{2+}. These basic building blocks are illustrated in Fig. 5.1.

The tetrahedra are joined together at their basal corners by means of shared oxygen atoms, in an hexagonal network which forms a flat sheet only 4.93 Å thick. This is illustrated in Fig. 5.2. The octahedra are similarly joined along their edges to form a triangular array as shown in Fig. 5.3. These sheets are about 5.05 Å thick.

The layered aluminosilicate clay minerals are of two main types, depending upon the ratios of tetrahedral to octahedral sheets, whether 1:1 or 2:1 (2:2 minerals are also recognized). In 1:1 minerals like kaolinite, an octahedral sheet is attached by the sharing of oxygens to a single tetrahedral sheet. In 2:1 minerals like montmorillonite, it is attached in the same way to two tetrahedral sheets, one on each side. This shown in Fig. 5.4.

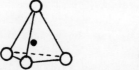

Fig. 5.1. The basic structural units of aliuminosilicate clay minerals: a tetrahedron of oxygen atoms surrounding a silicon ion (left), and an octahedron of oxygens or hydroxyls enclosing an aluminum ion (right).

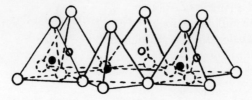

Fig. 5.2. Hexagonal network of tetrahedra forming a silica sheet.

B. Structure of Clay

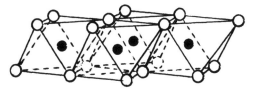

Fig. 5.3. Structural network of octahedra forming an alumina sheet.

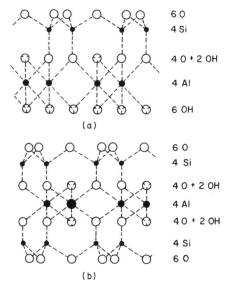

Fig. 5.4. Schematic representation of the structure of aluminosilicate minerals: (a) kaolinite and (b) montmorillonite.

A clay particle is composed of multiply stacked composite layers (or unit cells) of this sort, called *lamellae*.

The structure described is an idealized one. Typically, some substitutions of ions of approximately equal radii, called *isomorphous replacements*, occur during crystallization. In the tetrahedral sheets Al^{3+} may take the place of Si^{4+}, whereas in the octahedral layer Mg^{2+} can occasionally substitute for Al^{3+}. Consequently, internally unbalanced negative charges occur at different sites in the lamellae. Another source of unbalanced charge in clay crystals is the incomplete charge neutralization of terminal atoms on lattice edges. These unbalanced charges must be compensated externally by the adsorption of ions (mostly cations) which concentrate near the external surfaces of the clay particle and occasionally penetrate into the interlamellar spaces. The adsorbed cations, including Na^+, K^+, H^+, Mg^{2+}, Ca^{2+}, and Al^{3+}, are not an integral part of the lattice structure and can be replaced, or exchanged, by other cations in the soil solution. The cation exchange phenomenon is of great importance in the soil, as it affects the retention and release of nutrients and other salts, as well as the flocculation–dispersion processes of soil colloids.

C. The Principal Clay Minerals

Clay minerals are usually classified into two main groups, structured and amorphous. The structured clays are subclassified according to their internal crystal structure (i.e., the layered arrangement of tetrahedra and octahedra sheets) into two principal types 1:1 and 2:1 minerals. The 2:1 clay minerals are further divided into expanding and nonexpanding types. Finally, each of these types includes a number of specific minerals which can be identified on the basis of x ray, electron microscope, or thermal analysis techniques.

We shall now give a brief description of some of the prevalent clay minerals.

The most common mineral of the 1:1 type is *kaolinite*. Other minerals in the same group are *halloysite* and *dickite*. The basic layer in the crystal structure is a pair of silica–alumina sheets, and these are stacked in alternating fashion and held together by hydrogen bonding in a rigid, multilayered lattice which often forms an hexagonal platelet. Since water and ions cannot enter between the basic layers, these cannot ordinarily be split or separated. Moreover, since only the outer faces and edges of the platelets are exposed, kaolinite has a rather low specific surface. Kaolinite crystals generally range in planar diameter from 0.1 to 2 μm with a variable thickness in the range of about 0.02–0.05 μm. Owing to its relatively large particles and low specific surface, kaolinite exhibits less plasticity, cohesion, and swelling than most other clay minerals. The unit layer formula of kaolinite is $Al_4Si_4O_{10}(OH)_8$.

At the opposite end of the spectrum of aluminosilicate clay minerals is *montmorillonite*, a 2:1 mineral of the expanding type, which also includes *vermiculite* and *beidellite*. The lamellae of montmorillonite are stacked in loose assemblages called *tactoids*. Water and ions are drawn into the cleavage planes between the lamellae, and as the crystal expands like an accordion, it can readily be separated into flakelike thinner units and, ultimately, into individual lamellae, which are only 10 Å thick. As the montmorillonite crystals expand, their internal surfaces as well as external ones come into play, thus increasing the effective specific surface severalfold. Because of its tendency to expand and to disperse, montmorillonite exhibits pronounced swelling–shrinkage behavior, as well as high plasticity and cohesion. On drying, montmorillonitic soils, especially if dispersed, tend to crack and

C. The Principal Clay Minerals

form unusually hard clods. The unit layer formula of montmorillonite is $Al_{3.5}Mg_{0.5}Si_8O_{20}(OH)_4$.

A clay mineral intermediate in properties between kaolinite and montmorillonite is *illite*. It belongs to a group of clay minerals called *hydrous micas* which have a 2:1 silica–alumina ratio but are of the nonexpanding type. Isomorphous substitution of aluminum ions for silicon ions in the tetrahedral sheets (rather than substitutions in the octahedral sheets, as is the case in montmorillonite), to the extent of about 15%, accounts for the relatively high density of negative charges in these sheets. This, in turn, attracts potassium ions and "fixes" them tightly between adjacent lamellae. As a result, the layers are bound together so their separation, and hence expansion of the entire lattice, are effectively prevented. The unit layer formula of illite is $Al_4Si_7AlO_{20}(OH)_4K_{0.8}$, with the potassium occurring between the crystal units.

An example of a 2:2-type mineral is *chlorite*, in which magnesium rather than aluminum ions predominate in the octahedral sheets which are in combination with the tetrahedral silica sheets. Its unit layer formula is $Mg_6Si_6Al_2O_{20}(OH)_4$, with $Mg_6(OH)_{12}$ occurring between the layers. In behavior, chlorite resembles illite. An additional group of silicate clays, in which the lattice is continuous in only one direction, is known as *attapulgite* or *palygorskite*. The particles of this group are needlelike or tubelike, with microcavities providing internal surfaces.

Frequently, the various clay minerals do not occur separately but in a complex mixture. At times, even the internal structure is mixed or interstratified, giving rise to composite minerals which are somewhat loosely termed *bravaisite* (illite–montmorillonite, chlorite–illite, vermiculite–chlorite, etc.).

The clay fraction may also contain appreciable quantities of noncrystalline (amorphous) mineral colloids. *Allophanes*, for instance, are random combinations of poorly structured silica and alumina components expressible in the formula $Al_2O_3 \cdot 2SiO_2 \cdot H_2O$. The actual mole ratio of alumina to silica ranges in this group from 0.5 to 2.0. Phosphorus and iron oxides are frequently present in allophane. Notwithstanding its variable composition, allophane is sufficiently distinctive to be identified as a clay mineral. In behavior, this amorphous clay is similar to the crystalline clays in adsorption, ion exchange, and plasticity.

Still another constituent of the clay fraction is a group of hydrous oxides of iron and aluminum known as *sesquioxides*, which are prevalent mainly in the soils of tropical and subtropical regions and are responsible for the reddish or yellowish hue of these soils. Their composition can be formulated as $Fe_2O_3 \cdot nH_2O$ and $Al_2O_3 \cdot nH_2O$, in which the hydration ratio n is variable.

Table 5.1

TYPICAL PROPERTIES OF SELECTED CLAY MINERALS (APPROXIMATE VALUES)

Properties	Clay mineral				
	Kaolinite	Illite	Montmorillonite	Chlorite	Allophane
Planar diameter (μm)	0.1–4	0.1–2	0.01–1	0.1–2	
Basic layer thickness (Å)	7.2	10	10	14	
Particle thickness (Å)	500	50–300	10–100	100–1000	
Specific surface (m^2/gm)	5–20	80–120	700–800	80	
Cation exchange capacity (meq/100 gm)	3–15	15–40	80–100	20–40	40–70
Area per charge (Å^2)	25	50	100	50	120

Limonite and *goethite* are typical hydrated iron oxides, and *gibbsite* is a frequently encountered aluminum oxide. These sesquioxide clays are partly crystallized but often amorphous. Their electrostatic properties, adsorptive capacity, and plasticity are less pronounced than those of most of the silicate clays. Frequently, these oxides serve as cementing agents in the stabilization of soil aggregates, particularly in subtropical and tropical regions.

As this rather bewildering collection of mineral colloids suggests, the genesis of clay is a complex series of processes which differ from place to place. These processes range from a comparatively minor alteration of primary minerals (e.g., the transformation of muscovite mica into hydrous micas such as illite) to a thorough decomposition of primary minerals such as feldspar and subsequent recrystallization of kaolinite, for example.

Table 5.1 summarizes the properties of a few of the silicate clay minerals.

D. Humus: The Organic Constituent of Soil Colloids

Although the main topic of this chapter is the mineral colloidal matter known as clay, we are obliged to mention, at least briefly, the important fact that an entirely different kind of colloidal matter exists in soils, called *humus*. This generally dark-colored material, found mostly in the surface zone (the A horizon) of soils, is defined[1] as the more or less stable fraction of the soil organic matter remaining after the major portion of added plant and animal residues have decomposed. As such, humus does not include undecomposed or partially decomposed organic residues such as recent stubble or dead roots.

---

[1] "Glossary of Soil Science Terms," *Soil Sci. Soc. Am.*, Madison, Wisconsin (1970).

Like clay, humus particles are negatively charged. During hydration, each particle of humus forms a micelle and acts like a giant, composite anion, which adsorbs numerous cations. However, the cation exchange capacity of humus is much greater, per unit mass, than that of clay. Unlike most clay, moreover, humus is generally not crystalline but amorphous. As it is composed mostly of carbon, oxygen, and hydrogen, its charges are due not to isomorphous substitutions of cations but to the dissociation of carboxylic (—COOH) and phenolic (—O—OH) groups. Since the cation exchange process depends on replacement of the hydrogen in these groups, it is pH dependent, with the cation exchange capacity generally increasing at higher pH values.

Humus is not a single compound, nor does it have the same composition in different locations. Rather, it is a complex mixture of numerous compounds such as lignoproteins, polysaccharides, polyuronides, and others too varied to list. Furthermore, the organic colloids of humus, although more or less stable, are not nearly so stable as their mineral counterparts, and are in fact amenable to bacterial action, particularly if the soil's temperature, moisture, or aeration regimes are modified.

The content of humus in mineral soils varies from as high as 10% or even more in the top layer of *chernozem* (the typically black soils which occur in the American prairie and in the plains of Ukraine) down to nil in desert soils, and is of the order of 1–3% in many intermediate soils. The humus content generally decreases in depth through the B horizon, and becomes negligible at the bottom of the normal root zone, unless the soil is a deposit of alluvium with a high original content of humus. Organic soils such as peat or muck can contain considerably over 50% organic matter, though not all of that would fit the accepted definition of humus.

The importance of humus goes beyond its effect on cation adsorption or even plant nutrition. Humus often coagulates in association with clay and serves as a cementing agent, binding and stabilizing soil aggregates. This aspect will be discussed in Chapter 6 on soil structure.

E. The Electrostatic Double Layer

When a colloidal particle is more or less dry, the neutralizing counterions are attached to its surface. Upon wetting, however, some of the ions dissociate from the surface and enter into solution. A hydrated colloidal particle of clay or humus thus forms a *micelle*, in which the adsorbed ions are spatially separated, to a greater or lesser degree, from the negatively charged particle (Fig. 5.5). Together, the particle surface, acting as a multiple anion, and the "swarm" of cations hovering about it, form an *electrostatic double layer*.

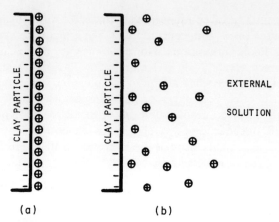

Fig. 5.5. Formation of a diffuse double layer in a hydrated micelle showing (a) the dry and (b) the hydrated state.

The swarm of adsorbed cations can be regarded as consisting of a layer more or less fixed in position at the immediate proximity of the particle surface and known as the *Stern layer* and of a diffuse assemblage of cations extending some distance away from the particle surface and gradually decreasing in concentration. This distribution is illustrated in Fig. 5.6. It results from an equilibrium between two opposing tendencies: (1) the electrostatic (*Coulombic*) attraction of the negatively charged surface for the positively charged ions, which tends to pull the cations inward so as to attain the minimum energy level, and (2) the kinetic (Brownian) motion of the liquid molecules, inducing the outward diffusion of the adsorbed cations in a tendency toward equalizing the concentration throughout the solution phase, thus maximizing entropy. The equilibrium distribution of cations is such as

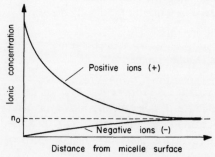

Fig. 5.6. The distribution of positive and negative ions in solution with distance from the surface of a clay micelle bearing net negative charge. Here n_0 is the ionic concentration in the bulk solution outside the electrical double layer.

E. The Electrostatic Double Layer

to minimize the *free energy* of the system as a whole. The actual concentration of cations inside the double layer can be 100 or even 1000 times greater than in the external, or intermicellar, solution (i.e., outside the range of influence of the particle).

As cations are adsorbed positively by the colloidal particle, anions are generally repelled, or adsorbed negatively, and thus are relegated from the micellar to the intermicellar solution. In some special cases, however, anions may also be attracted to specific sites on colloidal surfaces, but this phenomenon is much less prevalent than cation adsorption in soils.

Theoretical treatments of the electrostatic double layer have generally been based on the *Gouy–Chapman theory*, in which the negative charges are assumed to be constant and to be distributed uniformly over the particle surfaces (although in actual fact they originate within the crystal lattice). The strength of the surface charge is proportional to the charge density.

The effective thickness of the diffuse double layer can be estimated by means of the following equation:

$$z = \frac{1}{ev}\sqrt{\frac{\varepsilon kT}{8\pi n_0}} \quad (5.1)$$

in which z is the *characteristic length*, or extent, of the double layer, defined as the distance from the clay surface to where the ionic concentration is very nearly that of the external (intermicellar) solution; e is the elementary charge of an electron (4.77×10^{-10} esu); ε is the dielectric constant; k is the Boltzmann constant ($k = R/N$, where R is the gas constant and N is Avogadro's number); v is the valency of the ions in solution; n_0 is the concentration of the ions in the bulk solution (ions/cm^3); and T is the temperature (°K). A detailed mathematical analysis of the double layer was given by Overbeek (1952).

As the equation indicates, the diffuse double layer is compressed as the valency of the ions in solution is increased. For instance, if a solution of monovalent cations is replaced by a solution of divalent cations, the double layer becomes only half as thick (Fig. 5.7). The double-layer thickness is also affected by the solution's concentration (Fig. 5.8), as z of Eq. (5.1) is inversely proportional to the square root of n. Thus a tenfold increase of concentration will reduce the double layer to $1/\sqrt{10}$, i.e., to about $\frac{1}{3}$ of its previous thickness.

The above considerations do not account for interparticle interaction, i.e., the case where the diffuse ionic atmospheres of neighboring micelles intermingle. In this case, the concentration in the median plane, rather than in the external solution, is referred to, as calculated by Langmuir (1938), as

$$n_c = \pi^2/[v^2 B(d + x)^2 \times 10^{-16}] \quad (5.2)$$

Here n_c is cation concentration at the median plane (mole/liter), v is valence

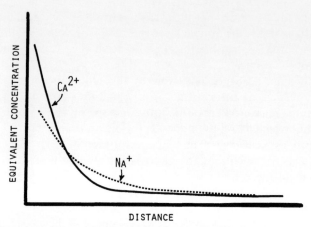

Fig. 5.7. Comparison of the distribution of strongly hydrated monovalent cation (Na) with that of a divalent cation (Ca) in the diffuse double layer. Because of its greater charge the divalent cation is attracted more strongly and held more closely to the particle surface. The area under each curve, representing milliequivalents of cations per unit surface, is equal.

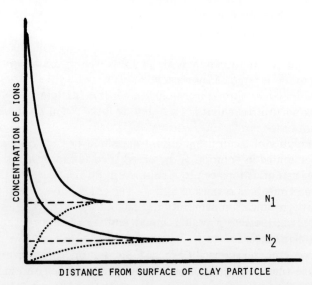

Fig. 5.8. Influence of ambient solution concentration (n_1, n_2) on the thickness of the diffuse double layer. The solid curves represent cations and the dotted curves anions. Increasing the concentration of the external solution is seen to compress the double layer.

F. Ion Exchange

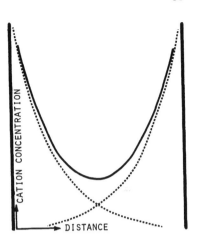

Fig. 5.9. Schematic illustration of the increased concentration of cations in the double layer due to interaction of adjacent particles. The dashed curves indicate the hypothetical distribution of cations for each of the two particles if it were suspended alone in the same ambient solution.

of the exchangeable cations, d is distance from either plate to the median plane (Å), x is a correction factor (1–4 Å), and B is a constant related to the temperature and dielectric constant (10^{15} cm/mmole). The distribution of cations between two charged plates is illustrated in Fig. 5.9.

F. Ion Exchange

The cations in the double layer can be replaced or exchanged by other cations introduced into the solution. Under chemically neutral conditions, the total number of exchangeable cation charges, expressed in terms of chemical equivalents per unit mass of soil particles, is nearly constant and independent of the species of cations present. It is thus considered to be an intrinsic property of the soil material, generally called the *cation exchange capacity* (or the *cation adsorption capacity*), and is commonly reported in milliequivalents of cations per 100 grams of soil. [The term "equivalent" is defined in chemistry as one gram atomic weight (mole) of hydrogen or the mass of any ion which can combine with or replace it in chemical reactions.] Thus expressed, the cation exchange capacity ranges from nil in sands to 100 meq/100 gm, or even more, in clays.

The cation exchange capacity depends not only on clay content but also on clay type (i.e., on specific surface and charge density), as indicated in Table 5.1. In the top layer of soils rich in organic matter, humus can account for as much as $\frac{1}{4}$ or $\frac{1}{3}$ of the exchange capacity. The cation exchange phenomenon affects the movement and retention of ions in the soil, which can be important in plant nutrition and in environmental processes involving the transport of pollutants. Cation exchange also affects the flocculation–dispersion processes of soil colloids, and hence the development and degradation of soil structure.

Because of differences in their valences, radii, and hydration properties, different cations are adsorbed with different degrees of tenacity or preference, and hence are more readily or less readily exchangeable. In general, the smaller the ionic radius and the greater the valence, the more closely and strongly is the ion adsorbed. On the other hand, the greater the ion's hydration, the farther it is from the adsorbing surface and the weaker its adsorption. Sodium ion, for example, has an atomic radius of only 0.98 Å when in the "naked" state, but it is strongly hydrated, and its effective radius increases eightfold when it is enveloped by water molecules. Monovalent cations, attracted by only a single charge, can be replaced more easily than divalent or trivalent cations. The order of preference in exchange reactions is as follows:

$$Al^{3+} \gg Ca^{2+} > Mg^{2+} \gg NH_4^+ > K^+ > H^+ > Na^+ > Li^+ \quad (5.3)$$

An example of an exchange reaction is the following

$$Na_2 \boxed{Clay} + Ca^{2+} \rightleftharpoons Ca \boxed{Clay} + 2Na^+ \quad (5.4)$$

In nature, soil material is seldom, if ever, adsorbed homogeneously with a single ionic species. Typically, the exchange capacity is taken up by several cations in varying proportions, all of which together constitute the so-called *exchange complex*. A typically heterogeneous exchange complex can be represented in the following way (with subscripts *a*, *b*, *c*, *d*, etc., indicating equivalent fractions of the cation exchange capacity, adding up to 100%):

In arid regions calcium and magnesium, and sometimes sodium, tend to predominate in the exchange complex. In humid regions, where soils are highly leached and often acid, hydrogen and aluminum ions, the latter often released from the clay crystal lattice under low pH conditions, play an important role. (We refer to hydrogen ions advisedly, knowing that H^+ as such may not exist independently but is probably associated with a water molecule, giving rise to the *hydronium ion* H_3O^+). The presence of exchangeable hydrogen is usually referred to as *base unsaturation* of the exchange complex, and its measure in equivalent terms for the soil mass of an agricul-

F. Ion Exchange

tural field is taken to be an indication of the amount of lime needed to neutralize the soil's reserve acidity.

The composition of the exchange complex depends on the concentration and ionic composition of the ambient solution. This dependence is expressed in the *Gapon equation*:

$$A_e/B_e = c([A_s]^{1/a}/[B_s]^{1/b}) \tag{5.5}$$

in which A and B are cations with valences of a and b, respectively, the subscript e indicates the concentration in the exchange complex, and the subscript s indicates the concentration in the ambient solution. The coefficient c depends somewhat on the nature of the charged surface and the cation adsorbed. This equation indicates that the adsorption mechanism favors cations of higher valency and that this preference increases as the solution becomes more dilute.

For the important case of calcium–sodium exchange, the *selectivity coefficient* (Shainberg, 1973), is given by

$$C_{Ca-Na} = n_0([Ca_e][Na_s]^2/[Na_e]^2[Ca_s]) \tag{5.6}$$

in which the subscript e denotes the ionic equivalent fractions of Ca and Na in the exchange complex, the subscript s denotes the mole fractions of these ions in the solution phase, and n_0 is the total molar concentration of the solution. The value of the selectivity coefficient for a wide range of exchangeable sodium fractions has been reported to be about 4 in a typical soil in Israel (Levy and Hillel, 1968). The use of this equation is illustrated in Sample Problem 4 at the end of this chapter.

For ordinarily encountered soil solution concentrations (in nonsaline soils) the high-valency cations predominate in the exchange complex if present in the soil solution in appreciable concentrations. In a mixed calcium–sodium system most of the calcium ions are usually adsorbed tightly in the Stern layer, whereas the sodium ions are relegated to the diffuse region of the double layer.

It should be noted that cation exchange reactions are rapid and reversible, and that therefore the composition of the exchange complex is highly dynamic as it responds to frequent changes in the composition of the soil solution. The composition of the soil's exchange complex in turn governs the soil's pH, as well as its swelling and flocculation–dispersion tendencies.

Anion exchange, as well as cation exchange, is known to occur in certain soils, and is usually attributed to the exposure of positive charges at the edges of clay crystals (as opposed to their faces), where the lattice bonds are broken. This has been noticed specifically in kaolinite. Anion exchange can be important in the retention of phosphate in soils. Organic matter (humus) can also exhibit anion exchange, particularly at low pH values.

G. Hydration and Swelling

In normal circumstances, clay particles are never completely dry. Even after being placed in an oven at 105°C for 24 hr, which is the standard for drying soil material, clay particles still retain appreciable amounts of adsorbed water, as shown in Fig. 5.10.

The strong affinity of clay surfaces for water is demonstrated by the *hygroscopic* nature of clay soils, i.e., their ability to adsorb and condense water vapor from the air. So-called air-dry soil is commonly found to have a mass wetness of several percent (relative to the oven-dry mass), the exact percentage depending of course upon the kind and quantity of clay present, as well as upon the humidity of the ambient air.

In the oven-dry state, the water associated with clay is so tightly held that it can be considered a part of the clay itself. As clay becomes increasingly hydrated, the water films surrounding each particle thicken and the water is more loosely held. The entire physical behavior of a clay-containing soil

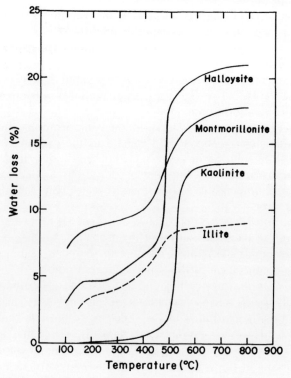

Fig. 5.10. Dehydration curves of clay minerals. (After Marshall, 1964.)

G. Hydration and Swelling

mass (including strength, consistency, plasticity, and the conduction of water and heat) is strongly influenced by the degree of hydration.

Water is attached to clay surfaces by a complex of mechanisms which include the electrostatic attraction of the dipolar, oriented water molecules to charged sites, as well as hydrogen bonding to exposed oxygen atoms on the clay crystal. Still another mechanism of hydration arises from the presence of adsorbed cations. As cations associated with the clay tend to hydrate themselves, they contribute to the overall hydration of the clay system. Quantitatively, this effect depends on the type of cations present and on the cation adsorption capacity.

The tenacity or strength of clay–water adsorption is clearly greatest for the first layer of water molecules. The second layer is attached to the first by hydrogen bonding, and the third to the second, and so forth, but the influence of the attractive force field of the clay surface diminishes with distance so that beyond a few molecular layers it apparently becomes vanishingly small. The physical condition and properties of the adsorbed water in the vicinity of the clay surface and its effective thickness have been a subject of controversy during the last two decades, with some investigators contending that it is quasi-crystalline and hence differs importantly from bulk water in viscosity, diffusivity to ions, dielectric constant, and density. Others claim that the difference in properties between adsorbed water and capillary water in the soil is scarcely detectable and in any case inconsequential. We shall say more about this when we deal with flow phenomena.

When a confined body of clay is allowed to sorb water, swelling pressures develop, which may be as high as several bars. The swelling pressures are related to the osmotic pressure difference between the double layer and the external solution. In a partially hydrated micelle, the thickness of the enveloping water is less than the potential thickness of the diffuse double layer. The double layer, thus truncated, will tend to expand to its full potential thickness and dilution by the osmotic absorption of additional water, if available. As each micelle expands its negatively charged swarm of cations repels those of the adjacent micelle, and thus the micelles tend to push each other apart. This has the internal effect of closing the large pores (with consequent effect on the sytem's permeability) and the external effect of causing the system as a whole to swell.

We have already shown that the concentration of ions between two associated clay particles is greater than in the external solution. The actual concentration difference depends on the interparticle distance (i.e., on the degree of hydration) and on the potential extent of the diffuse double layer (which, in turn, depends upon the valence and concentration of the adsorbed cations). The osmotic attraction for external water is generally twice as high with monovalent than with divalent cations, since there are normally twice as many of the former. Hence swelling and repulsion will be greatest with monovalent cations such as sodium, and with distilled water as the external solution. With calcium as the predominant cation in the exchange complex, swelling is greatly reduced. A similar restraining effect is caused at low pH values by the presence of trivalent aluminum. High salinity of the soil solution will also suppress swelling. However, if a saline soil in which sodium salts predominate is leached of excess salts without concurrent addition of calcium ions (e.g., in the form of gypsum) strong swelling may result from the predominance of sodium ions in the adsorbed phase.

The time-dependent volume increase of clays and clay–sand mixtures in the process of hydration is illustrated in Fig. 5.11. This time dependence is due to the low permeability of clay systems. The eventual swelling is seen to depend on the amount and nature of the clay present. In general, swelling, increases with increasing specific surface area. It is also affected by the arrangement or orientation of soil particles and by the possible presence of interparticle cementation by such materials as iron or aluminum oxides, carbonates, or humus, which might restrict expansion of the soil matrix.

Attempts have been made to calculate swelling pressures on the basis of double-layer theory. The *van't Hoff equation* for monovalent cations is

$$P = RT(n_c - 2n_0) \qquad (5.7)$$

where P is the calculated swelling pressure resulting from the osmotic pressure differential between the interparticle midplane [with an ionic concen-

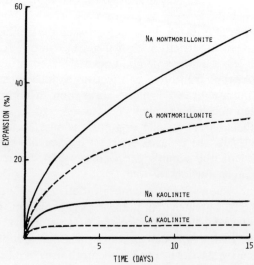

Fig. 5.11. Volume changes of montmorillonite and kaolinite clays during the process of water absorption.

tration n_c as given in Eq. (5.2)] and the ambient intermicellar solution (with a concentration n_0), R is the gas constant, and T is the absolute temperature. Calculations based on this idealized model, which assumes uniform spacing and parallel orientation of clay platelets, have failed to predict actual swelling pressures under real conditions where particles are arranged randomly and where tactoid formation, flocculation, and the presence of several cations and clay minerals, as well as of silt and sand particles and large pores, can influence soil behavior during swelling.

We have already mentioned that swelling can cause the clogging of soil pores and hence a reduction of permeability. This is generally an undesirable effect, but it may be useful in the special situations in which it is necessary to make the soil impervious, as in reservoirs and canal linings.

When a hydrated body of clay is dried, a process opposite to swelling occurs, namely *shrinkage*. In the field, the shrinkage which typically begins at the surface often causes the formation of numerous cracks, which break the soil mass into fragments of various sizes, from small aggregates to large blocks. An extreme example of this can be observed in *vertisols*, which are soils rich in expansive clay (e.g., montmorillonite). When subject to alternating wetting and drying, as in a semiarid region, such soils tend to heave and settle and to form wide, deep cracks and slanted sheer planes extending deeply into the soil profile. Vertisols can be problematic both in agriculture, where they are exceedingly difficult to till because they tend to puddle and then desiccate and harden, and in engineering, where their successive heaving and subsidence tends to warp road pavements and to damage buildings.

H. Flocculation and Dispersion

As clay particles interact with each other, forces both of repulsion and of attraction can come into play. The one or the other type of force may predominate, depending on physicochemical conditions. When the repulsive forces are dominant, the particles separate and remain apart from each other, and the clay is said to be *dispersed* or *peptized*. When the attractive forces prevail, the clay becomes *flocculated*, a phenomenon analogous to the *coagulation* of organic colloids, as the particles associate in packets or *flocs*.

These phenomena can be observed quite readily in dilute aqueous suspensions of clay. For the clay particles to enter into and remain in a state of suspension, they must be dispersed, usually by the addition of a sodic dispersing agent and by mechanical agitation, in the manner described in connection with mechanical analysis. The dispersed suspension is typically turbid, and remains so as long as the suspension is stable. This state can be changed rather dramatically by the addition of salt or polyvalent cations. The turbid suspension suddenly clarifies, as the clay particles flocculate and the flocs settle to the bottom.

The major repulsive force between clay micelles derives from the like charges of the ionic swarms surrounding each particle, as manifested by the swelling tendency of clay–water systems. An attractive force may result, however, if two clay platelets are brought close enough together (within 15 Å or so) so that their counterions intermingle to form a unified layer of positive charges which then attracts the negatively charged particles on both sides of it. This process, sometimes called *plate condensation*, brings about the formation of *tactoids*, or packets of parallel-oriented platelets which remain in more or less stable association (Fig. 5.12).

Another type of electrostatic attraction between particles can occur whenever the plate edges develop positive charges, as they tend to do at low pH values. If the repulsion normally due to the diffuse double layer is not so great as to prevent clay particles from coming together, the positive charges on the edge of one particle may form bonds with the negative charges on the face of another. With this attachment being repeated for many particles, a card-house structure of flocs can develop, as shown in Fig. 5.13.

When particles approach each other very closely, as during drying or consolidation, still another attractive factor appears, the *London–van der Waals force*. Finally, when the particles are pushed still closer together and begin impinging on each other, a resistance once again arises, called the *Born force*, which, however, is only active over extremely short distances.

The important fact about these various attractive and repulsive forces is that they operate with different intensities and over different ranges. The combined force field, shown schematically in Fig. 5.14, consists of regions within which net attraction prevails, and regions over which repulsion predominates. For example, Coulombic (electrostatic) forces are inversely proportional to distance squared, whereas London–van der Waals forces are inversely proportional to the seventh power of the distance, so that the

Fig. 5.12. Schematic representation of montmorillonite tactoid.

H. Flocculation and Dispersion

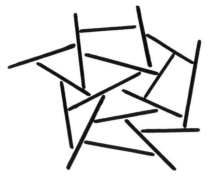

Fig. 5.13. Schematic representation of "cardhouse" structure resulting from edge-to-face electrostatic bonding.

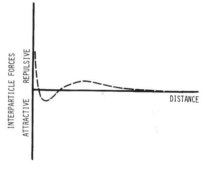

Fig. 5.14. Schematic representation of the combined force field surrounding a hydrated clay particle.

latter are effective within a narrow space of only a few angstroms whereas the former extend to distances at least ten times as great.

Starting from a dilute stable suspension in which the micelles are far apart so that the repulsive forces prevail, flocculation can only be achieved if these repulsive forces are suppressed sufficiently to allow colliding particles to coalesce or clump together, rather than bounce apart, as they approach within the range of the attractive forces. Repulsion is maximized, favoring dispersion, when the double layer is fully extended, as when the ambient solution is very dilute and of high pH (thus preventing positive charges from forming at the particle edges), when the dominant cation is monovalent, and when the soil is fully hydrated. Mechanical agitation also inhibits flocculation. Repulsion is minimized, allowing flocculation, whenever the solution concentration is high and the monovalent cations are replaced by polyvalent ones.

The balance of repulsive versus attractive forces is reversible, and under certain circumstances can easily shift either way. A soil can thus disperse, flocculate, redisperse, reflocculate, and so forth several times over, as, for example, when it is irrigated with water of varying salinity and ionic composition.

We conclude that cation exchange, swelling, cohesion, and particle arrangement are all closely related dynamic phenomena which are strongly influenced by the basic properties and transient conditions of the electrostatic double layer of the colloidal clay.

Sample Problems

1. The specific surface of clay mineral X is 100 m^2/gm and the cation exchange capacity is 25 meq/100 gm. The corresponding values for mineral Y are 20 m^2/gm and 10 meq/100 gm. Calculate the adsorptive charge densities of these minerals.

Recall that the number of charges in an equivalent equals Avogadro's number, $N = 6.02 \times 10^{23}$ (i.e., 6.02×10^{20}/meq). For mineral X,

$$\begin{aligned} \text{no. of charges per cm}^2 &= (.25 \times 6.02 \times 10^{20} \text{ charge/gm}) \\ &\quad \div (100 \text{ m}^2/\text{gm} \times 10^4 \text{ cm}^2/\text{m}^2) \\ &= 1.505 \times 10^{14} \end{aligned}$$

$$\begin{aligned} \text{area per charge} &= (1.505 \times 10^{14} \text{ charge/cm}^2)^{-1} \\ &= 6.64 \times 10^{-15} \text{ cm}^2/\text{charge} \\ &= 66.4 \text{ Å}^2/\text{charge} \end{aligned}$$

For mineral Y,

$$\begin{aligned} \text{no. of charges per cm}^2 &= (.10 \times 6.02 \times 10^{20} \text{ charge/gm}) \\ &\quad \div (20 \text{ m}^2/\text{gm} \times 10^4 \text{ cm}^2/\text{m}^2) \\ &= 3.01 \times 10^{14} \end{aligned}$$

$$\begin{aligned} \text{area per charge} &= (3.01 \times 10^{14} \text{ charge/cm}^2)^{-1} \\ &= 3.32 \times 10^{-15} \text{ cm}^2/\text{charge} \\ &= 33.2 \text{ Å}^2/\text{charge} \end{aligned}$$

Note: The mineral with the lesser specific surface and exchange capacity has the higher charge density.

2. If the cation exchange capacity of a soil is 40 meq/100 gm, and if sodium ions constitute 25% of the exchange complex, what is the minimum amount of gypsum required per hectare to replace sodium with calcium in the upper 20 cm layer of the soil? Assume a bulk density of 1.2 gm/cm^3.

The chemical formula for gypsum is $CaSO_4 \cdot 2H_2O$. Hence the molecular weight is 172 and the equivalent weight is 86.

Mass of soil in upper 20 cm layer
$= 10^4$ m²/hectare $\times$ (2 $\times$ 10⁻¹ m depth) $\times$ 10⁶ cm³/m³ $\times$ 1.2 gm/cm³
$= 2.4 \times 10^9$ gm.
Equivalents of sodium to be removed from upper 20 cm layer
$= 2.4 \times 10^9$ gm $\times$ (0.40 $\times$ 0.25 $\times$ 10⁻³ eq/gm) $= 2.4 \times 10^5$ eq.
Mass of gypsum needed $= 2.4 \times 10^5$ eq $\times$ 86 gm/eq $\cong 2 \times 10^7$ gm
$= 20$ metric tons.

3. The specific surface of a given clay is 100 m²/gm and the bulk density is 1.25 gm/cm³. What is the average thickness of the water enveloping the particles if the volumetric wetness θ of the saturated clay–water system is 50%? Assuming parallel orientation of clay platelets, estimate the median-plane concentration (mole/liter) of sodium and calcium.
Recall that

$$\text{bulk density} = \text{mass of grains/bulk volume. Hence:}$$

The mass of solids per cubic centimeter of bulk volume $= 1$ cm³ $\times$ 1.25 gm/cm³ $= 1.25$ g.
Volume of water per cubic centimeter of bulk volume $= 1$ cm³ $\times$ 0.5 $= 0.5$ cm³.
Surface area of grains per gram $= 100$ m² $\times$ 10,000 cm²/m² $= 10^6$ cm².
Surface area of grains per cubic centimeter $= 1.25 \times 10^6$ cm².
Thickness of water film $=$ (volume of water)/(surface area of grains) $= (0.5 \text{ cm}^3)/(1.25 \times 10^6 \text{ cm}^2) = 4 \times 10^{-7}$ cm $= 40$ Å.

If the clay–water system is saturated, and the clay platelets are in parallel orientation, the average interplate distance is 80 Å. Using Eq. (5.2),

$$n_c = \frac{\pi^2}{v^2 B(d+x)^2 10^{-16}}$$

For calcium,

$$n_c \cong \frac{10}{4 \times 10^{15} \times (40+2)^2 \times 10^{-16}} \cong 1.42 \times 10^{-2} \text{ mole/liter}$$

For sodium,

$$n_c \cong \frac{10}{1 \times 10^{15} \times 42^2 \times 10^{-16}} \cong 5.67 \times 10^{-2} \text{ mole/liter}$$

4. Calculate the fractional composition of the exchange complex of a soil which is equilibrated with a solution of $0.01 M$ in which the molar frac-

tions of Ca^{2+} and Na^+ are equal (i.e., 0.5). Recalculate for solution concentrations of $0.1M$ and $0.001M$.

Referring to Eq. (5.6), with the selectivity coefficient C_{Ca-Na} taken to be 4.0, we can write

$$[Ca_e]/[Na_e]^2 = (4.0/0.01)[0.5/(0.5)^2] = 800 \quad \text{or} \quad [Na_e]^2 = 1/800$$

Hence the equivalent fraction of exchangeable sodium is $[Na_e] = 1/\sqrt{800} \approx 0.035$. Thus, sodium constitutes 3.5% of the exchange complex, with calcium constituting the remaining 96.5%. If the solution concentration is increased tenfold to $0.1M$, we get

$$[Ca_e]/[Na_e]^2 = (4.0/0.1)(0.5/0.25) = 80$$

Hence $[Na_e] = 1/\sqrt{80} = 0.106 = 10.6\%$, which reduces the exchangeable calcium to 89.4% of the exchange complex. From the same equation, it is easy to show that, on the other hand, if the concentration of the ambient solution were decreased to $0.001M$, the calcium ion would be adsorbed preferentially to the extent of 98.9% of the exchange capacity.

> Get used to thinking that there is nothing
> Mother Nature loves so well
> as to change existing forms
> and to make new ones like them.
> "Meditations II"
> Marcus Aurelius Antonius
> 121–180 C.E.

6 Soil Structure and Aggregation

A. Introduction

Bricks thrown haphazardly atop one another become an unsightly heap. The same bricks, only differently arranged and mutually bonded, can form a home or a factory. Similarly, a soil can be merely a loose and unstable assemblage of random particles, or it can consist of a distinctly structured pattern of interbonded particles associated into aggregates having regular sizes and shapes. Hence it is not enough to study the properties of individual soil particles. To understand how the soil behaves as a composite body, we must consider the manner in which the various particles are packed and held together in a continuous spatial network which is commonly called the *soil matrix*, or *fabric*.

The arrangement and organization of the particles in the soil is called *soil structure*. Since soil particles differ in shape, size, and orientation, and can be variously associated and interlinked, the mass of them can form complex and irregular configurations which are in general exceedingly difficult if not impossible to characterize in exact geometric terms. A further complication is the inherently unstable nature of soil structure and hence its inconstancy in time, as well as its nonuniformity in space. Soil structure is strongly affected by changes in climate, biological activity, and soil management practices, and it is vulnerable to destructive forces of a mechanical and physicochemical nature. For these various reasons, we have no truly objective or universally applicable method to measure soil structure per se, and the term soil structure therefore denotes a qualitative concept rather than a directly quantifiable property. The numerous methods which have been proposed for characterization of soil structure are in fact indirect methods which measure one soil attribute or another which are supposed to be dependent upon structure, rather than the structure itself, and many of these methods are at best specific to the purpose for which they were devised and at worst completely arbitrary.

The complexity and difficulty associated with soil structure notwithstanding, we can readily perceive and appreciate its critical importance, inasmuch as it determines the total porosity as well as the shapes of individual pores and their size distribution. Hence, soil structure affects the retention and transmission of fluids in the soil, including infiltration and aeration. Moreover, as soil structure influences the mechanical properties of the soil, it may also affect such disparate phenomena as germination, root growth, tillage, overland traffic, and erosion. Agriculturists are usually interested in having the soil, at least in its surface zone, in a loose and highly porous and permeable condition. Engineers, on the other hand, desire a dense and rigid structure so as to provide maximal stability and resistance to shear and minimal permeability. In either case, knowledge of soil structure relationships is essential for efficient management.

B. Types of Soil Structure

In general, we can recognize three broad categories of soil structure— *single grained*, *massive*, and *aggregated*. When particles are entirely unattached to each other, the structure is completely loose, as it is in the case of coarse granular soils or unconsolidated deposits of desert dust. Such soils were labeled structureless in the older literature of soil physics, but, since even a loose arrangement is a structure of sorts, we prefer the term single grained structure. On the other hand, when the soil is tightly packed in large cohesive blocks, as is sometimes the case with dried clay, the structure can be called massive. Between these two extremes, we can recognize an intermediate condition in which the soil particles are associated in quasi-stable small clods known as *aggregates* or *peds*. This last type of structure, called aggregated, is generally the most desirable condition for plant growth, especially in the critical early stages of germination and seedling establishment. The formation and maintenance of stable aggregates is the essential feature of soil *tilth*, a qualitative term used by agronomists to describe that highly desirable, yet unfortunately elusive, physical condition in which the soil is an optimally loose, friable, and porous assemblage of aggregates permitting free movement of water and air, easy cultivation and planting, and unobstructed germination and root growth.

C. Structure of Granular Soils

The structure of most coarse-grained, or granular, soils is single grained, as there is little tendency for the grains to adhere to each other and to form aggregates. The actual arrangement and internal mode of packing of the grains depends upon the distribution of grain sizes and shapes, as well as upon the manner in which the material has been deposited or formed in place.

Hypothetically speaking, the two extreme cases of possible packing arrangements are, on the one hand, a system of uniform spherical grains in a state of open packing (and hence minimal density) and, on the other hand, a gradual distribution of grain sizes in which progressively smaller grains fill the voids between larger ones in an "ideal" succession which provides maximal density. A system of uniform grains is called *monodisperse*, and one which includes various grain sizes is *polydisperse*.

Let us consider, for a moment, the hypothetical case of a system consisting entirely of uniform spheres. Although not realistic, this is a useful exercise, even apart from its didactic value, inasmuch as it can help us to establish theoretical limits of porosity by which we can later evaluate real systems. With monodisperse spheres, the minimal density and hence maximal porosity is obtained in the case of cubic (open) packing. In this mode of packing, each grain touches six neighbors on opposite sides of three orthogonal axes and hence is said to have a *coordination number* of six. The bulk density is then

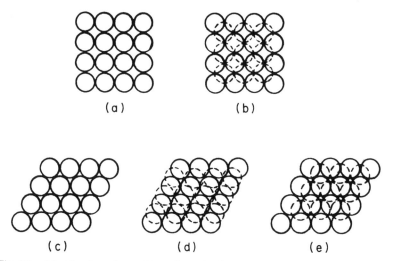

Fig. 6.1. Models of regular packing of equal spheres. (a) Simple cubic, (b) cubical tetrahedral, (c) tetragonal sphenoidal, (d) pyramidal, and (e) tetrahedral. From Deresiewicz, 1958, Fig. 1.

$\pi\rho_s/6$ (where ρ_s is the density of the solid particles), and the porosity is 47.6%, regardless of the diameter of the spheres. If $\rho_s = 2.65$ gm/cm^3, we obtain a bulk density of about 1.39 gm/cm^3. The densest possible packing of uniform spheres is obtained with either the tetrahedral or the octahedral arrangement, both of which have a coordination number of 12, a bulk density of $\pi/3\sqrt{2}$ (about 1.97 gm/cm^3), and a porosity of 25.9%.

A schematic representation of these modes of packing, as well as of two intermediate arrangements with coordination numbers of 8 and 10, are given in Fig. 6.1. A quantitative summary is given in Table 6.1.

The analysis of polydisperse systems is obviously more complex, even if we continue to assume spherical shape. A mixture of two grain sizes depends on the relative concentration of the components. For any particular ratio of grain sizes, there exists an optimal composition such that will result in minimal porosity. In principle, the density of a polydisperse system, consisting of many particle sizes, can be greater than that of a monodisperse system, as smaller particles can fit into the spaces between larger ones, and so ad infinitum (Fig. 6.2). Such systems were analyzed by Wise (1952) and Deresiewicz (1958). Depending on the array of relative particle radii, coordination numbers as high as 30 or more, and porosities lower than 20% are possible.

Natural granular soils resemble collections of spheres in the sense that the particles are often rounded (though deposits of angular particles also occur). The actual porosities of natural sediments range within the theoretically derivable limits for ideal packings of monodisperse and polydisperse spheres; i.e., they lie between 25 and 50%, in general.

Table 6.1

PACKING OF SPHERES[a]

Type of packing	Coordination number	Spacing of layers	Volume of unit prism	Density	Porosity (%)
Simple cubic	6	$2R$	$8R^3$	$\pi/6$ (0.5236)	47.64
Cubic tetrahedral	8	$2R$	$4\sqrt{3}R^3$	$\pi/3\sqrt{3}$ (0.6046)	39.54
Tetragonal sphenoidal	10	$R\sqrt{3}$	$6R^3$	$2\pi/9$ (0.6981)	30.19
Pyramidal	12	$R\sqrt{2}$	$4\sqrt{2}R^3$	$\pi/3\sqrt{2}$ (0.7405)	25.95
Tetrahedral	12	$2R\sqrt{2/3}$	$4\sqrt{2}R^3$	$\pi/3\sqrt{2}$	25.95

[a] From Deresiewicz (1958, Table 1).

D. Structure of Aggregated Soils

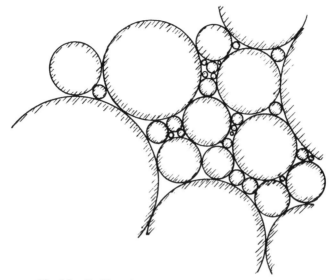

Fig. 6.2. Packing of polydisperse particles (hypothetical).

Loose deposits of granular material cannot be compacted very effectively by the application of static pressure (which, however, can be very effective in compacting unsaturated, clayey soils), unless, of course, the pressure exceeds the crushing strength of the individual grains. Short of that, however, static pressure is merely transmitted along the soil skeleton and borne by the high frictional resistance to the mutual sliding of the grains. Granular material can, however, be compacted by the application of vibratory action. The vibration pulsates the grains and allows smaller ones to enter between larger ones, thus increasing the packing density. The vibration of granular materials is effective in the dry and in the saturated states, whereas at unsaturated moisture contents surface tension forces of the water menisci wedged between the particles lend cohesiveness and hence greater rigidity to the matrix.

The structure of nonideal granular materials in which the grains are not spherical but oblong or angular is much more difficult to formulate theoretically, as the orientations of particles must be taken into account.

D. Structure of Aggregated Soils

In soils with an appreciable content of clay, the primary particles tend, under favorable circumstances, to group themselves into structural units known as *secondary particles*, or *aggregates*. Such aggregates are not characterized by any universally fixed size, nor are they necessarily stable. The visible aggregates, which are generally of the order of several millimeters to several centimeters in diameter, are often called *peds*, or *macroaggregates*. These are usually assemblages of smaller groupings, or microaggregates, which themselves are associations of the ultimate structural units, being the flocs, clusters, or packets of clay particles. Bundles of the latter units attach themselves to, and sometimes engulf, the much larger primary particles of sand and silt. The internal organization of these various groupings can now be studied by means of electron microscopy, particularly with the use of newly developed scanning techniques.

A prerequisite for aggregation is that the clay be flocculated. However, flocculation is a necessary but not a sufficient condition for aggregation. As stated by Richard Bradfield of Cornell University some 40 years ago, "aggregation is flocculation—plus!" That "plus" is *cementation*.

The complex interrelationship of physical, biological, and chemical reactions involved in the formation and degradation of soil aggregates was reviewed by Harris *et al*. (1965). An important role is played by the extensive networks of roots which permeate the soil and tend to enmesh soil aggregates. Roots exert pressures which compress aggregates and separate between adjacent ones. Water uptake by roots causes differential dehydration, shrinkage, and the opening of numerous small cracks. Moreover, root exudations and the continual death of roots and particularly of root hairs promote microbial activity which results in the production of humic cements. Since these binding substances are transitory, as they are susceptible to further microbial decomposition, organic matter must be replenished and supplied continually if aggregate stability is to be maintained in the long run.

Active humus is accumulated and soil aggregates are stabilized most effectively under perennial sod-forming herbage. Annual cropping systems, on the other hand, hasten the decomposition of humus and the destruction of soil aggregates. The foliage of close-growing vegetation, and its residues, also protect surface soil aggregates against slaking by water, particularly under raindrop impact, to which aggregates become especially vulnerable if exposed and desiccated in the absence of a protective cover.

The influence of a cropping system on soil aggregation is seen to be a function of root activity (density and depth of rooting and the rate of root proliferation), density and continuity of surface cover, and the mode and frequency of cultivation and traffic. Crops or cropping systems which develop sparse roots, provide little vegetative cover, and require intensive mechanical cultivation of the soil are the least likely to maintain optimal tilth. Moreover, soil wetness at the time of cultivation has a great bearing on whether aggregated structure is maintained or destroyed. Cultivation of excessively wet soil causes puddling (plastic remolding), whereas cultivation of ex-

D. Structure of Aggregated Soils

cessively dry soil is likely to result in grinding or pulverizing the soil into dust. In between lies a rather narrow range of wetness considered optimal for cultivation, at which the soil will readily break up into clods of the desired size range with the least amount of effort. The precise value of "optimal wetness" must be determined for each soil specifically, and perhaps for each type of tillage implement as well.[1]

When we speak of *microbial activity* as affecting soil aggregation we are obliged to point out that this catch-all phrase in reality refers to the time-variable activity of numerous microorganisms, including thousands of species of bacteria, fungi, actinomycetes, etc. Especially important are rhizospheric bacteria, which flourish in direct association with roots of specific plants, as well as fungi, which often form extensive adhesive networks of fine filaments known as mycelia or hyphae. The composition of the soil microfauna and microflora depends on the thermal and moisture regimes, on soil pH and oxidation–reduction potential, the nutrient status of the soil substrate, and the type and quantity of organic matter present (Alexander, 1977).

Soil microorganisms bind aggregates by a complex of mechanisms, such as adsorption, physical entanglement and envelopment, and cementation by excreted mucilagenous products. Prominent among the many microbial products capable of binding soil aggregates are polysaccharides, hemicelluloses or uronides, levans, as well as numerous other natural polymers. Such materials are attached to clay surfaces by means of cation bridges, hydrogen bonding, van der Waals forces, and anion adsorption mechanisms. Polysaccharides, in particular, consist of large, linear, and flexible molecules capable of forming multiple bonds with several particles at once. In some cases, organic polymers hardly penetrate between the individual clay particles but form a protective capsule around soil aggregates. In other cases, solutions of active organic agents penetrate into soil aggregates and then precipitate more or less irreversibly as insoluble (though still biologically decomposable) cements.

In addition to increasing the strength and stability of intra-aggregate bonding, organic products may further promote aggregate stability by reducing wettability and swelling. Some of the organic materials are inherently hydrophobic, or become so as they dehydrate, so that the organo–clay complex may have a reduced affinity for water.

Some inorganic materials can also serve as cementing agents. The importance of clay, and its state of flocculation, should be obvious by now.

---

[1] A problem often encountered in field practice is that the wetness range which is optimal for tillage may also be "optimal" for compaction by traffic, which is generally an undersirable consequence of tillage. This problem will be discussed in Chapter 14.

Table 6.2

POSSIBLE AGGREGATE BONDING MECHANISMS[a]

I. *Clay Domain-Clay Domain*[a]

A. Domain face-Domain face
Cation[b] bridge between negative faces. Mechanism similar to that for orientation of clay platelets into domains

$$\text{Face}^{-}\text{---}M^{n+}\text{---}^{-}\text{Face}$$

B. Domain edge-Domain face
Positive edge site to negative face

$$\text{Edge Al—OH}_2^{+}\text{---}^{-}\text{Face}$$

II. *Clay Domain-Organic Polymer*[c]*-Clay Domain*

A. Domain edge-Organic polymer-(Domain)
  1. Anion exchange: Positive edge site to polymer carboxyl

$$\text{Edge Al—OH}_2^{+}\text{---}^{-}\text{OOC—R—COO}^{-}\text{---}$$

  2. Hydrogen bonding between edge hydroxyl and polymer carbonyl or amide

$$\text{Edge—OH---O}{=}\text{C—R—}\overset{O}{\underset{H}{\overset{\|}{C}}}\text{—N---HO—Edge}$$

  3. Cation bridge between negative edge site and polymer carboxyl.

$$\text{Edge—O}^{-}\text{---}M^{n+}\text{---}^{-}\text{OOC—R—COO}^{-}\text{---}$$

  4. Van der Waals attraction between edge and polymer

B. Domain face-Organic polymer-(Domain)
  1. Hydrogen bonding between polymer hydroxyl and external or internal (expanding lattice minerals) face silicate oxygens

$$\text{Face Si—O---HO—R—OH---}$$

  2. Cation bridge between domain external face and polymer carboxyl or other polarizable group

$$\text{External Face}^{-}\text{---}M^{n+}\text{---}^{-}\text{OOC—R—COO}^{-}\text{---}$$

  3. Van der Waals attraction between face and polymer

III. *Quartz-(Silt, Inorganic and Organic Colloids)-Quartz*

A. Chemical bonds established between quartz surface gels of hydrated alumino silicates and active groups of other aggregate constituents
B. Quartz grains held in a matrix of silt and clay stabilized primarily by:
  1. Oriented clay particles
  2. Irreversibly dehydrated silicates   sesquioxides   or humic-sesquioxide complexes
  3. Irreversibly dehydrated humic materials
  4. Silt-size microaggregates stabilized by iron humates
  5. Organic colloids and clay domains bonded by mechanisms cited under I and II.

---

[a] After Harris *et al.* (1965).

[b] Clay domain defined as a group of clay crystals oriented sufficiently close together by cations or hydrogen bonds between the crystal faces to behave as a single unit.

[c] M^{n+} = Free cation or positively charged metal oxide or hydroxide.

[d] R = Organic polymer with axis horizontal or perpendicular to clay domain.

E. Additional Factors Affecting Aggregation

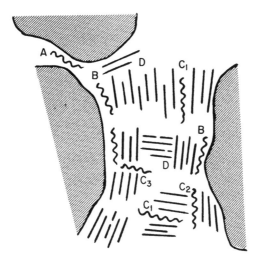

Fig. 6.3. Possible arrangements of quartz particles, clay domains, and organic matter in a soil aggregate. (A) Quartz–organic colloid–quartz, (B) quartz–organic colloid–clay domain, (C) clay domain–organic colloid–clay domain, (C_1) face–face, (C_2) edge–face, (C_3) edge–edge, (D) clay domain edge–clay domain face. (After Emerson, 1959.)

Cohesiveness between clay particles is, in fact, the ultimate internal binding force within microaggregates. Calcium carbonate, as well as iron and aluminum oxides, can impart considerable stability to otherwise weak soil aggregates. The latter, in particular, are responsible for the significant stability of aggregates in tropical soils which often contain little organic matter.

A summary of possible aggregate bonding mechanisms was compiled by Harris *et al.* (1965), and is presented in Table 6.2. An interesting model of the internal bonding forms which can constitute a soil aggregate was provided by Emerson (1959) and is shown in Fig. 6.3. Emerson's model is composed of sand or silt-size quartz particles and of "domains" of oriented clay bonded by electrostatic forces. Stability of the aggregate is enhanced by linkage of organic polymers between the quartz particles and the faces or edges of clay crystals.

E. Additional Factors Affecting Aggregation

Prominent among the macrobiological factors which influence soil structure are earthworms, which have aroused the interest of such diverse observers as Aristotle and Darwin. The role of earthworms in the soil was described most vividly and aptly two centuries ago by George White (as quoted by Russel, 1912):

> Worms seem to be the great promoters of vegetation, which would proceed but lamely without them, by boring, perforating, and loosening the soil, and rendering it pervious to rains and fibers of plants, by drawing straws and stalks of leaves and twigs into it, and, most of all, by throwing up such infinite numbers of lumps of earth called wormcasts, which being their excrement, is a fine manure of grain and grass .... The earth without worms would soon become cold, hardbound, and void of fermentation, and consequently sterile.

A population of several million earthworms (of various species) per acre is not uncommon in regions where the supply of moisture and fresh organic matter is adequate. Such a population of earthworms can digest and expel as "casts" many tons of soil per acre per year.

In addition to its indirect influence on soil structure via its effect on biological activity in the soil, the climate can play a direct role in the formation and destruction of aggregates. We have already mentioned freezing and thawing processes, which can induce uneven stresses and strains in the soil body. The formation of ice-filled cavities, which tend to be larger as the freezing process is slower, affects pore-size distribution as the expansion of water upon freezing compresses and separates aggregates.

Quite apart from freezing and thawing, alternate wetting and drying,[2] causing differential expansion and shrinkage, may enhance aggregation in some cases while contributing to aggregate breakdown in others. A particularly destructive condition may result when thoroughly desiccated aggregates are suddenly submerged in water. The water drawn into each aggregate over its entire periphery may trap and compress the air originally present in the dry aggregate. As the cohesive strength of the outer part of the clod is reduced by swelling, and as the pressure of the entrapped air builds up in proportion to its compression, sooner or later the latter exceeds the former and the clod may actually explode. More typically, however, a series of small explosions, each marked by the escape of a bubble of air, shatters the clod into fragments. This destructive process is known as *air slaking* (Fig. 6.4).

Another process, already mentioned, by which water can destroy aggregates is the hammering impact of falling raindrops, as well as the scouring action of flowing water in surface runoff. On the other hand, drying processes, causing shrinkage, can (in themselves) increase aggregate stability by making

---

[2] Freezing, incidentally, is itself analogous to drying, in that it causes the extraction of liquid water from regions of the soil toward the freezing sites (ice lenses).

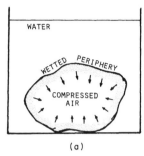
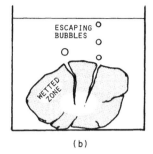

Fig. 6.4. Air slaking of an initially dry aggregate suddenly submerged in water. (a). *Early stage:* The periphery of the aggregate is wetted and water moves into the aggregate compressing the air ahead of it. (b) *Bursting stage:* As the wetted zone is weakened by swelling and the pressure of entrapped air increases in proportion to its compression, the point is reached at which the aggregate is shattered and air bubbles out. This point can be quite abrupt. The final result is the overall collapse of the shattered aggregate.

the aggregates more dense and cohesive, as well as by causing gluelike organic gums and gels to "set" irreversibly so as to serve as stable cementing agents.

With so many factors active simultaneously, aggregate stability is obviously not an absolute attribute but a function of the relative strength of the intraaggregate bonds versus the stresses induced by swelling, scouring, and air entrapment. Accordingly, the task of maintaining aggregation involves strengthening the internal bonding mechanisms while at the same time reducing the destructive forces, e.g., avoiding sudden submergence and direct exposure of surface aggregates to desiccation, raindrop impact, and running water, as well as to heavy compressive traffic and excessively pulverizing tillage.

F. Characterization of Soil Structure

Soil structure, or fabric, can be studied directly by microscopic observation of thin slices under polarized light. The structural associations of clay can be examined by means of electron microscopy, using either transmission or scanning techniques. The structure of single-grained soils, as well as of aggregated soils, can be considered quantitatively in terms of the total porosity and of the pore-size distribution. The specific structure of aggregated soils can, furthermore, be characterized qualitatively by specifying the typical shapes of aggregates found in various horizons within the soil profile, or quantitatively by measuring their sizes. Additional methods of characterizing soil structure are based on measuring mechanical properties and permeability to various fluids. None of these methods has been accepted universally. In each case, the choice of the method to be used depends upon the problem, the soil, the equipment available, and, not the least, upon the soil physicist.

Total porosity f of a soil sample is usually computed on the basis of measured bulk density ρ_b, using the following equation (see Chapter 2):

$$f = 1 - \rho_b/\rho_s \tag{6.1}$$

where ρ_s is the average particle density.

Bulk density is generally measured by means of a *core sampler* designed to extract "undisturbed" samples of known volume from various depths in the profile. An alternative is to measure the volumes and masses of individual clods (not including interclod cavities) by *immersion in mercury* or by coating with paraffin wax prior to *immersion in water*. Still other methods are the *sand-funnel* and *balloon techniques* used in engineering, and *gamma-ray attenuation densitometry* (Blake, 1965).

Pore-size distribution measurements can be made in coarse-grained soils by means of the *pressure-intrusion method* (Diamond, 1970), in which a nonwetting liquid, generally mercury, is forced into the pores of a predried sample. The pressure is applied incrementally, and the volume pentrated by the liquid is measured for each pressure step, equivalent (by the capillary theory) to a range of pore diameters. In the case of fine-grained soils, capillary condensation methods, or, more commonly, *desorption methods* are used (Vomocil, 1965). In the desorption method, a presaturated sample is subjected to a stepwise series of incremental suctions, and the capillary theory is used to obtain the equivalent pore-size distribution. Water is commonly used as the permeating liquid, though nonpolar liquids have also been tried in an attempt to assess, by the comparison, the possible effect of water saturation and desorption in modifying soil structure. Where the aggregates are fairly distinct, it is sometimes possible to divide pore-size distribution into two distinguishable ranges, namely *macropores* and *micropores*. The macropores are mostly the interaggregate cavities which serve as the principal avenues for the infiltration and drainage of water and for aeration. The micropores are the intraaggregate capillaries responsible for the retention of water and solutes. However, the demarcation is seldom truly distinct, and the separation between macropores and micropores is often arbitrary.

The shapes of aggregates observable in the field (illustrated in Fig. 6.5) can be classified as follows:

(1) *Platy:* Horizontally layered, thin and flat aggregates resembling wafers. Such structures occur, for example, in recently deposited clay soils.

(2) *Prismatic* or *columnar:* Vertically oriented pillars, often six sided,

F. Characterization of Soil Structure

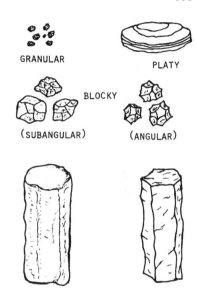

Fig. 6.5. Observable forms of soil aggregation.

up to 15 cm in diameter. Such structures are common in the B horizon of clayey soils, particularly in semiarid regions. Where the tops are flat, these vertical aggregates are called prismatic, and where rounded—columnar.

(3) *Blocky:* Cubelike blocks of soil, up to 10 cm in size, sometimes angular with well-defined planar faces. These structures occur most commonly in the upper part of the B horizon.

(4) *Spherical:* Rounded aggregates, generally not much larger than 2 cm in diameter, often found in a loose condition in the A horizon. Such units are called granules, and where particularly porous, crumbs.

The shapes, sizes, and densities of aggregates generally vary within the profile. As the overburden pressure increases with depth, and inasmuch as the deeper layers do not experience such extreme fluctuations in moisture content (as does the alternately saturated and desiccated surface layer), the decrease of swelling and shrinkage activity causes the deeper aggregates to be larger. A typical structural profile in semiarid regions consists of a granulated A horizon underlain by a prismatic B horizon, whereas in humid temperate regions a granulated A horizon may occur with a platy or blocky B horizon. The number of variations found in nature are, however, legion. A detailed classification of aggregate shapes is given in Table 6.3.

Aggregate size distribution is a determinant of pore-size distribution and has a bearing on the erodibility of the surface, particularly by wind. In the field, adjacent particles often adhere to each other, though of course not as

Table 6.3

CLASSIFICATION OF SOIL STRUCTURE ACCORDING TO SOIL SURVEY STAFF (1951)

A Type: Shape and arrangement of peds

	Platelike. Horizontal axes longer than vertical. Arranged around a horizontal plane.	Prismlike. Horizontal axes shorter than vertical. Arranged around vertical line. Vertices angular.		Blocklike–polyhedral-spheroidal. Three approximately equal dimensions arranged around a point.			Spheroidal–polyhedral. Plane or curved surfaces not accommodated to faces of surrounding peds.	
				Blocklike–polyhedral. Plane or curved surfaces accommodated to faces of surrounding peds.				
		Without rounded caps	With rounded caps	Faces flattened; vertices sharply angular		Mixed rounded, flattened faces; many rounded vertices	Relatively nonporous peds	Porous peds
						Subangular		
	Platy	Prismatic	Columnar	Blocky		blocky	Granular	Crumb
B class: Size of peds								
1. Very fine or very thin	<1 mm	<10 mm	<10 mm	<5 mm		<5 mm	<1 mm	1 mm
2. Fine or thin	1–2 mm	10–20 mm	10–20 mm	5–10 mm		5–10 mm	1–2 mm	1–2 mm
3. Medium	2–5 mm	20–50 mm	20–50 mm	10–20 mm		10–20 mm	2–5 mm	2–5 mm
4. Coarse or thick	5–10 mm	50–100 mm	50–100 mm	20–50 mm		20–50 mm	5–10 mm	
5. Very coarse or very thick	>10 mm	>100 mm	>100 mm	>50 mm		>50 mm	>10 mm	

C grade: Durability of peds

0. Structureless   No aggregation or orderly arrangement.
1. Weak   Poorly formed, nondurable, indistinct peds that break into a mixture of a few entire and many broken peds and much unaggregated material.
2. Moderate   Well-formed, moderately durable peds, indistinct in undisturbed soil, that break into many entire and some broken peds but little unaggregated material.
3. Strong   Well-formed, durable, distinct peds, weakly attached to each other, that break almost completely into entire peds.

tenaciously as do the particles within each aggregate. Separating and classifying soil aggregates necessarily involves a disruption of the original, in situ structural arrangement. The application of too great a force may break up the aggregates themselves. Hence determination of aggregate size distribution depends on the mechanical means employed to separate the aggregates.

Aggregate screening methods were reviewed by Kemper and Chepil (1965). Screening through flat sieves is difficult to standardize and entails frequent clogging of the sieve openings. Chepil (1962) presented a detailed plan for a rotary sieve machine equipped with up to 14 concentrically nested sieves of various aperture sizes. The operation of this machine can be standardized, thus minimizing the arbitrary personal factor, and clogging is virtually eliminated. Samples for *dry sieving analysis* should be taken when the soil is reasonably dry, and care must be taken to avoid change of structure during handling The electrically rotated sieves are slanted downward so that the classified aggregates gradually tumble into separate bins for weighing.

Various indexes have been proposed for expressing the distribution of aggregate sizes. If a single characteristic parameter is desired, such as might allow correlation with various factors (e.g., erosion, infiltration, evaporation, or aeration), a method must be adopted for assigning an appropriate weighting factor to each size range of aggregates. A widely used index is the *mean weight diameter* (Van Bavel, 1949; Youker and McGuinness, 1956), based on weighting the masses of aggregates of the various size classes according to their respective sizes. The mean weight diameter X is thus defined by the following equation:

$$X = \sum_{i=1}^{n} x_i w_i \quad (6.2)$$

where x_i is the mean diameter of any particular size range of aggregates separated by sieving, and w_i is the weight of the aggregates in that size range as a fraction of the total dry weight of the sample analyzed. The summation accounts for all size ranges, including the group of aggregates smaller than the openings of the finest sieve.

An alternative index of aggregate size distribution is the *geometric mean diameter* Y calculated according to the equation (Mazurak, 1950):

$$Y = \exp\left[\left(\sum_{i=1}^{n} w_i \log x_i\right) \bigg/ \left(\sum_{i=1}^{n} w_i\right)\right] \quad (6.3)$$

wherein w_i is the weight of aggregates in a size class of average diameter x_i, and $\sum_{i=1}^{n} w_i$ is the total weight of the sample.

Gardner (1956) reported that the aggregates of many soils exhibit a logarithmic–normal distribution which can be characterized in terms of two parameters, namely the *geometric mean diameter* and the *log standard deviation*. Other indexes proposed for characterizing aggregate size distribution are the so-called *coefficient of aggregation* (Retzer and Russell, 1941) and the *weighted mean diameter* and standard deviation (Puri and Puri, 1949). No universal prescription can be offered on which of these alternative indexes are to be preferred.

G. Aggregate Stability

Determining the momentary state of aggregation of a soil at any particular time might not suffice to portray the soil's true structural characteristics as they may vary dynamically over a period of time. By any measure, the degree of aggregation is indeed a time-variable property, as aggregates form, disintegrate, and reform periodically. For instance, a newly cultivated field may for a time exhibit a nearly optimal array of aggregate sizes, with large interaggregate pores favoring high infiltration rates and unrestricted aeration. This blissful state often proves to be ephemeral, however, as many farmers have repeatedly discovered to their great chagrin. Soil structure may begin to deteriorate quite visibly and rapidly, as the soil is subject to destructive forces resulting from intermittent rainfall (causing slaking, swelling, shrinkage, and erosion) followed by dry spells (causing the soil to be vulnerable to deflation by wind). Repeated traffic, particularly by heavy machinery, furthermore tends to crush the aggregates remaining at the surface, and to compact the soil to some depth below the surface.

Soils vary, of course, in the degree to which they are vulnerable to externally imposed destructive forces. *Aggregate stability* is a measure of this vulnerability. More specifically, it expresses the resistance of aggregates to breakdown when subjected to potentially disruptive processes. Since the reaction of a soil to forces acting on it depends not only on the soil itself but also to a large degree upon the nature of the forces and the manner they are applied, aggregate stability is not measurable in absolute terms. Rather, it is a relative, and partly even subjective, concept. This, however, does not detract from its importance.

To test aggregate stability, soil physicists generally subject samples of aggregates to artificially induced forces designed to simulate phenomena which are likely to occur in the field. The nature of the forces applied during such testing depends on the investigator's perception of the natural phenomenon which he wishes to simulate, as well as on the equipment available and the mode of its employment. The degree of stability is then assessed by determining the fraction of the original sample's mass which has withstood destruction and retained its physical integrity, or, conversely, the fraction

G. Aggregate Stability

which appears, by some arbitrary but reproducible criterion, to have disintegrated.

If an indication of mechanical stability is sought, measurements can be made of the resistance of aggregates to prolonged dry sieving, or to crushing forces. The latter can be applied statically, as by compression between two plates; or dynamically, by dropping the aggregates from a given height onto a hard surface to observe the readiness with which they tend to shatter under repeated collisions. If the objective is to study the resistance to erosion by wind, the aggregates can be placed in a wind tunnel and tested for deflation under specified wind conditions (Chepil, 1958).

Most frequently, however, the concept of aggregate stability is applied in relation to the destructive action of water. Although mentioned before, it bears repeating that the very wetting of aggregates may cause their collapse, as the bonding substances dissolve or weaken and as the clay swells and possibly disperses. If wetting is nonuniform, one part of the aggregate will swell more than another, and the resulting stress, compounded during subsequent shrinkage, may fracture the aggregate. Aggregates are more vulnerable to sudden than to gradual wetting, owing to the air occlusion effect. Raindrops and flowing water provide the energy to detach particles and transport them away. Abrasion by particles carried as suspended matter in runoff water scours the surface and contributes to the overall breakdown of aggregated structure at the soil surface.

The classical and still most prevalent procedure for testing the water stability of soil aggregates is the *wet sieving method* (Tiulin, 1928; Yoder, 1936; De Boodt and De Leenheer, 1958). A representative sample of air-dry aggregates is placed on the uppermost of a set of graduated sieves and immersed in water to simulate flooding. The sieves are then oscillated vertically and rhythmically, so that water is made to flow up and down through the screens and the assemblage of aggregates. In this manner, the action of flowing water is simulated. At the end of a specified period of sieving (e.g., 20 min) the nest of sieves is removed from the water and the oven-dry weight of material left on each sieve is determined. As pointed out by Kemper (1965), the results should be corrected for the coarse primary particles retained on each sieve to avoid designating them falsely as aggregates. This is done by dispersing the material collected from each sieve, using a mechanical stirrer and a sodic dispersing agent, then washing the material back through the same sieve. The weight of sand retained after the second sieving is then subtracted from the total weight of undispersed material retained after the first sieving, and the percentage of stable aggregates %SA is given by

$$\%SA = 100 \times \frac{(\text{weight retained}) - (\text{weight of sand})}{(\text{total sample weight}) - (\text{weight of sand})} \quad (6.4)$$

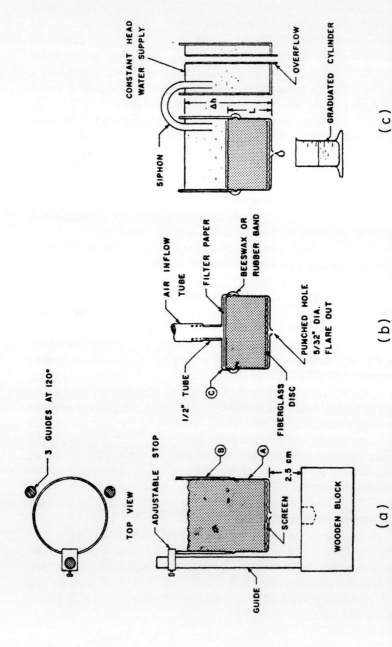

Fig. 6.6. Equipment for air and water permeability of fragmented soils: (a) The soil is compacted by dropping the soil-filled container on a wooden block. Three parallel upright rods guide the container. (b) For air-permeability measurements, the cylinder extension B is removed, the soil in can A is struck off level, and lid C with air-inlet tube is sealed to the can with beeswax or a rubber band. (c) For water-permeability measurements, lid C is replaced with cylinder extension E and water is supplied to the container with a siphon from a constant level water supply tank. (After Reeve, 1965.)

G. Aggregate Stability

Obviously, the value of %SA depends upon the period of time the aggregates are shaken in water. Russell and Feng (1947) reported the following relationship:

$$\log(\%SA) = a - b \log t \tag{6.5}$$

wherein t is time, a is the logarithm of the original weight of the sample, and b is the slope of the $\log(\%SA)$ versus $\log t$ curve. The difference in mean weight–diameter between dry-sieved and wet-sieved samples has also been used as an indicator of stability (De Leenheer and De Boodt, 1954).

The crucial step in this procedure is the initial wetting. Abrupt immersion of dry aggregates may result in rapid disintegration by entrapped air. Gradual wetting is achieved by capillary rise under tension if the aggregates are placed over the water surface for a time, or, even better, by wetting under vacuum with deaired water, which precludes the entrapment of air bubbles entirely (Kemper, 1965).

An alternative approach is to subject soil aggregates to simulated rain. In the *drop method* (McCalla, 1944), individual aggregates are bombarded with drops of water in a standardized manner. The number of drops needed for total dissipation of the aggregate, or the fractional mass of the aggregate remaining after a given time, can be determined. A better way is to subject an aggregated soil surface in the field to periods of simulated rainstorms of controllable raindrop sizes and velocities (Morin *et al.*, 1967; Amerman *et al.*, 1970). The condition of the soil surface can then be compared to the initial condition, and the degree of aggregate stability thereby assessed.

To determine the stability of microaggregates, comparative *sedimentation analysis* can be carried out with dispersed and undispersed samples. An index indicating the fractional amount of clay associated in microaggregates can then be calculated. Still another index of soil structural stability is obtained by comparing the *permeability* of the soil to an inert fluid with the permeability toward water (Reeve, 1965). The permeability of an appropriately packed soil sample is first measured by using air, a fluid that is presumed to have no effect on structure, and then by using water. The latter, being a polar liquid, reacts with the soil, modifying the structure and usually reducing the permeability. A ratio of unity, if such were to occur, would indicate perfect stability. Values greater than unity indicate a scale of increasing instability. The experimental technique is illustrated in Fig. 6.6. Finally, it is possible to assess structural stability by comparing the retention of water in a presaturated sample as a function of increasing suction with the retention of a nonpolar liquid such as toluene.

H. Soil Crusting

As we have already mentioned, it is the aggregates exposed at the soil surface which are most vulnerable to destructive forces. The surface aggregates which collapse and slake down during wetting may form a slick layer of dispersed mud, sometimes several centimeters thick, which clogs the surface macropores and thus tends to inhibit the infiltration of water into the soil and the exchange of gases between the soil and the atmosphere (aeration). Such a layer is often called a *surface seal*. Upon drying, this dispersed layer shrinks to become a dense, hard crust which impedes seedling emergence by its hardness and tears seedling roots as it cracks, forming a characteristic polygonal pattern.

The effect of soil crusting on seedlings depends on crust thickness and strength, as well as on the size of the seeds and vigor of the seedlings. Particularly sensitive are small seeded grasses and vegetables. In soils which exhibit a strong crusting tendency one can often observe that seedling emergence occurs only through the crust's cracks, while numerous unfortunate seedlings lie smothered under the hard crust fragments between cracks. Such an occurrence can doom an entire crop from the very outset.

Attempts have been made to characterize soil crusting, particularly with respect to its effect on seedling emergence, in terms of the resistance of the dry crust to the penetration of a probe (Parker and Taylor, 1965), as well as in terms of its strength as exhibited in the *modulus of rupture* test. (See Chapter 13.) These tests were designed to imitate the process by which a seedling forces its way upward by penetrating and rupturing the crust. Using the modulus of rupture technique, Richards (1953) reported that the emergence of bean seedlings in a fine sandy loam was reduced from 100 to 0% when crust strength increased from 108 to 273 mbar, whereas Allison (1956) found that the emergence of sweet corn was prevented only when crust strength exceeded 1200 mbar. However, the critical crust strength which prevents emergence obviously depends on crust thickness and soil wetness as well as on plant species and depth of seed placement (Hillel, 1972).

Crust strength increases as the rate of drying decreases and as the degree of colloidal dispersion increases. As water evaporates, the soil surface often becomes charged with a relatively high concentration of sodic salts and, consequently, with a high exchangeable sodium percentage. With subsequent infiltration of rain or irrigation water, the salts are leached but the exchangeable sodium percentage (ESP) remains high. The resulting combination of high ESP and low salt concentration induces colloidal dispersion, which contributes to the formation of a dense crust.

I. Soil Conditioners

We have already described the beneficial effect on soil aggregation of various natural polymers, products of the microbial decay of organic matter in the soil. Such substances as polysaccharides and polyuronides promote aggregate stability by gluing particles together within aggregates as well as by coating aggregate surfaces. As the soil dries, the gel-like glues sometimes undergo practically irreversible dehydration, becoming more or less stable cementing agents which bind clay flocs to each other as well as to silt and sand grains.

Considerable work has been done to develop and apply synthetic compounds capable of duplicating the effect of natural polymers. Where natural aggregation or aggregate stability is found lacking, such synthetic polymers, called *soil conditioners*, can help in the stabilization of artificially formed aggregates. Materials have indeed been produced whick are effective in relatively small quantities (e.g., 0.1% of the treated soil mass) and can produce a dramatic improvement of soil structure, with consequent beneficial effect upon infiltration, aeration, and the prevention of crusting and erosion.

The first soil conditioners, introduced commercially in 1951 under the trade name krilium, were a hydrolized polyacrylonitrile (HPAN) and a copolymer of vinyl acetate and maleic acid (VAMA). Subsequently, numerous additional substances have been offered, including polyvinyl acetate (PVAc), polyvinyl alcohol (PVA), polyacrylic acid (PAA), and polyacrylamide (PAM), to mention just a few. Many more formulations, often based on industrial byproducts, have been marketed as soil conditioners even without specification of their chemical composition or their exact physical effects. Some of the substances could be applied in water-soluble form and others in an emulsifiable form: some acted as polyanions, others as polycations, and still others as nonionic binders. The mechanisms by which the various polymeric formulations can stabilize aggregates were at first poorly understood but have since been shown to include electrostatic or exchange reactions, hydrogen bonding, and van der Waals forces (Greenland, 1965). The activity of soil conditioning polymers has been attributed to a quantifiable property called *functionality*, being the number of active groups (e.g., carboxyl amide or sulfonic groups) present per unit mass of the polymer. The effectiveness is also related to molecular weight or polymer chain length (Schamp, 1971), with optimal molecular weights reported to be as high as 10^6.

The introduction of soil conditioners in the early 1950s was greeted initially with great, and probably excessive, enthusiasm. Exaggerated claims were made by a quickly proliferating legion of enterprises bent on a too rapid commercial exploitation of the new discovery. Some (though far from all) of the myriad products introduced into the market were indeed effective in stabilizing aggregates, but the costs, relative to the provable benefits, were

Fig. 6.7. Plow-zone clods: (a) Highly porous and friable and (b) compact and dense. Note the profusion of roots in (a) and their almost total absence in (b). (Courtesy of Dr. A. C. Trouse, National Tillage Machinery Laboratory, U.S. Department of Agriculture, Auburn, Alabama.)

exorbitant. Within a few years, therefore, soil conditioners lost all credibility and were abandoned. The disenchantment may have been as exaggerated, however, as the acclaim which had preceded it. The basic idea of soil conditioning was, and remains, essentially sound. Lately, interest in them has been revived, albeit in a more cautious way and with due consideration of the mechanisms and costs involved (DeBoodt, 1972a,b).

The effect of clod structure on root growth is illustrated in Fig. 6.7.

J. Hydrophobization of Soil Aggregates

A method has recently been developed (Hillel, 1977b) to treat the soil surface zone mechanically and chemically so as to form a top layer of water-repellent clods which seems to improve the water economy of agricultural soils by maximizing the infiltration of water while minimizing evaporation, weed germination, and erosion. Natural mineral surfaces are typically hydrophilic (wettable) in that they exhibit an acute angle of contact with water. Hence a porous assemblage of dry mineral particles normally sorbs water spontaneously by capillary attraction. On the other hand, particle surfaces which adsorb a water-repellent coating are characterized by an obtuse wetting angle, and therefore narrow pores between such particles exhibit capillary repulsion and resist the entry of water (see Fig. 6.8). Hydrophobic materials have been used to help seal a precompacted soil surface against infiltration of rain in order to induce runoff from barren slopes in desert regions, a practice called *water harvesting*. It has been found that the same materials can be used to produce the opposite effect of increasing infiltration in semiarid agricultural regions if the soil surface is first broken into clods and the clods are treated to make them individually water repellent.

Rainwater reaching hydrophobic clods trickles off each clod and flows downward through the open spaces, thus penetrating directly into the deeper soil layers. The bypassed clods themselves remain dry even while the cloddy layer as a whole conducts water. If rain intensity exceeds the soil's maximal absorption rate, then the excess can accrue and remain temporarily in the interclod cavities of the surface zone rather than run off, as it would if the soil surface were smooth and devoid of the cloddy mulch.

The effect of clod hydrophobization can become even more important during the dry spells which generally follow rainfall events in semiarid and arid regions. The evaporation phase normally begins when the soil top layer is in a state of near saturation, so radiation and wind can extract a considerable fraction of moisture contributed by the preceding rain. On the other hand, when a top layer of water-repellent dry clods is present, capillary conduction from the underlying moist soil is inhibited. The layer of dry clods can constitute a barrier through which soil moisture can escape only by the relatively slow process of vapor diffusion.

In addition to runoff and direct evaporation, a major cause of water loss and poor yields is the infestation of weeds, which compete with crop plants

Fig. 6.8. Comparison between natural (hydrophilic) and silicone treated (hydrophobic) clods (a) during wetting and (b) after drying. Unstable natural clods on the right imbibe water and slake down to a muddy suspension which cakes upon drying to form a dense crust. The treated clods on the left remain discrete and stable when water is added or evaporated. (After Hillel, 1977.)

for nutrients, space, and light, as well as for soil moisture. Hydrophobization of top layer clods can inhibit germination of weed seeds in the soil surface zone, since this layer remains dry through repeated cycles of rain and shine. On the other hand, germination of crop seeds can be ensured by a method of planting which cuts a furrow through the top layer and inserts the seeds into the underlying moist soil.

The approach described differs from the soil conditioning approach in that no attempt is made to cement the clods internally but merely to make their surfaces water repellent by treating them with a hydrophobic agent. Of the various chemical agents tested, the organosilicone group appear to be most suitable. Since only monomolecular adsorption is required to transform mineral particle surfaces from a hydrophilic to a hydrophobic state, the desired effect can be achieved by a thin spraying over the peripheries of the clods without total impregnation. However, extensive screening and testing of the soil in place are still required to establish the most persistent and economical chemical formulation and mode of application for different soil and climatic conditions. Indeed, the whole issue of soil structure management remains a vital and challenging topic of research, requiring the cooperation of chemists, physicists, microbiologists, and agronomists.

Sample Problems

1. Calculate the overall porosity and bulk density values for an assemblage of uniform spherical aggregates in open (cubic) and dense (tetrahedral) packing. Assume that each aggregate has a bulk density of 1.80 gm/cm^3. How much of the porosity is due to macro (interaggregate) pores, and how much to micro (intraaggregate) pores in each packing mode?

To determine porosity:

For open packing, the volume of a sphere = $\frac{4}{3}\pi r^3 = (\pi/6)d^3$, where r is radius and d diameter. Assuming the diameter to be of unit length, the volume of the sphere = $\pi/6 = 0.5236$. In cubic packing, each such sphere occupies a cube of unit volume ($d^3 = 1 \times 1 \times 1 = 1$). Hence the macro (interaggregate) porosity = $1 - \pi/6 = 0.4764$. Recall that porosity = $1 - (\rho_b/\rho_s)$, where ρ_b is bulk density and ρ_s particle density. The porosity of each aggregate = $1 - (\rho_b/\rho_s) = 1 - (1.8/2.65) = 0.32$. As a fraction of the unit cube volume, this micro porosity = $0.5236 \times 0.32 = 0.1676$. Therefore, the total porosity = $0.4764 + 0.1676 = 0.644 = 64.4\%$.

For dense packing, in tetrahedral packing, the fractional volume of the spheres = 0.7405 (see Table 6.1). The macro porosity = $1 - 0.7405 = 0.2595$. The porosity of each aggregate, as before = 0.32. As a fraction of total volume, the micro porosity = 0.7405 × 0.32 = 0.237. Therefore, total porosity = 0.2595 + 0.237 = 0.496.

To compute bulk densities (ρ_b): Since porosity, $f = 1 - (\rho_b/\rho_s)$. Bulk density, $\rho_b = \rho_s(1 - f) = 2.65(1 - 0.644) = 0.934$ gm/cm^3 for open packing = $2.65(1 - 0.496) = 1.3396$ gm/cm^3 for dense packing.

Note: Even though the porosity of the aggregates themselves did not change, the overall macroporosity and microporosity (and hence the total porosity and bulk density) changed greatly with the change in packing arrangement.

2. Calculate the mean weight diameters of the assemblages of aggregates given in the following tabulation. The percentages refer to the mass fractions of dry soil in each diameter range.

Aggregate diameter range (mm)	Dry sieving		Wet sieving	
	Virgin soil	Cultivated soil	Virgin soil	Cultivated soil
0.0–0.5	10%	25%	30%	50%
0.5–1.0	10%	25%	15%	25%
1–2	15%	15%	15%	15%
2–5	15%	15%	15%	5%
5–10	20%	10%	15%	4%
10–20	20%	7%	5%	1%
20–50	10%	3%	5%	0%

First, we determine the mean diameters of the seven aggregate diameter ranges. These are

Range: 0–0.5, 0.5–1, 1–2, 2–5, 5–10, 10–20, 20–50 mm
Mean: 0.25, 0.75, 1.5, 3.5, 7.5, 15, 35 mm

Recall that the mean weight diameter X is defined by Eq. (6.2):

$$X = \sum_{i=1}^{i=n} x_i w_i$$

Hence, for the dry-sieved virgin soil,

$X = (0.25 \times 0.1) + (0.75 \times 0.1) + (1.5 \times 0.15) + (3.5 \times 0.15)$
$\quad + (7.5 \times 0.2) + (15 \times 0.2) + (35 \times 0.1) = 8.85$ mm

For the dry-sieved cultivated soil,

$X = (0.25 \times 0.25) + (0.75 \times 0.25) + (1.5 \times 0.15) + (3.5 \times 0.15)$
$\quad + (7.5 \times 0.1) + (15 \times 0.07) + (35 \times 0.03) = 4.30$ mm

Sample Problems

For the wet-sieved virgin soil,

$$X = (0.25 \times 0.3) + (0.75 \times 0.15) + (1.5 \times 0.15) + (3.5 \times 0.15) \\ + (7.5 \times 0.15) + (15 \times 0.05) + (35 \times 0.05) = 4.56 \text{ mm}$$

For the wet-sieved cultivated soil,

$$X = (0.25 \times 0.5) + (0.75 \times 0.25) + (1.5 \times 0.15) + (3.5 \times 0.05) \\ + (7.5 \times 0.04) + (15 \times 0.01) + (35 \times 0.0) = 1.16 \text{ mm}$$

Note: Wet sieving reduced the mean weight diameter from 8.85 to 4.56 mm in the virgin soil and from 4.30 to 1.16 mm in the cultivated soil. This indicates the degree of instability of the various aggregates under the slaking effect of immersion in water. The influence of cultivation is generally to reduce the water stability of soil aggregates and hence to render the soil more vulnerable to crusting and erosion processes.

Part III:

THE LIQUID PHASE

> Water spilt on the ground ...
> cannot be gathered up again ....
> Samuel II, 14:14.

7 Soil Water: Content and Potential

A. Introduction

The variable amount of water contained in a unit mass or volume of soil, and the energy state of water in the soil are important factors affecting the growth of plants. Numerous other soil properties depend very strongly upon water content. Included among these are mechanical properties such as consistency, plasticity, strength, compactibility, penetrability, stickiness, and trafficability. In clayey soils, swelling and shrinkage associated with addition or extraction of water change the overall specific volume (or bulk density) of the soil as well as its pore-size distribution. Soil water content also governs the air content and gas exchange of the soil, thus affecting the respiration of roots, the activity of microorganisms, and the chemical state of the soil (e.g., oxidation–reduction potential).

The per mass or per volume fraction of water in the soil can be characterized in terms of *soil wetness*. The physicochemical condition or state of soil water is characterized in terms of its free energy per unit mass, termed the *potential*. Of the various components of this potential, it is the *pressure* or *matric* potential which characterizes the tenacity with which soil water is held by the soil matrix. Wetness and matric potential are functionally related to each other, and the graphical representation of this relationship is termed the *soil-moisture characteristic* curve. The relationship is not unique, however, as it is affected by direction and rate of change of soil moisture and is sensitive to changes in soil volume and structure. Both wetness and matric potential vary widely in space and time as the soil is wetted by rain, drained by gravity, and dried by evaporation and root extraction.

The wettest possible condition of a soil is that of *saturation*, defined as the condition in which all the soil pores are filled with water. The saturation valu is relatively easy to define in the case of nonswelling (e.g., sandy) soils. It can be difficult or even impossible to define in the case of swelling soils, as such soils may continue to imbibe water and swell even after all pores have been filled with water. The lowest wetness we are likely to encounter in nature is a variable state called *air dryness*, and in the laboratory it is an arbitrary state known as the *oven-dry* condition.

B. The Soil-Water Content (Wetness)[1]

The fractional content of water in the soil can be expressed in terms of either mass or volume ratios. As given in Chapter 2

$$w = M_w/M_s \tag{7.1}$$

$$\theta = V_w/V_t = V_w/(V_s + V_w + V_a) \tag{7.2}$$

where w, the mass wetness, is the dimensionless ratio of water mass M_w to dry soil mass M_s, whereas θ, the volume wetness, is the ratio of water volume V_w to total (bulk) soil volume V_t. The latter is equal to the sum of the volumes of solids (V_s), water (V_w), and air (V_a). Both θ and w are usually multiplied by 100 and reported as percentages by volume or mass.

The two expressions can be related to each other by means of the bulk density ρ_b and the density of water:

$$\theta = w(\rho_b/\rho_w) = w\Gamma_b \tag{7.3}$$

where Γ_b is the *bulk specific gravity* of the soil (a dimensionless ratio which usually lies in the range between 1.1 and 1.7). The conversion is relatively simple for nonswelling soils in which bulk density, and hence bulk specific gravity, are constant regardless of wetness, but it can be difficult in the case of swelling soils as the bulk density must be known as a function of mass wetness. The relation of bulk density (or, rather, of its reciprocal, known as *bulk specific volume*) to mass wetness is illustrated schematically in Fig. 7.1.

In many cases, it is useful to express the water content of a soil profile

---

[1] The author prefers the term *wetness* to *soil water content* not only for reasons of verbal economy but also because wetness implies intensity whereas content implies extensity. One can speak of the water content when referring to the total amount of water in a bucket or in a defined volume of soil but the degree of wetness obviously pertains to the *relative* concentration rather than the absolute amount of water in a porous body independent of the body's size.

C. Measurement of Soil Wetness

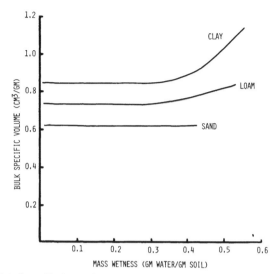

Fig. 7.1. Relation of bulk specific volume (reciprocal of bulk density) to mass wetness for nonexpansive and expansive soils (schematic).

in terms of *depth*, i.e., as the volume of water contained in a specified total depth of soil d_t per unit area of land. This indicates the equivalent depth d_w soil water would have if it were extracted and then ponded over the soil surface. Thus

$$d_w = \theta d_t = w\Gamma_b d_t \tag{7.4}$$

Usually d_w is given in millimeters, as are rainfall and evaporation.

Another expression of soil wetness is the *liquid ratio* ϑ, defined as the volume of water per unit volume of the solid phase:

$$\vartheta = w\frac{\rho_s}{\rho_w} = \theta\frac{\rho_w}{\rho_b}\frac{\rho_s}{\rho_w} = \theta\frac{\rho_s}{\rho_b} \tag{7.5}$$

This expression is useful, along with the *void ratio* (defined in Chapter 2), in connection with soils having a nonrigid (swelling and shrinking) matrix.

C. Measurement of Soil Wetness

The need to determine the amount of water contained in the soil arises frequently in many agronomic, ecological, and hydrological investigations aimed at understanding the soil's chemical, mechanical, hydrological, and biological relationships. There are direct and indirect methods to measure soil moisture (Gardner, 1965), and, as we have already pointed out, there are several alternative ways to express it quantitatively. As yet we have no universally recognized standard method of measurement and no uniform way to compute and present the results of soil moisture measurements. We shall proceed to describe, briefly, some of the most prevalent methods for this determination.

1. SAMPLING AND DRYING

The traditional (gravimetric) method of measuring mass wetness consists of removing a sample by augering into the soil and then determining its moist and dry weights. the *moist weight* is determined by weighing the sample as it is at the time of sampling, and the *dry weight* is obtained after drying the sample to a constant weight in an oven. The more or less standard method of drying is to place the sample in an oven at 105°C for 24 hr. An alternative method of drying, suitable for field use, is to impregnate the sample in a heat-resistant container with alcohol, which is then burned off, thus vaporizing the water (Bouyoucos, 1937). The mass wetness, also called *gravimetric wetness*, is the ratio of the weight loss in drying to the dry weight of the sample (mass and weight being proportional):

$$w = \frac{\text{(wet weight)} - \text{(dry weight)}}{\text{dry weight}} = \frac{\text{weight loss in drying}}{\text{weight of dried sample}} \quad (7.6)$$

The gravimetric method, depending as it does on sampling, transporting, and repeated weighings, entails practically inevitable errors. It is also laborious and time consuming, since a period of at least 24 hr is usually considered necessary for complete oven drying. The standard method of oven drying is itself arbitrary. Some clays may still contain appreciable amounts of adsorbed water (Nutting, 1943) even at 105°C. On the other hand, some organic matter may oxidize and decompose at this temperature so that the weight loss may not be due entirely to the vaporation of water.

The errors of the gravimetric method can be reduced by increasing the sizes and number of samples. However, the sampling method is destructive and may disturb an observation or experimental plot sufficiently to distort the results. For these reasons, many workers prefer indirect methods, which permit making frequent or continuous measurements at the same points, and, once the equipment is installed and calibrated, with much less time, labor, and soil disturbance.

2. ELECTRICAL RESISTANCE

The electrical resistance of a soil volume depends not only upon its water content, but also upon its composition, texture, and soluble-salt concen-

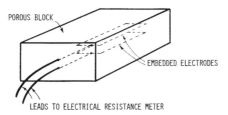

Fig. 7.2. An electrical resistance block (schematic). The embedded electrodes may be plates, screens, or wires in a parallel or in a concentric arrangement.

tration. On the other hand, the electrical resistance of porous bodies placed in the soil and left to equilibrate with soil moisture can sometimes be calibrated against soil wetness. Such units (generally called *electrical resistance blocks*) generally contain a pair of electrodes embedded in gypsum (Bouyoucos and Mick, 1940), nylon, or fiberglass (Colman and Hendrix, 1949). (See Fig. 7.2.)

Porous blocks placed in the soil tend to equilibrate with the soil moisture (matric) suction, rather than with the soil moisture per se. Different soils can have greatly differing wetness versus suction relationships (e.g., a sandy soil may retain less than 5% moisture at, say, 15-bar suctions, whereas a clayey soil may retain three or four times as much). Hence, calibration of porous blocks against suction (tension) is basically preferable to calibration against soil wetness, particularly when the soil used for calibration is a disturbed sample differing in structure from the soil in situ.

The equilibrium of porous blocks with soil moisture may be affected by *hysteresis*, i.e., by the direction of change of soil moisture (whether increasing or decreasing). Furthermore, the hydraulic properties of the blocks (or inadequate contact with the soil) may prevent the rapid attainment of equilibrium and cause a time lag between the state of water in the soil and the state of water being measured in the block. This effect, as well as the sensitivity of the block, may not be constant over the entire range of variation in soil wetness which we may desire to measure. Thus, gypsum blocks are more responsive in the dry range, whereas porous nylon blocks, because of their larger pore sizes, are more sensitive in the wet range of soil moisture variation.

The electrical conductivity of moist porous blocks is due primarily to the permeating fluid rather than to the solid matrix. Thus it depends upon the electrolytic solutes present in the fluid as well as upon the volume content of the fluid. Blocks made of such inert materials as fiberglass, for instance, are highly sensitive to even small variations in salinity of the soil solution. On the other hand, blocks made of plaster of Paris (gypsum) maintain a nearly constant electrolyte concentration corresponding primarily to that of a saturated solution of calcium sulfate. This tends to mask, or buffer, the effect of small or even moderate variations in the soil solution (such as those due to fertilization or low levels of salinity). However, an undesirable consequence of the solubility of gypsum is that these blocks eventually deteriorate in the soil. Hence the relationship between electrical resistance and moisture suction varies not only from block to block but also for each block as a function of time, since the gradual dissolution of the gypsum changes the internal porosity and pore-size distribution of the blocks.

For these and other reasons (e.g., temperature sensitivity) the evaluation of soil wetness by means of electrical resistance blocks is likely to be of limited accuracy. On the other hand, an advantage of these blocks is that they can be connected to a recorder to obtain a continuous indication of soil moisture changes in situ.

3. Neutron Scattering

First developed in the 1950s, this method has gained widespread acceptance as an efficient and reliable technique for monitoring soil moisture in the field (Holmes, 1956; van Bavel, 1963). Its principal advantages over the gravimetric method are that it allows less laborious, more rapid, nondestructive, and periodically repeatable measurements, in the same locations and depths, of the volumetric wetness of a representative volume of soil. The method is practically independent of temperature and pressure. Its main disadvantages, however, are the high initial cost of the instrument, low degree of spatial resolution, difficulty of measuring moisture in the soil surface zone, and the health hazard associated with exposure to neutron and gamma radiation.

The instrument, known as a *neutron moisture meter* (Fig. 7.3) consists of two main components: (a) a *probe*, which is lowered into an access tube[2] inserted vertically into the soil, and which contains a *source of fast neutrons* and a *detector of slow neutrons*; (b) a *scaler* or ratemeter (usually battery powered and portable) to monitor the flux of slow neutrons scattered by the soil.

A source of fast neutrons is generally obtained by mixing a radioactive emitter of *alpha particles* (helium nuclei) with beryllium. Frequently used is a 2–5 millicurie pelletized mixture of radium and beryllium. An Ra–Be source emits about 16,000 neutrons per second per milligram (or millicurie) of radium. The energies of the neutrons emitted by this source vary from 1 to 15 MeV (million electron volts), with a preponderant energy range

---

[2] The purpose of the access tube is both to maintain the bore hole into which the probe is lowered and to standardize measuring conditions. Aluminum tubing is usually the preferred material for access tubes since it is nearly transparent to a neutron flux.

C. Measurement of Soil Wetness

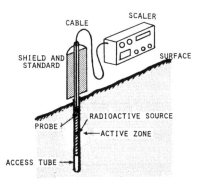

Fig. 7.3. Components of a portable neutron soil-moisture meter, including a probe (with a source of fast neutrons and a detector of slow neutrons) lowered from a shield containing hydrogenous material (e.g., paraffin, polyethylene, etc.) into the soil via an access tube. A scaler–ratemeter is shown alongside the probe. Recent models incorporate the scaler into the shield body and the integrated unit weights no more than 8 kgm.

of 2–4 MeV and an average speed of about 1600 km/sec. Hence, they are called *fast neutrons*. An alternative source of fast neutrons is a mixture of americium and beryllium. Both radium and americium incidentally also emit gamma radiation, but that of the americium is lower in energy and hence less hazardous than that of the radium. The source materials are chosen for their longevity (e.g., radium–beryllium has a half-life of about 1620 yr) so that they can be used for a number of years without an appreciable change in radiation flux.

The fast neutrons are emitted radially into the soil, where they encounter and collide elastically (as do billiard balls) with various atomic nuclei. Through repeated collisions, the neutrons are deflected and "scattered," and they gradually lose some of their kinetic energy. As the speed of the initially fast neutrons diminishes, it approaches a speed which is characteristic for particles at the ambient temperature. For neutrons this is about 2.7 km/sec, corresponding to an energy of about 0.03 eV. Neutrons that have been slowed to such a speed are said to be *thermalized* and are called *slow neutrons*. Such slow neutrons continue to interact with the soil and are eventually absorbed by the nuclei present.

The effectiveness of various nuclei present in the soil in moderating or thermalizing fast neutrons varies widely. The average loss of energy is maximal for collisions between particles of approximately the same mass. Of all nuclei encountered in the soil, the ones most nearly equal in mass to neutrons are the nuclei of hydrogen, which are therefore the most effective fast neutron moderators of all soil constituents. Thus, the average number of collisions required to slow a neutron from, say, 2 MeV to thermal energies is 18 for hydrogen, 114 for carbon, 150 for oxygen, and $9N + 6$ for nuclei of larger mass number N. If the soil contains an appreciable concentration of hydrogen, the emitted fast neutrons are thermalized before they get very far from the source, and the slow neutrons thus produced scatter randomly in the soil, quickly forming a swarm or cloud of constant density around the probe. The equilibrium density of the slow neutron cloud is determined by the rate of emission by the source and the rates of thermalization and absorption by the medium (i.e., soil) and is established within a small fraction of a second. Certain elements which might be present in the soil, incidentally, exhibit a high absorption capacity for slow neutrons (e.g., boron, cadmium, and chlorine), and their presence in nonnegligible concentrations might tend to decrease the density of slow neutrons. By and large, however, the density of slow neutrons formed around the probe is nearly proportional to the concentration of hydrogen in the soil, and therefore more or less proportional to the volume fraction of water present in the soil. Thus

$$N_w = m\theta + b \quad \text{and} \quad N_w/N_s = y\theta \tag{7.7}$$

in which N_w is slow neutron count rate in wet soil, N_s is count rate in water or in a standard absorber (i.e., the shield of the probe), θ is volumetric wetness, and m and b are the slope and x intercept, respectively, of the line indicating N_w as function of θ.

As the thermalized neutrons repeatedly collide and bounce about randomly, a number of them (proportional to the density of neutrons thus thermalized and scattered, and hence approximately linearly related to the concentration of soil moisture) return to the probe, where they are counted by the detector of slow neutrons. The detector cell is usually filled with BF_3 gas. When a thermalized neutron encounters a ^{10}B nucleus and is absorbed, an alpha particle is emitted, creating an electrical pulse on a charged wire. The number of pulses over a measured time interval is counted by a scaler, or indicated by a ratemeter.

The effective volume of soil in which the water content is measured depends upon the energy of emitted neutrons as well as upon the concentration of hydrogen nuclei; i.e., for a given source and soil, it depends by and large upon the volume concentration of soil moisture. If the soil is dry rather than wet, the cloud of slow neutrons surrounding the probe will be less dense and extend farther from the source, and vice versa for wet soil. With the commonly used radium–beryllium or americium–beryllium sources, the so-called *sphere of influence*, or effective volume of measurement, varies with a radius of less than 10 cm in a wet soil to 25 cm or more in a dry soil. This low and variable degree of spatial resolution makes the neutron moisture meter unsuitable for the detection of moisture profile discontinuities (e.g., wetting fronts or boundaries between distinct layers in the soil). Moreover, measurements close to the surface (say, within 20 cm of the surface, depending on soil wetness) are precluded[3] because of the escape of fast neutrons through the surface. However, the relatively large volume monitored can be an advantage in water balance studies, for instance, as such a volume is more truly representative of field soil than the very much smaller samples generally taken for the gravimetric measurement of soil moisture.

Although attempts have been made to predict the relation between count rate and soil wetness from theoretical considerations, such are the anomalies and variability among different soils that it is best to calibrate the neutron moisture meter specifically for each soil type and set of circumstances in which it is to be used. The "universal" calibration curves provided by manufacturers may not be directly applicable to various field conditions. Calibration can be carried out in soil-filled bins of predetermined moisture content, or directly in the field by making numerous measurements of soil moisture with the neutron and the gravimetric methods concurrently. Since in most soils a nearly linear relation between count rate and volumetric wetness is obtained, the neutron method can be said to provide uniform precision throughout the entire range of variation of soil moisture. In reality, calibration curves often deviate from a perfectly straight line, and the degree of deviation depends both on the soil and on the probe used (i.e., the positioning of the detector relative to the source, etc.).

It is often assumed that neutron thermalization in the soil is due solely to the presence of water. In actual situations, appreciable amounts of hydrogen are present in clay and, particularly, in organic matter. Moreover, in certain rare cases the boron content of the soil may be sufficiently high to cause an appreciable lowering of slow neutron density, since boron is a particularly effective absorber of slow neutrons. To a lesser degree, this is also true for chlorine. It has been estimated that a boron concentration exceeding 10 ppm (parts per million) and a total chlorine concentration exceeding 1000 ppm could affect the calibration curve to a significant degree.

For the sake of safety, and also to provide a convenient means of making standard readings, the probe containing the fast neutron source is normally carried inside a protective *shield*, which is a cylindrical container filled with lead and some hydrogenous material such as paraffin or polyethylene designed to prevent the escape of fast neutrons. The shield should also protect users of the neutron soil moisture meter against emitted gamma radiation. Improper or excessive use of the equipment can be hazardous. The danger from exposure to radiation depends upon the strength of the source, the quality of the shield, the distance from source to operator, and the duration of contact. With strict observance of safety rules, the equipment can be

---

[3] A special surface probe is available commercially to allow measurement of the average moisture in the soil's top layer. It is used mostly in engineering practice in connection with soil compaction.

used without undue risk. However, it is altogether too easy to become complacent and careless, since the radiation can be neither seen nor felt. A recent analysis of the radiation exposure hazard was given by Gee et al. (1976).

4. Gamma-Ray Absorption

The *gamma-ray scanner* for measuring soil moisture generally consists of two spatially separated units, or probes: (1) a *source*, usually containing a pellet of radioactive cesium (^{137}Cs emitting gamma radiation with an energy of 0.661 MeV); and (2) a *detector*, normally consisting of a scintillation counter (e.g., sodium iodide crystal or a synthetic scintillator) connected to a photomultiplier and preamplifier. If the emission of radiation is monoenergetic and radial, and if the space between the source and the detector is empty and the two units are a constant distance apart, then the fraction of the emitted radiation received by the detector will depend only on the angular section intercepted, that is, on the distance of separation and the size of the scintillation unit. On the other hand, should the space between the units be filled with some material, a fraction of the original radiation which would otherwise be detected will be absorbed, depending upon the interposing mass, i.e., upon the thickness and density of the material. In the event the material placed between source and detector is a body of soil of constant bulk density, the intensity of the transmitted radiation will vary only with changes in water content. In fact, it will be an exponential function of soil wetness as follows (Gurr, 1962); Ferguson and Gardner, 1962):

$$N_w/N_d = \exp(-\theta_m \mu_w x) \tag{7.8}$$

where N_w/N_d is the ratio of the monoenergetic radiation flux transmitted through wet soil (N_w) to that transmitted through dry soil (N_d), μ_w is the mass attenuation coefficient for water, x is the thickness of the transmitting soil body, and θ_m is the mass of water per unit bulk volume of soil. Equation (7.8) can be transformed to give soil wetness as an explicit function of the relative transmission rate:

$$\theta_m = -\frac{\ln(N_w/N_d)}{\mu_w x} = -\frac{\log_{10}(N_w/N_d)}{0.4343 \mu_w x} \tag{7.9}$$

where $\ln(N_w/N_d)$ is the natural logarithm of the count ratio for wet to dry soil, and $\log_{10}$ is the common logarithm for the same ratio.

In the above equations, the attenuation due to air in the soil sample is assumed to be negligible (owing to the low density of air) as is the attenuation due to the material constituting the generally thin walls separating the soil from the source and detector units. Since the equations pertain only to the transmission of a primary beam of monoenergetic radiation, all secondary radiation resulting from the interactions of the primary rays in the soil must be screened out of the measurement. This is done by means of a discriminator or pulse height analyzer connected to the scaler.

The gamma-ray absorption method is used mostly in the laboratory, where the dimensions and density of the soil sample, as well as the ambient temperature, can be precisely controlled. A high degree of spatial resolution (e.g., 2 mm or so) can be obtained by collimation of the radiation, i.e., by drilling a narrow hole or slot into the lead wall shielding the source (as well, perhaps, as into a second shield placed in front of the detector), thus allowing passage of only a very narrow beam. Since the absorption of radiation is dependent upon the entire mass intervening between source and detector, the readings can only be related uniquely to changing soil moisture if soil bulk density is constant or if its change is monitored simultaneously. To permit simultaneous measurement of bulk density and soil moisture changes in swelling or shrinking soils, dual-source gamma-ray scanners have been developed, in which both cesium 137 and americium 241 are used. Analysis of the concurrent transmission of the two beams can allow separation of the change in total attenuating mass between that due to soil bulk density and that due to soil moisture (Gardner, 1965).

The double-probe gamma-ray method has also been adapted to field use (Fig. 7.4; Vomocil, 1954). In principle, this technique offers several advantages over the neutron moisture meter in that it allows much better depth

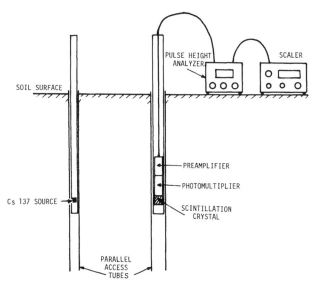

Fig. 7.4. Double-probe gamma-ray apparatus for monitoring soil moisture or density.

resolution in measurement of soil moisture profiles (i.e., about 1 cm in effective measurement width), sufficient to detect discontinuities between profile layers as well as movement of wetting fronts and conditions prevailing near the soil surface. However, the field device is still too cumbersome for general usage. Not the least of the problems is the accurate installation and alignment of two access tubes which must be strictly parallel, and the accurate determination of soil bulk density as it might vary in depth and time. Problems of temperature sensitivity of the electronic device which plagued early designs can apparently be solved, but field calibration with the high degree of depth resolution required remains a difficulty.[4]

5. OTHER METHODS

Additional approaches to the measurement of soil wetness include techniques based on the dependence of soil thermal properties upon water content, and the use of ultrasonic waves, radar waves, and dielectric properties. Some of these and other methods have been tried in connection with the remote sensing of land areas from aircraft or satellites. However, most of the methods currently under development are not yet practical for routine use in the field.

D. Energy State of Soil Water

Soil water, like other bodies in nature, can contain energy in different quantities and forms. Classical physics recognizes two principal forms of energy, *kinetic* and *potential*. Since the movement of water in the soil is quite slow, its kinetic energy, which is proportional to the velocity squared, is generally considered to be negligible. On the other hand, the potential energy, which is due to position or internal condition, is of primary importance in determining the state and movement of water in the soil.

The potential energy of soil water varies over a very wide range. Differences in potential energy of water between one point and another give rise to the tendency of water to flow within the soil. The spontaneous and universal tendency of all matter in nature is to move from where the potential energy is higher to where it is lower and for each parcel of matter to equilibrate with its surroundings. Soil water obeys the same universal pursuit of that elusive state known as *equilibrium*, definable as a condition of uniform potential energy throughout. In the soil, water moves constantly in the

---

[4] The health hazard associated with use of gamma-ray equipment is similar in principle to that discussed in connection with the neutron moisture meter. The equipment is considered safe only if strict attention is paid to all the safety rules.

D. Energy State of Soil Water

direction of decreasing potential energy. The rate of decrease of potential energy with distance is in fact the moving force causing flow. A knowledge of the relative potential energy state of soil water at each point within the soil can allow us to evaluate the forces acting on soil water in all directions, and to determine how far the water in a soil system is from equilibrium. This is analagous to the well-known fact that an object will tend to fall spontaneously from a higher to a lower elevation, but that lifting it requires work. Since potential energy is a measure of the amount of work a body can perform by virtue of the energy stored in it, knowing the potential energy state of water in the soil and in the plant growing in that soil can help us to estimate how much work the plant must expend to extract a unit amount of water.

Clearly, it is not the absolute amount of potential energy "contained" in the water which is important in itself, but rather the relative level of that energy in different regions within the soil. The concept of *soil-water potential*[5] is a criterion, or yardstick, for this energy. It expresses the specific potential energy of soil water relative to that of water in a standard reference state. The standard state generally used is that of a hypothetical reservoir of pure water, at atmospheric pressure, at the same temperature as that of soil water (or at any other specified temperature), and at a given and constant elevation. Since the elevation of this hypothetical reservoir can be set at will, it follows that the potential which is determined by comparison with this standard is not absolute, but by employing even so arbitrary a criterion we can determine the relative magnitude of the specific potential energy of water at different locations or times within the soil.

Just as an energy increment can be viewed as the product of a force and a distance increment, so the ratio of a potential energy increment to a distance increment can be viewed as constituting a force. Accordingly, the force acting on soil water, directed from a zone of higher to a zone of lower potential, is equal to the negative *potential gradient* ($-d\phi/dx$), which is the change of energy potential with distance x. The negative sign indicates that the force acts in the direction of decreasing potential.

The concept of soil-water potential is of great fundamental importance. This concept replaces the arbitrary categorizations which prevailed in the early stages of the development of soil physics and which purported to recognize and classify different forms of soil water: e.g., gravitational water, capillary water, hygroscopic water. The fact is that all of soil water, not merely a part of it, is affected by the earth's gravitational field, so that in

---

[5] The potential concept was first applied to soil water by Buckingham, in his classic and still-pertinent paper on the "capillary" potential (1907). Gardner (1920) showed how this potential is dependent upon the water content. Richards (1931) developed the tensiometer for measuring it in situ.

effect it is all gravitational. Furthermore, the laws of capillarity do not begin or cease at certain values of wetness or pore sizes.

In what way, then, does water differ from place to place and from time to time within the soil? Not in form, but in potential energy. The possible values of soil-water potential are continuous, and do not exhibit any abrupt discontinuities or changes from one condition to another (except in changes of phase). Rather than attempt to classify soil water, our task therefore is to obtain the measure of its potential energy state.

When the soil is saturated and its water is at a hydrostatic pressure greater than the atmospheric pressure (as, for instance, under a water table) the potential energy level of that water may be greater than that of the reference state reservoir described, and water will tend to move spontaneously from the soil into such a reservoir. If, on the other hand, the soil is moist but unsaturated, its water will no longer be free to flow out toward a reservoir at atmospheric pressure. On the contrary, the spontaneous tendency will be for the soil to draw water from such a reservoir if placed in contact with it, much as a blotter draws ink.

Under hydrostatic pressures greater than atmospheric, the potential of soil water (in the absence of osmotic effects) is greater than that of the reference state and therefore can be considered positive. In an unsaturated soil, the water is constrained by capillary and adsorptive forces, hence its energy potential is generally negative, since its equivalent hydrostatic pressure is less than that of the reference state. The potential energy of soil water is also reduced by the presence of solutes, i.e., by the osmotic effect.

Under normal conditions in the field, the soil is generally unsaturated and the soil-water potential is negative. Its magnitude at any point depends not only on hydrostatic pressure but also upon such additional physical factors as elevation (relative to that of the reference elevation), concentration of solutes, and temperature.

E. Total Soil-Water Potential

We have already described the energy potential of soil water in a qualitative way. Thermodynamically, this energy potential can be regarded in terms of the difference in partial specific free energy between soil water and standard water. More explicitly, a soil physics terminology committee of the International Soil Science Society (Aslyng, 1963) defined the *total potential* of soil water as "the amount of work that must be done per unit quantity of pure water in order to transport reversibly and isothermally an infinitesimal quantity of water from a pool of pure water at a specified

E. Total Soil-Water Potential

elevation at atmospheric pressure to the soil water (at the point under consideration)."

This is merely a formal definition, since in actual practice the potential is not measured by transporting water as per the definition, but by measuring some other property related to the potential in some known way (e.g., hydrostatic pressure, vapor pressure, elevation, etc.). The definition specifies transporting an infinitesimal quantity, in any case, to ensure that the determination procedure does not change either the reference state (i.e., the pool of pure, free water) or the soil-water potential being measured. It should be recognized that this definition provides a conceptual rather than an actual working tool. It can be argued (in view of the hysteresis phenomenon to be discussed in Section I) that no change in soil wetness can in practice be carried out reversibly, or that the total potential need not be restricted to isothermal conditions. A most serious difficulty is encountered in attempting to allocate the total potential among the various components or mechanisms comprising it, since these may not be mutually independent.

The above definition is based upon the specific differential Gibbs free energy function. The differential form provides a criterion of equilibrium and of the direction in which changes can be expected to occur in nonequilibrium systems. Philip (1960) introduced the integral form of the thermodynamic function,[6] to provide a criterion for the complete potential energy of a soil–water system.

Soil water is subject to a number of force fields, which cause its potential to differ from that of pure, free water. Such force fields result from the attraction of the solid matrix for water, as well as from the presence of solutes and the action of external gas pressure and gravitation. Accordingly, the *total potential* of soil water can be thought of as the sum of the separate contributions of these various factors, as follows:

$$\phi_t = \phi_g + \phi_p + \phi_o + \cdots \quad (7.10)$$

where ϕ_t is the total potential, ϕ_g the gravitational potential, ϕ_p the pressure (or matric) potential, ϕ_o the osmotic potential, and the ellipses signify that additional terms are theoretically possible.

Not all of the separate potentials given in Eq. (7.10) act in the same way, and their separate gradients may not always be equally effective in causing flow (for example, the osmotic potential gradient requires a semipermeable membrane to induce liquid flow). The main advantage of the total-potential concept is that it provides a unified measure by which the state of water can

---

[6] Defined for a small element of soil of a given wetness as minus the work required, isothermally and reversible, per unit quantity of water, to *completely* remove the water from the soil and to transform it into pure, free water at some datum level.

be evaluated at any time and everywhere within the soil–plant–atmosphere continuum.

F. Thermodynamic Basis of the Potential Concept

It might be useful at this point to digress from our topic of soil water in order to clarify the thermodynamic background of the potential concept.

Over the past few decades, numerous attempts have been made to apply the principles and terminology of thermodynamics to the retention and movement of water in the soil–plant system. An early and comprehensive effort in this direction was made by Edlefsen and Anderson (1943). Classical thermodynamics (Guggenheim, 1959) deals with equilibrium states and reversible processes and can thus serve to describe the forces acting on water and its energy of retention. However, equilibrium states occur only rarely in nature, and spontaneous processes tend to be irreversible. To describe such processes, a branch of thermodynamics known as nonequilibrium or irreversible thermodynamics, has been developed in recent decades (Prigogine, 1961; de Groot, 1963; Katchalsky and Curran, 1965). The application of irreversible thermodynamics to soil-water phenomena will be mentioned in another chapter, while the present one is based on classical thermodynamic relations.

The *soil-water potential* concept depends ultimately upon the first and second laws of classical thermodynamics. The first law is merely the well-known energy conservation law, which states that energy can be converted from one form to another, but can be neither created nor destroyed. In equation form;

$$dQ = dU + P\,dV + dW \tag{7.11}$$

where dQ is heat added to the system, dU is change in internal energy U of the system, $P\,dV$ is the work of expansion done by the system (P pressure, V volume), and dW is all the other work done by the system on its surroundings.

The second law of thermodynamics specifies that the direction of change in an isolated system is always toward equilibrium. This law has subtle and far-reaching implications. It has been stated verbally in several different ways,[7] none of which appears to convey its complete meaning to the layman.

---

[7] For instance (Reid, 1960): "Heat can be conveyed from a lower to a higher temperature only by expenditure of work," or "it is impossible to devise a machine whose only net effect is to remove heat from a reservoir and to lift a weight," or "all natural or spontaneous processes are irreversible," or "the total mount of entropy in nature is increasing."

F. Thermodynamic Basis of the Potential Concept

Mathematically, the second law rests on the definition of two properties, the *absolute temperature* T (being always positive) and the entropy S, such that

$$dQ = T\,dS \quad \text{for reversible processes}$$
$$dQ < T\,dS \quad \text{for irreversible processes} \quad (7.12)$$

where dQ is, as before, the heat input into the system and dS is the change in entropy. The intensive property[8] of temperature is familiar and hence intuitively understandable. The meaning of entropy, however, is not readily apparent. It is a measure of the internal disorder or randomness of a system. The change in entropy is equal to the ratio of the heat input to the temperature of the system, i.e., $dS = dQ/T$. In irreversible processes, dS in the system is greater than zero, and thus the entropy tends to increase spontaneously.[9]

The second law of thermodynamics can now be stated as

$$dU = T\,dS - P\,dV \quad (7.13)$$

where dU is, again, the change in internal energy U of the system.

In a system of variable composition, the total differential of the internal energy can be expressed as a function of S, V, and n_i, where n_i is the number of moles of a given component in the system (Guggenheim, 1959; Slatyer, 1967):

$$dU = (\partial U/\partial S)_{V,n_i}\,dS + (\partial U/\partial V)_{S,n_i}\,dV + (\partial U/\partial n_i)_{S,V,n_j}\,dn_i \quad (7.14)$$

A very useful thermodynamic quantity is the *Gibbs free energy* G defined as

$$G = U + PV - TS \quad (7.15)$$

The *chemical potential* μ_i of a component in a system of variable composition is defined as the partial molal Gibbs free energy of that component, $\bar{G}_i$. The change in the free energy of the system with change in the concentration of the component considered is equivalent to

$$\bar{G}_i = (\partial G/\partial n_i)_{T,P,n_j} = \mu_i \quad (7.16)$$

---

[8] *Intensive properties* (e.g., temperature, pressure, concentration) are independent of the size of the system, whereas *extensive properties* (e.g., mass, volume) are defined by the system as a whole.

[9] Values for the entropy of water at different temperatures were reported by de Jong (1968):

Temperature (°K):	0	250	273	273	298	298
State:	—	Solid	Solid	Liquid	Liquid	Vapor
Entropy (cal/mole °K):	0	9.0	9.8	15.1	16.7	45.1

The total differential of the chemical potential is

$$d\mu_i = \left(\frac{\partial \mu_i}{\partial T}\right)_{P,n_i} dT - \left(\frac{\partial \mu_i}{\partial P}\right)_{T,n_i} dP - \left(\frac{\partial \mu_i}{\partial n_i}\right)_{T,P,n_j} dn_i \quad (7.17)$$

Thus the chemical potential is an expression of the potential energy state of a component in a mixed system in the absence of external forces, i.e., when temperature, pressure, and composition are the only effective variables. The chemical potential thus excludes the effects of gravitational, centrifugal, or externally imposed electrical force fields. The chemical potential is constant in the system when the intensive parameters of temperature, pressure, and concentration are constant and a state of equilibrium exists. In an unequilibrated system, a difference in chemical potential of a component between two locations determines the direction (but not necessarily the rate) in which the component will tend to move spontaneously within the system.

How these relationships apply in the case of soil water is still a matter of some controversy. A comprehensive and critical review of this subject was given by Bolt and Frissel (1960). The major difficulty encountered in the formulation of a working equation for the potential of soil water is the selection of the independent variables to ensure that they do not overlap (i.e., that the various effects are separated in the terms of the summation equation). In particular, this difficulty arises from the complex nature of the forces of interaction between water and the solid matrix (including adsorption, exchangeable ion, and capillary effects).

The difference in chemical potential between water in the soil and pure free water at the same temperature has been called the *moisture potential* of the soil. The *total potential* given in the preceding section includes a gravitational term as well.

We shall now proceed to describe the various components of the total potential of soil water separately.

G. Gravitational Potential

Every body on the earth's surface is attracted toward the earth's center by a gravitational force equal to the weight of the body, that weight being the product of the mass of the body and the gravitational acceleration. To raise a body against this attraction, work must be expended, and this work is stored by the raised body in the form of *gravitational potential energy*. The amount of this energy depends on the body's position in the gravitational force field.

H. Pressure Potential

The gravitational potential of soil water at each point is determined by the elevation of the point relative to some arbitrary reference level. For the sake of convenience, it is customary to set the reference level at the elevation of a pertinent point within the soil, or below the soil profile being considered, so that the gravitational potential can always be taken as positive or zero. On the other hand, if the soil surface is chosen as the reference level, as is often done, then the gravitational potential for all points below the surface is negative with respect to that reference level.

At a height z above a reference, the gravitational potential energy E_g of a mass M of water occupying a volume V is

$$E_g = Mgz = \rho_w V g z \tag{7.18}$$

where ρ_w is the density of water and g the acceleration of gravity. Accordingly, the gravitational potential in terms of the potential energy per unit mass is

$$\phi_g = gz \tag{7.19}$$

and in terms of potential energy per unit volume is

$$\phi_{g,v} = \rho_w g z \tag{7.20}$$

The gravitational potential is independent of the chemical and pressure conditions of soil water, and dependent only on relative elevation.

H. Pressure Potential

When soil water is at hydrostatic pressure greater than atmospheric, its pressure potential is considered positive. When it is at a pressure lower than atmospheric (a subpressure commonly known as *tension* or *suction*) the pressure potential is considered negative. Thus, water under a free-water surface is at positive pressure potential, while water at such a surface is at zero pressure potential, and water which has risen in a capillary tube above that surface is characterized by a negative pressure potential. This principle is illustrated in Fig. 7.5.

The positive pressure potential which occurs below the ground water level has been termed the *submergence potential*. The hydrostatic pressure P of water with reference to the atmospheric pressure is

$$P = \rho g h \tag{7.21}$$

where h is the submergence depth below the free-water surface (called the *piezometric head*.

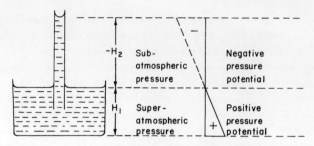

Fig. 7.5. Superatmospheric and subatmospheric pressures below and above a free-water surface.

The potential energy[10] of this water is then

$$E = P\,dV \tag{7.22}$$

and thus the submergence potential, taken as the potential energy per unit volume, is equal to the hydrostatic pressure, P:

$$\phi_{ps} = P \tag{7.23}$$

A negative pressure potential has often been termed *capillary potential*, and more recently, *matric potential*.[11] This potential of soil water results from the capillary and adsorptive forces due to the soil matrix. These forces attract and bind water in the soil and lower its potential energy below that of bulk water. As shown in Chapter 3, capillarity results from the surface tension of water and its contact angle with the solid particles. In an unsaturated (three-phase) soil system, curved menisci form which obey the equation of capillarity

$$P_0 - P_c = \Delta P = \gamma(1/R_1 + 1/R_2) \tag{7.24}$$

where P_0 is the atmospheric pressure, conventionally taken as zero, P_c the

---

[10] To describe swelling and compressible soils, Philip (1969) postulated the existence of an "overburden potential" in addition to the other manifestations of a pressure potential. Where any addition of water to the medium demands a local increase in bulk volume, the overburden potential is the work performed (per unit weight of added water) against gravity and the external load.

[11] The terms matric potential, matric suction, and soil-water suction have been used interchangeably. According to the I.S.S.S. committee cited in Section E, the matric suction is defined as "the negative gauge pressure, relative to the external gas pressure on the soil water, to which a solution identical in composition with the soil solution must be subjected in order to be in equilibrium through a porous membrane wall with the water in the soil." This definition implies the use of either a *tensiometer*, relative to the prevailing atmospheric pressure (conventionally taken to be zero) of the gas phase, or of a *pressure plate extraction apparatus*, in which the gas phase is pressurized sufficiently to bring the water phase to atmospheric pressure. These techniques are described in Section N.

H. Pressure Potential

pressure of soil water, which can be smaller than atmospheric, ΔP is the pressure deficit, or subpressure, of soil water, γ the surface tension of water, and R_1 and R_2 are the principal radii of curvature of a point on the meniscus.

If the soil were like a simple bundle of capillary tubes, the equations of capillarity might by themselves suffice to describe the relation of the negative pressure potential, or tension, to the radii of the soil pores in which the menisci are contained. However, in addition to the capillary phenomenon, the soil also exhibits adsorption, which forms hydration envelopes over the particle surfaces. These two mechanisms of soil–water interaction are illustrated in Fig. 7.6.

The presence of water in films as well as under concave menisci is most important in clayey soil and at high suctions, and it is influenced by the electric double layer and the exchangeable cations present. In sandy soils, adsorption is relatively unimportant and the capillary effect predominates. In general, however, the negative pressure potential results from the combined effect of the two mechanisms, which cannot easily be separated, since the capillary "wedges" are at a state of internal equilibrium with the adsorption "films," and the ones cannot be changed without affecting the others. Hence, the older term capillary potential is inadequate and the better term is *matric potential*, as it denotes the total effect resulting from the affinity of the water to the whole matrix of the soil, including its pores and particle surfaces together.

Some soil physicists prefer to separate the positive pressure potential from the matric potential, assuming the two to be mutually exclusive. Accordingly, soil water may exhibit either of the two potentials, but not both simultaneously. Unsaturated soil has no pressure potential, only a matric potential, expressible in negative pressure units. This, however, is really a matter of formality. There is an advantage in unifying the positive pressure poten-

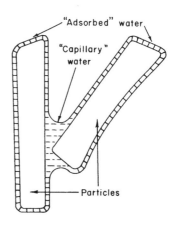

Fig. 7.6. Water in an unsaturated soil is subject to capillarity and adsorption, which combine to produce a matric suction.

tial and the matric potential (with the latter considered merely as a negative pressure potential) in that this unified concept allows one to consider the entire moisture profile in the field in terms of a single continuous potential extending from the saturated region into the unsaturated region, below and above the water table.

An additional factor which may affect the pressure of soil water is a possible change in the pressure of the ambient air. In general, this effect is negligible, as the atmospheric pressure remains nearly constant, small barometric pressure fluctuations notwithstanding. However, in the laboratory, the application of air pressure to change soil-water pressure or suction is a common practice. Hence, this effect has been recognized and termed the *pneumatic potential*. In an unsaturated soil ϕ_p can be taken as equal to the sum of the matric (ϕ_m) and pneumatic (ϕ_a) potentials.

In the absence of solutes, the liquid and vapor phases in an unsaturated porous medium are related at equilibrium by

$$\phi_m = RT \ln(p/p_0) \qquad (7.25)$$

where p/p_0 is the relative humidity, R the gas constant for water vapor, and T the absolute temperature. The term p/p_0 represents the ratio of the partial pressure of vapor in the air phase relative to the partial pressure at saturation.

I. Osmotic Potential

The presence of solutes in soil water affects its thermodynamic properties and lowers its potential energy. In particular, solutes lower the vapor pressure of soil water. While this phenomenon may not affect liquid flow in the soil significantly, it does come into play whenever a membrane or diffusion barrier is present which transmits water more readily than salts. The osmotic effect is important in the interaction between plant roots and soil, as well as in processes involving vapor diffusion. A more complete discussion of the osmotic effect was given in Chapter 3. A discussion of solute transport will be given in Chapter 10.

J. Revised Terminology

The foregoing terminology pertaining to soil water potential was based largely upon recommendations of an International Soil Science Society (I.S.S.S.) Soil Physics Committee (A. Aslying, 1963). More recently, this terminology was reexamined by another committee of the I.S.S.S. (Bolt,

K. Quantitative Expression of Soil-Water Potential

1976). It provided a more complete analysis of soil-water phenomena and recommended that the earlier terminology be modified somewhat. First, the committee recommended adoption by soil physicists of SI units as part of the global effort by physical scientists to standardize a system of units based on the mks (meter–kilogram–second) rather than on the heretofore more popular cgs (centimeter–gram–second) system.[12] Second, in contrast to the 1963 report, the term "water" was limited to the chemical constituent H_2O, whereas "liquid phase" was used to indicate the aqueous solution of the soil. In our usage, that aqueous solution is called "soil water" or the "soil solution." The committee accepted our earlier use of the term "wetness" for the mass of water lost from the soil upon drying at 105°C, relative to the mass of dry soil. However, instead of our term "volumetric wetness" the committee recommended "volume fraction of liquid," which seems cumbersome. The terms "total potential," "gravitational potential," "osmotic potential," and "pressure potential" were left more or less intact, but the terms "wetness potential" or "tensiometric potential" were used in lieu of "matric potential," to encompass swelling soils as well as rigid matrix soils. Neither of the new terms seems preferable to matric potential, and the term tensiometric potential seems especially inadequate, as it refers to a particular instrument capable of measuring only a small part of the total range of the potential in question (namely, the matric potential, which itself is part of the pressure potential). Finally, the committee specified three components of the pressure potential: "wetness potential" as already mentioned, "envelope-pressure potential," which was found preferable to "overburden potential," and "penumatic potential," referring to the excess gas pressure made to act on the soil's liquid phase.[13]

K. Quantitative Expression of Soil-Water Potential

The soil-water potential (Table 7.1) is expressible physically in at least three ways.

1. *Energy per unit mass:* This is often taken to be the fundamental expression of potential, using units of ergs per gram or joules per kilogram. The dimensions of energy per unit mass are L^2T^{-2}.

---

[12] In our book, we continue to use the cgs. system, not wanting to be at variance with the main body of the soil physics literature, which still uses cgs. almost universally.

[13] Lest this interminable terminological debate deter newcomers from entering the fray of soil physics, they might take comfort from the fact that their confusion is fully shared by the professionals. Viewed more positively, however, the lack of a consensus on terminology can be taken as the characteristic of a vigorous and challenging young science . . . .

Table 7.1

ENERGY LEVELS OF SOIL WATER

Soil-water potential				Soil-water suction[a]		Vapor pressure (torr) 20°C (%)	Relative humidity[b] at 20°C (%)
Per unit mass		Per unit volume		Pressure (bar)	Head (cm H$_2$O)		
erg/gm	joule/kg	bar	cm H$_2$O				
0	0	0	0	0	0	17.5350	100.00
-1×10^4	-1	-0.01	-10.2	0.01	10.2	17.5349	100.00
-5×10^4	-5	-0.05	-51.0	0.05	51.0	17.5344	99.997
-1×10^5	-10	-0.1	-102.0	0.1	102.0	17.5337	99.993
-2×10^5	-20	-0.02	-204.0	0.2	204.0	17.5324	99.985
-3×10^5	-30	-0.3	-306.0	0.3	306.0	17.5312	99.978
-4×10^5	-40	-0.4	-408.0	0.4	408.0	17.5299	99.971
-5×10^5	-50	-0.5	-510.0	0.5	510.0	17.5286	99.964
-6×10^5	-60	-0.6	-612.0	0.6	612.0	17.5273	99.965
-7×10^5	-70	-0.7	-714.0	0.7	714.0	17.5260	99.949
-8×10^5	-80	-0.8	-816.0	0.8	816.0	17.5247	99.941
-9×10^5	-90	-0.9	-918.0	0.9	918.0	17.5234	99.934
-1×10^6	-100	-1.0	-1020	1.0	1020	17.5222	99.927
-2×10^6	-200	-2	-1040	2	1040	17.5089	99.851
-3×10^6	-300	-3	-3060	3	3060	17.4961	99.778
-4×10^6	-400	-4	-4080	4	4080	17.4833	99.705
-5×10^6	-500	-5	-5100	5	5100	17.4704	99.637
-6×10^6	-600	-6	-6120	6	6120	17.4572	99.556

[a] In the absence of osmotic effects (soluble salts), soil-water suction equals matric suction; otherwise, it is the sum of matric and osmotic suctions.

[b] Relative humidity of air in equilibrium with the soil at different suction values.

2. *Energy per unit volume:* Since water is a practically incompressible liquid, its density is almost independent of potential. Hence, there is a direct proportion between the expression of the potential as energy per unit mass and its expression as energy per unit volume. The latter expression yields the dimensions of pressure (for, just as energy can be expressed as the product of pressure and volume, so the ratio of energy to volume gives a pressure). This equivalent pressure can be measured in terms of dynes per square centimeter, Newtons per square meter, bars, or atmospheres. The basic dimensions are those of force per unit area: $ML^{-1}T^{-2}$. This method of expression is convenient for the osmotic and pressure potentials, but is seldom used for the gravitational potential.

3. *Energy per unit weight* (*hydraulic head*): Whatever can be expressed in units of hydrostatic pressure can also be expressed in terms of an equivalent head of water, which is the height of a liquid column corresponding

K. Quantitative Expression of Soil-Water Potential

to the given pressure. For example, a pressure of 1 atm is equivalent to a vertical water column, or hydraulic head, of 1033 cm, and to a mercury head of 76 cm. This method of expression is certainly simpler, and often more convenient, than the previous methods. Hence, it is common to characterize the state of soil water in terms of the total potential head, the gravitational potential head, and the pressure potential head, which are usually expressible in centimeters of water. Accordingly, instead of

$$\phi = \phi_g + \phi_p \tag{7.26}$$

one could write

$$H = H_g + H_p$$

which reads: The total potential head of soil water (H) is the sum of the gravitational (H_g) and pressure (H_p) potential heads. H is commonly called, simply, the *hydraulic head*.

In attempting to express the negative pressure potential of soil water (relative to atmospheric pressure) in terms of an equivalent hydraulic head, we must contend with the fact that this head may be of the order of $-10,000$ or even $-100,000$ cm of water. To avoid the use of such cumbersomely large numbers, Schofield (1935) suggested the use of pF (by analogy with the pH acidity scale) which he defined as the logarithm of the negative pressure (tension, or suction) head in centimeters of water. A pF of 1 is, thus, a tension head of 10 cm H_2O, a pF of 3 is a tension head of 1000 cm H_2O, and so forth.

The use of various alternative methods for expressing the soil-water potential can be perplexing to the uninitiated. It should be understood that these alternative expressions are in fact equivalent, and each method of expression can be translated directly into any of the other methods. If we use ϕ to designate the potential in terms of energy per mass, P for the potential in terms of pressure, and H for the potential head, then

$$\phi = P/\rho_w \tag{7.27}$$

$$H = P/\rho_w g = \phi/g \tag{7.28}$$

where ρ_w is the density of liquid water and g the acceleration of gravity.

A remark is in order concerning the use of the synonymous terms tension and suction in lieu of negative or subatmospheric pressure. Tension and suction are merely semantic devices to avoid the use of the unesthetic negative sign which generally characterizes the pressure of soil water, and to allow us to speak of the osmotic and matric potentials in positive terms. These two potentials, separately and in combination, are illustrated in Fig. 7.7.

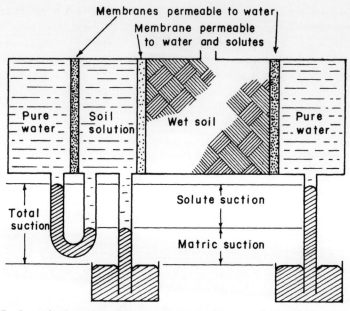

Fig. 7.7. In an isothermal equilibrium system, matric suction is the pressure difference across a membrane separating soil solution in situ from the same solution in bulk, the membrane being permeable to solution but not to solid particles or air; osmotic suction is the pressure difference across a semipermeable membrane separating bulk phases of pure water and the soil solution; and total suction is the sum of the matric and osmotic suction values, and is thus the pressure difference across a semipermeable membrane separating pure water from a soil that contains solution. An ideal semipermeable membrane is permeable to water only. (After Richards, 1965.)

L. Soil-Moisture Characteristic Curve

In a saturated soil at equilibrium with free water at the same elevation, the actual pressure is atmospheric, and hence the hydrostatic pressure and the suction (or tension) are zero.

If a slight suction, i.e., a water pressure slightly subatmospheric, is applied to water in a saturated soil, no outflow may occur until, as suction is increased, a certain critical value is exceeded at which the largest pore of entry begins to empty. This critical suction is called the *air-entry suction*. Its value is generally small in coarse-textured and in well-aggregated soils. However, since in coarse-textured soils the pores are often more nearly uniform in size, these soils may exhibit critical air-entry phenomena more distinctly and sharply than do fine-textured soils.

As suction is further increased, more water is drawn out of the soil and

L. Soil-Moisture Characteristic Curve

more of the relatively large pores, which cannot retain water against the suction applied, will empty out. Recalling the capillary equation[14] ($-P = 2\gamma/r$), we can readily predict that a gradual increase in suction will result in the emptying of progressively smaller pores, until, at high suction values, only the very narrow pores retain water. Similarly, an increase in soil-water suction is associated with a decreasing thickness of the hydration envelopes covering the soil-particle surfaces. Increasing suction is thus associated with decreasing soil wetness. The amount of water remaining in the soil at equilibrium is a function of the sizes and volumes of the water-filled pores and hence it is a function of the matric suction. This function is usually measured experimentally, and it is represented graphically by a curve known as the soil-moisture retention curve, or the soil-moisture characteristic (Childs, 1940).

As yet, no satisfactory theory exists for the prediction of the matric suction versus wetness relationship from basic soil properties. The adsorption and pore-geometry effects are often too complex to be described by a simple model. Several empirical equations have been proposed which apparently describe the soil-moisture characteristic for some soils and within limited suction ranges. One such equation was advanced by Visser (1966):

$$\psi = a(f - \theta)^b/\theta^c \tag{7.29}$$

where ψ is the matric suction, f is the porosity, θ is the volumetric wetness, and a, b, and c are constants. The actual use of this equation is hampered by the difficulty of evaluating its constants. Visser found that b varied between 0 and 10, a between 0 and 3, and f between 0.4 and 0.6.

Other equations to describe the relationship between wetness and matric suction have been proposed by Laliberte (1969), White *et al.* (1970), Su and Brooks (1975), and van Genuchten (1978). An equation presented by Brooks and Corey (1966) is

$$(\theta - \theta_r)/(\theta_m - \theta_r) = (\psi_e/\psi)^\lambda \tag{7.30}$$

where suction values greater than the air-entry suction ψ_e. The exponent λ has been termed the *pore-size distribution index*. In this equation, θ is the volume wetness (a function of the suction ψ), θ_m is the maximum wetness (saturation or near saturation), and θ_r a "residual" wetness considered to be confined to small pores which do not form a continuous network (e.g., intraaggregate pores).

Gardner *et al.* (1970) proposed the empirical relation

$$\psi = a\theta^{-b} \tag{7.31}$$

---

[14] The use of this equation assumes that the contact angle is zero and that soil pores are approximately cylindrical.

Constant b was found to have a value of 4.3 for a fine sandy loam. This relation fits only a limited range of the characteristic curve, but it may be useful in analyzing processes in which the water-content range is narrow (e.g., redistribution or internal drainage).

The amount of water retained at relatively low values of matric suction (say, between 0 and 1 bar of suction) depends primarily upon the capillary effect and the pore-size distribution, and hence is strongly affected by the structure of the soil. On the other hand, water retention in the higher suction range is due increasingly to adsorption and is thus influenced less by the structure and more by the texture and specific surface of the soil material. According to Gardner (1968), the water content at a suction of 15 bar (often taken to be the lower limit of soil moisture availability to plants) is fairly well correlated with the surface area of a soil and would represent, roughly, about 10 molecular layers of water if it were distributed uniformly over the particle surfaces.

It should be obvious from the foregoing that the soil-moisture characteristic curve is strongly affected by soil texture. The greater the clay content, in general, the greater the water retention at any particular suction, and the more gradual the slope of the curve. In a sandy soil, most of the pores are relatively large, and once these large pores are emptied at a given suction, only a small amount of water remains. In a clayey soil, the pore-size distribution is more uniform, and more of the water is adsorbed, so that increasing the matric suction causes a more gradual decrease in water content (Fig. 7.8).

Soil structure also affects the shape of the soil-moisture characteristic curve, particularly in the low-suction range. The effect of compaction upon

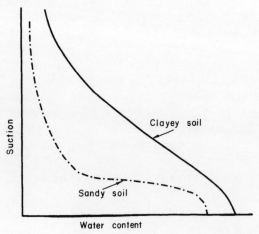

Fig. 7.8. The effect of texture on soil-water retention.

L. Soil-Moisture Characteristic Curve

a soil is to decrease the total porosity, and, especially, to decrease the volume of the large interaggregate pores. This means that the saturation water content and the initial decrease of water content with the application of low suction are reduced. On the other hand, the volume of intermediate-size pores is likely to be somewhat greater in a compact soil (as some of the originally large pores have been squeezed into intermediate size by compaction), while the intraaggregate micropores remain unaffected and thus the curves for the compacted and uncompacted soil may be nearly identical in the high suction range (Fig. 7.9). In the very high suction range, the predominant mechanism of water retention is adsorptive rather than capillary, and hence the retention capacity becomes more of a textural than a soil-structural attribute.

If two soil bodies differing in texture or structure are brought into direct physical contact, they will tend toward a state of potential-energy equilibrium in which the water potential would become equal throughout, but each of the two bodies will retain an amount of water determined by its own soil-moisture characteristic. Two soil bodies or layers can thus attain equilibrium and yet exhibit a marked nonuniformity, or discontinuity, in wetness. We can easily envision a situation in which a drier soil layer will contribute water to a wetter one simply because the water potential of the former is higher than that of the latter, owing to textural, structural, elevational, osmotic, or thermal differences.

In a nonshrinking soil, the soil-moisture characteristic curve, once obtained, allows calculation of the effective pore-size distribution (i.e., the

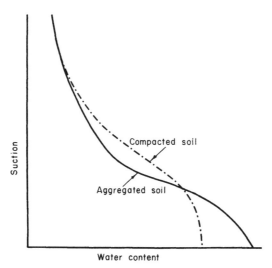

Fig. 7.9. The effect of soil structure on soil-water retention.

volumes of different classes of pore sizes). If an increase in matric suction from ψ_1 to ψ_2 results in the release of a certain volume of water, then that volume is evidently equal to the volume of pores having a range of effective radii between r_1 and r_2, where ψ_1 and r_1, and ψ_2 and r_2, are related by the equation of capillarity, namely $\psi = 2\gamma/r$.

An important fact which deserves to be stressed once again is that water in an unsaturated soil, being at subatmospheric pressure, will not spontaneously seep into the atmosphere. To flow out of the soil and into the atmosphere, soil-water pressure must exceed atmospheric pressure. Similarly, for a fine-textured soil to drain into the large pores of an initially dry, coarse-textured soil water must be at a sufficiently low tension, close to atmospheric pressure (at which the tension, or suction, is zero).

The slope of the soil-moisture characteristic curve, which is the change of water content per unit change of matric potential, is generally termed the differential (or specific) *water capacity*[15] c_θ:

$$c_\theta = d\theta/d\phi_p \quad \text{or} \quad c_\theta = -d\theta/d\psi \tag{7.32}$$

This is an important property in relation to soil-moisture storage and availability to plants. The actual value of c depends upon the wetness range, the texture, and the hysteresis effect.

An important and as yet incompletely understood phenomenon is the possible change in soil-moisture characteristics caused by the swelling and shrinkage of clay, which in turn is affected by the composition and concentration of the soil solution (Russell, 1941). Swelling is generally suppressed when the soil solution is fairly concentrated with electrolytes, particularly in the presence of a preponderance of divalent cations such as calcium. On the other hand, swelling—and hence water retention at any suction value— can be much more pronounced when the soil solution is dilute and with a preponderance of monovalent cations such as sodium (Dane and Klute, 1977). Other factors that affect the soil-moisture characteristic are the entrapment and persistence of air bubbles (Peck, 1969) and the change in soil structure resulting from sudden wetting, as well as from prolonged saturation (Hillel and Mottes, 1966).

M. Hysteresis

The relation between matric potential and soil wetness can be obtained in two ways: (1) in *desorption*, by taking an initially saturated sample and

---

[15] This by analogy with the differential heat capacity, which is the change in heat content per unit change in the thermal potential (temperature). However, while the latter is fairly constant with temperature for many materials, the differential water capacity in soils is strongly dependent on the matric potential.

M. Hysteresis

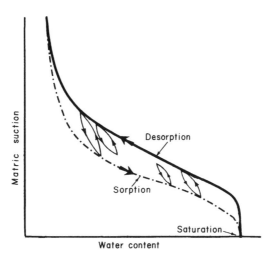

Fig. 7.10. The suction–water content curves in sorption and desorption. The intermediate loops are *scanning curves*, indicating transitions between the main branches.

applying increasing suction to gradually dry the soil while taking successive measurements of wetness versus suction; and (2) in *sorption*, by gradually wetting up an initially dry soil sample while reducing the suction. Each of these two methods yields a continuous curve, but the two curves will in general not be identical. The equilibrium soil wetness at a given suction is greater in desorption (drying) than in sorption (wetting). This dependence of the equilibrium content and state of soil water upon the direction of the process leading up to it is called *hysteresis*[16] (Haines, 1930; Miller and Miller, 1955a,b, 1956; Philip, 1964; Topp and Miller, 1966; Bomba, 1968; Topp, 1969).

Figure 7.10 shows a typical soil-moisture characteristic curve and illustrates the hysteresis effect in the soil–water equilibrium relationship.

The hysteresis effect may be attributed to several causes:

(1) the geometric nonuniformity of the individual pores (which are generally irregularly shaped voids interconnected by smaller passages), resulting in the "inkbottle" effect, illustrated in Fig. 7.11;

---

[16] A description of hysteresis in more general physical terms was offered by Poulovassilis (1962): If a physical property Y depends upon an independent variable X, then it may occur that the relationship between Y and X is unique, and, in particular, independent of whether X is increasing or decreasing. Such a relationship is reversible. Many physical properties, are, however, irreversible, and even when the changes of X are made very slowly the curve for increasing X does not coincide with that for decreasing X. This phenomenon, called hysteresis, is commonly observed in magnetism, for instance.

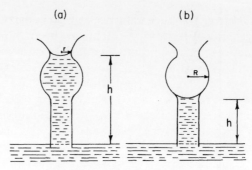

Fig. 7.11. "Ink-bottle" effect determines equilibrium height of water in a variable-width pore: (a) in capillary drainage (desorption) and (b) in capillary rise (sorption).

(2) the contact-angle effect, mentioned in Chapter 3, by which the contact angle is greater and hence the radius of curvature is greater, in an advancing meniscus than in the case of a receding one.[17] A given water content will tend therefore to exhibit greater suction in desorption than in sorption.

(3) entrapped air, which further decreases the water content of newly wetted soil. Failure to attain true equilibrium can accentuate the hysteresis effect.

(4) swelling, shrinking, or aging phenomena, which result in differential changes of soil structure, depending on the wetting and drying history of the sample (Hillel and Mottes, 1966). The gradual solution of air, or the release of dissolved air from soil water, can also have a differential effect upon the suction–wetness relationship in wetting and drying systems.

Of particular interest is the ink bottle effect. Consider the hypothetical pore shown in Fig. 7.11. This pore consists of a relatively wide void of radius R, bounded by narrow channels of radius r. If initially saturated, this pore will drain abruptly the moment the suction exceeds ψ_r, where $\psi_r = 2\gamma/r$. For this pore to rewet, however, the suction must decrease to below ψ_R, where $\psi_R = 2\gamma/R$, whereupon the pore abruptly fills. Since $R > r$, it follows that $\psi_r > \psi_R$. Desorption depends on the narrow radii of the connecting channels, whereas sorption depends on the maximum diameter of the large pores. These discontinuous spurts of water, called *Haines jumps*, can be observed readily in coarse sands. The hysteresis effect is in general more pronounced in coarse-textured soils in the low-suction range, where pores may empty at an appreciably larger suction than that at which they fill.

[17] Contact-angle hysteresis can arise because of surface roughness, the presence and distribution of adsorbed impurities on the solid surface, and the mechanism by which liquid molecules adsorb or desorb when the interface is displaced.

In the past, hysteresis was generally disregarded in the theory and practice of soil physics. This may be justifiable in the treatment of processes entailing monotonic wetting (e.g., infiltration) or drying (e.g., evaporation). But the hysteresis effect may be important in cases of composite processes in which wetting and drying occur simultaneously or sequentially in various parts of the soil profile (e.g., redistribution). It is possible to have two soil layers of identical texture and structure at equilibrium with each other (i.e., at identical energy states) and yet they may differ in wetness or water content if their wetting histories have been different. Furthermore, hysteresis can affect the dynamic, as well as the static, properties of the soil (i.e., hydraulic conductivity and flow phenomena).

The two complete characteristic curves, from saturation to dryness and vice versa, are called the *main branches* of the hysteretic soil moisture characteristic. When a partially wetted soil commences to drain, or when a partially desorbed soil is rewetted, the relation of suction to moisture content follows some intermediate curve as it moves from one main branch to the other. Such intermediate spurs are called *scanning curves*. Cyclic changes often entail wetting and drying scanning curves, which may form loops between the main branches (Fig. 7.11). The $\psi-\theta$ relationship can thus become very complicated. Because of its complexity, the hysteresis phenomenon is too often ignored, and the soil moisture characteristic which is generally reported is the *desorption curve*, also known as the soil-moisture release curve. The *sorption curve*, which is equally important but more difficult to determine, is seldom even attempted.

N. Measurement of Soil-Moisture Potential

The measurement of soil wetness, described earlier in this chapter, though essential in many soil physical and engineering investigations, is obviously not sufficient to provide a description of the state of soil water. To obtain such a description, evaluation of the energy status of soil water (soil-moisture potential, or suction) is necessary. In general, the twin variables, wetness and potential, should each be measured directly, as the translation of one to the other on the basis of calibration curves of soil samples is too often unreliable.

Total soil-moisture potential is often thought of as the sum of matric and osmotic (solute) potentials and is a useful index for characterizing the energy status of soil water with respect to plant water uptake. The sum of the matric and gravitational (elevation) heads is generally called the hydraulic head (or hydraulic potential) and is useful in evaluating the directions and magnitudes of the water-moving forces throughout the soil profile. Methods are available for measuring matric potential as well as total soil moisture potential, separately or together (Black, 1965). A schematic representation of the components of the soil-moisture potential is shown in Fig. 7.7. To measure matric potential in the field, an instrument known as the tensiometer is used, whereas in the laboratory use is often made of tension plates and of air-pressure extraction cells. Total soil moisture potential can be obtained by measuring the equilibrium vapor pressure of soil water by means of thermocouple psychrometers.

We shall now describe the tensiometer, which has won widespread acceptance as a practical device for the in situ measurement of matric suction, hydraulic head, and hydraulic gradients.

1. THE TENSIOMETER

The essential parts of a tensiometer are shown in Fig. 7.12. The tensiometer consists of a porous cup, generally of ceramic material, connected through a tube to a manometer, with all parts filled with water. When the cup is placed in the soil where the suction measurement is to be made, the bulk water inside the cup comes into hydraulic contact and tends to equilibrate with soil water through the pores in the ceramic walls. When initially placed in the soil, the water contained in the tensiometer is generally at atmospheric pressure. Soil water, being generally at subatmospheric pressure, exercises a suction which draws out a certain amount of water from the rigid and airtight tensiometer,

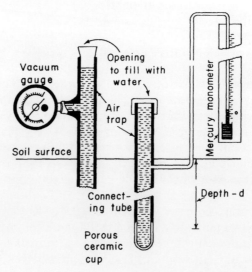

Fig. 7.12. Schematic illustration of the essential parts of a tensiometer. (After S. J. Richards, 1965.)

thus causing a drop in its hydrostatic pressure. This pressure is indicated by a manometer, which may be a simple water- or mercury-filled U tube, a vacuum gauge, or an electrical transducer.

A tensiometer left in the soil for a long period of time tends to follow the changes in the matric suction of soil water. As soil moisture is depleted by drainage or plant uptake, or as it is replenished by rainfall or irrigation, corresponding readings on the tensiometer gauge occur. Owing to the hydraulic resistance of the cup and the surrounding soil, or of the contact zone between the cup and the soil, the tensiometer response may lag behind suction changes in the soil. This lag time can be minimized by the use of a null-type device or of a transducer-type manometer with rigid tubing, so that practically no flow of water need take place as the tensiometer adjusts to changes in the soil matric suction.

Since the porous cup walls of the tensiometer are permeable to both water and solutes, the water inside the tensiometer tends to assume the same solute composition and concentration as soil water, and the instrument does not indicate the osmotic suction of soil water (unless equipped with some type of an auxiliary salt sensor).

Suction measurements by tensiometry are generally limited to matric suction values of below 1 atm. This is due to the fact that the vacuum gauge or manometer measures a partial vacuum relative to the external atmospheric pressure, as well as to the general failure of water columns in macroscopic systems to withstand tensions exceeding 1 bar.[18] Furthermore, as the ceramic material is generally made of the most permeable and porous material possible, too high a suction may cause air entry into the cup, which would equalize the internal pressure to the atmospheric. Under such conditions, soil suction will continue to increase even though the tensiometer fails to show it.

In practice, the useful limit of most tensiometers is at about 0.8 bar of maximal suction. To measure higher suctions, the use of an osmometer with a semipermeable membrane at the wall has been proposed, but this instrument is still in the experimental stage. However, the limited range of suction measurable by the tensiometer is not as serious a problem as it may seem at first sight. Though the suction range of 0–0.8 bar is but a small part of the total range of suction variation encountered in the field, it generally encompasses the greater part of the soil wetness range. In many agricultural soils the tensiometer range accounts for more than 50% (and in coarse-textured soils 75% or more) of the amount of soil water taken up by plants.

---

[18] It is interesting that water in capillary systems can maintain continuity at tensions of many bars. Witness, for example, the continuity of liquid water in the xylem vessels of tall trees. The ultimate tensile strength of water is apparently equivalent to several hundred bars.

Thus, where soil management (particularly in irrigation) is aimed at maintaining low-suction conditions which are most favorable for plant growth, tensiometers are definitely useful.

Despite their many shortcomings, tensiometers are practical instruments, available commercially, and, when operated and maintained by a skilled worker, are capable of providing reliable data on the in situ state of soil-moisture profiles and their changes with time.

Tensiometers have long been used in guiding the timing of irrigation of field and orchard crops, as well as of potted plants (Richards and Weaver, 1944). A general practice is to place tensiometers at one or more soil depths representing the root zone, and to irrigate when the tensiometers indicate that the matric suction has reached some prescribed value. The use of several tensiometers at different depths can indicate the amount of water needed in irrigation, and can also allow calculation of the hydraulic gradients in the soil profile. If $\psi_1, \psi_2, \psi_3, \ldots, \psi_n$ are the matric suction values in centimeters of water head ($\cong$ millibars) at depths $d_1, d_2, d_3, \ldots, d_n$ measured in centimeters below the surface, the average hydraulic gradient i between depths d_n and d_{n+1} is

$$i = [(\psi_{n+1} + d_{n+1}) - (\psi_n + d_n)]/(d_{n+1} - d_n) \qquad (7.33)$$

Measurement of the hydraulic gradient is particularly important in the region below the root zone, where the direction and magnitude of water movement cannot easily be ascertained otherwise.

The still considerable cost of tensiometers may limit the number of instruments used below the number needed for characterizing the often highly variable distribution of moisture and hence the pattern of suction in heterogeneous soils. Air diffusion through the porous cup into the vacuum gauging system requires frequent purging with deaired water. Tensiometers are also sensitive to temperature gradients between their various parts. Hence the above-ground parts should preferably be shielded from direct exposure to the sun. When installing a tensiometer, it is important that good contact be made between the cup and the soil so that equilibration is not hindered by contact-zone impedance to flow.

2. The Thermocouple Psychrometer

At equilibrium, the potential of soil moisture is equal to the potential of the water vapor in the ambient air. If thermal equilibrium is assured and the gravitational effect is neglected, the vapor potential can be taken to be equal to the sum of the matric and osmotic potentials, since air acts as an ideal semipermeable membrane in allowing only water molecules to pass (provided that the solutes are nonvolatile).

N. Measurement of Soil-Moisture Potential

A *psychrometer* is an instrument designed to indicate the *relative humidity*[19] of the atmosphere in which it is placed, generally by measuring the difference between the temperatures registered by a wet bulb and a dry bulb thermometer. The *dry bulb thermometer* indicates the temperature of a nonevaporating surface in thermal equilibrium with the ambient air. The *wet bulb thermometer* indicates the generally lower temperature of an evaporating surface, where latent heat is adsorbed in proportion to the rate of evaporation. If the atmosphere has a low relative humidity (i.e., is relatively dry) its evaporative demand and hence the evaporation rate will be higher, resulting in a greater depression of wet bulb temperature relative to dry bulb temperature, and vice versa. The relative humidity of a body of air in equilibrium with a moist porous body will depend upon the temperature, as well as upon the state of water in the porous system—i.e., on the constraining effects of adsorption, capillarity, and solutes, all of which act to reduce the evaporability of water relative to that of pure, free water at the same temperature. Hence the relative humidity of an unsaturated soil's atmosphere will generally be under 100%, and the deficit to "saturation" will depend upon the soil moisture potential, or suction, due to the combined effects of the osmotic and matric potentials. However, throughout most of the range of variation of soil moisture, this deficit to saturation is found to be very small, generally less than 2%. If we are to attempt to measure soil-moisture potential in terms of its effect upon the equilibrium relative humidity of the ambient air, we must have a very accurate psychrometer indeed.

Fortunately, recent years have witnessed the development of highly precise, miniaturized *thermocouple psychrometers* which indeed make possible the in situ measurement of soil moisture potential (Dalton and Rawlins, 1968; Brown, 1970). A thermocouple is a double junction of two dissimilar metals. If the two junctions are subjected to different temperatures, they will generate a voltage difference. If, on the other hand, an electromotive force (emf) is applied between the junctions, a difference in temperature will result; depending on which way a direct current is applied, one junction can be heated while the other is cooled, and vice versa. The *soil psychrometer* (Fig. 7.13) consists of a fine wire thermocouple, one junction of which is equilibrated with the soil atmosphere by placing it inside a hollow porous cup embedded in the soil, while the other junction is kept in an insulated medium to provide a temperature lag. During operation, an emf is applied so that the junction exposed to the soil atmosphere is cooled to a temperature below the dew point of that atmosphere, at which point a droplet of water

---

[19] Relative humidity is merely the ratio of the partial pressure p of water vapor in the air to the equilibrium partial pressure p_0 of vapor in a vapor-saturated air at the same temperature, multiplied by 100 to express relative humidity in percent units. Thus, $RH = 100(p/p_0)$.

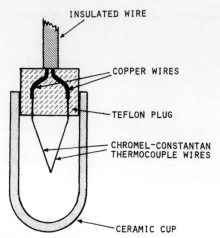

Fig. 7.13. Cross section of a thermocouple psychrometer contained in an air-filled ceramic cup.

condenses on the junction, allowing it to become, in effect, a wet bulb thermometer. This is a consequence of the so-called *Peltier effect* (see, for example, Yavorsky and Detlaf, 1972). The cooling is then stopped, and as the water from the droplet reevaporates the junction attains a wet bulb temperature which remains nearly constant until the junction dries out, after which it returns to the ambient soil temperature. While evaporation takes place, the difference in temperature between the wet bulb and the insulated junction serving as dry bulb generates an emf which is indicative of the soil moisture potential. The relative humidity (i.e., the vapor pressure depression relative to that of pure, free water) is related to the soil-water potential according to

$$\phi = \bar{R}T \ln(p/p_0) \tag{7.34}$$

where p is the vapor pressure of soil water, p_0 the vapor pressure of pure, free water at the same temperature and air pressure, and $\bar{R}$ is the specific gas constant for water vapor.

The measurement of vapor pressure or relative humidity is obviously highly sensitive to temperature changes. Hence the need for very accurate temperature control and monitoring. Under field conditions, the accuracy claimed is of the order of about 0.5 bar of soil moisture potential. The instrument is thus not practical at low soil moisture suction values, but can be quite useful considerably beyond the suction range of the tensiometer (i.e., from 2 to 50 bar). It is used mostly in research, and is now manufactured commercially. Applications to the measurement of plant-water potential are also being investigated, as are applications in engineering practice.

3. MEASUREMENT OF THE SOIL-MOISTURE CHARACTERISTIC CURVES

The fundamental relation between soil wetness and matric suction is often determined by means of a tension plate assembly (Fig. 7.14) in the low suction (<1 bar) range, and by means of a pressure plate or pressure membrane apparatus (Fig. 7.15) in the higher suction range. These instruments allow the application of successive suction values and the repeated measurement of the equilibrium soil wetness at each suction.

The maximum suction value obtainable by porous-plate devices is limited to 1 bar if the soil air is kept at atmospheric pressure and the pressure difference across the plate is controlled either by vacuum or by a hanging water column. Matric suction values considerably greater than 1 bar (say, 20 bar or even more) can be obtained by increasing the pressure of the air phase. This requires placing the porous-plate assembly inside a pressure chamber, as shown in Fig. 7.15. The limit of matric suction obtainable with such a

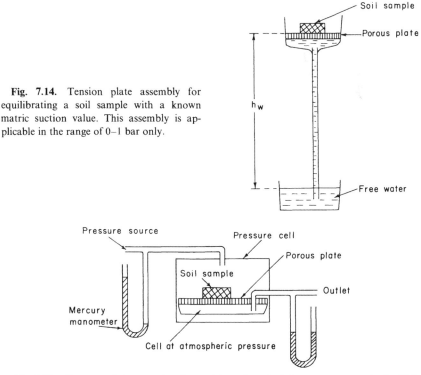

Fig. 7.14. Tension plate assembly for equilibrating a soil sample with a known matric suction value. This assembly is applicable in the range of 0–1 bar only.

Fig. 7.15. Pressure plate apparatus for moisture characteristic measurements in the high-suction range. The lower side of the porous plate is in contact with water at atmospheric pressure. Air pressure is used to extract water from initially saturated soil samples.

device is determined by the design of the chamber (i.e., its safe working pressure) and by the maximal air-pressure difference the saturated porous plate can bear without allowing air to bubble through its pores. Ceramic plates generally do not hold pressures greater than about 20 bar, but cellulose acetate membranes can be used with pressures exceeding 100 bar.

Soil moisture retention in the low-suction range (0–1 bar) is strongly influenced by soil structure and pore-size distribution. Hence, measurements made with disturbed samples (e.g., dried, screened, and artificially packed samples) cannot be expected to represent field conditions. The use of undisturbed soil cores is therefore preferable. Even better, in principle, is the in situ determination of the soil moisture characteristic by making simultaneous measurements of wetness (e.g., with the neutron moisture meter) and suction (using tensiometers) in the field. Unfortunately, this approach has often been frustrated by soil heterogeneity and by uncertainties over hysteretic phenomena as they occur in the field.

We have already mentioned that the soil-moisture characteristic is hysteretic. Ordinarily, the desorption curve is measured by gradually and monotonically extracting water from initially saturated samples. The resulting curve, often called the soil-moisture release curve, is applicable to processes involving drainage, evaporation, or plant extraction of soil moisture. On the other hand, the sorption curve is needed whenever infiltration or wetting processes are studied. Modified apparatus is required for the measurement of wetness versus suction during sorption (Tanner and Elrick, 1958). Both primary curves and knowledge of scanning patterns in transition from wetting to drying soil (and vice versa) are needed for a complete description.

Sample Problems

1. The accompanying data were obtained by gravimetric sampling just prior to and two days following an irrigation.

Sampling time	Sample number	Depth (cm)	Bulk density (gm/cm^3)	Gross weight of sample plus container		Weight container (gm)
				Wet sample	Dried sample	
Before irrigation	1	0–40	1.2	160	150	50
	2	60–100	1.5	145	130	50
After irrigation	3	0–40	1.2	230	200	50
	4	40–100	1.5	206	170	50

Sample Problems

From these data, calculate the mass and volume wetness values of each layer before and after the irrigation, and determine the amount of water (in millimeters) added to each layer and to the profile as a whole.

Using Eq. (7.6), we obtained the following mass wetness values:

$$w_1 = \frac{160 - 150}{150 - 50} = \frac{10 \text{ gm}}{100 \text{ gm}} = 0.1, \quad w_2 = \frac{146 - 130}{130 - 50} = \frac{16 \text{ gm}}{80 \text{ gm}} = 0.2$$

$$w_3 = \frac{230 - 200}{200 - 50} = \frac{30 \text{ gm}}{150 \text{ gm}} = 0.2, \quad w_4 = \frac{206 - 170}{170 - 50} = \frac{36 \text{ gm}}{120 \text{ gm}} = 0.3$$

Using Eq. (7.3), we obtain the following volume wetness values:

$$\theta_1 = 1.2 \times 0.1 = 0.12, \quad \theta_2 = 1.5 \times 0.2 = 0.30$$
$$\theta_3 = 1.2 \times 0.2 = 0.24, \quad \theta_4 = 1.5 \times 0.3 = 0.45$$

Using Eq. (7.4), we obtain the following water depths per layer;

$$d_{w_1} = 0.12 \times 400 \text{ mm} = 48 \text{ mm}, \quad d_{w_2} = 0.30 \times 600 \text{ mm} = 180 \text{ mm}$$
$$d_{w_3} = 0.24 \times 400 \text{ mm} = 96 \text{ mm}, \quad d_{w_4} = 0.45 \times 600 \text{ mm} = 270 \text{ mm}$$

Depth of water in profile before irrigation = 48 + 180 = 228 mm.
Depth of water in profile after irrigation = 96 + 270 = 366 mm.
Depth of water added to top layer = 96 − 48 = 48 mm.
Depth of water added to bottom layer = 270 − 180 = 90 mm.
Depth of water added to entire profile = 48 + 90 = 138 mm.

2. From calibration of a neutron probe we know that when a soil's volumetric wetness is 15% we get a reading of 24,000 cpm (counts per minute), and at a wetness of 40% we get 44,000 cpm. Find the equation of the straight line defining the calibration curve (in the form of $Y = mX + b$, where Y is counts per minute, X is volumetric wetness, m is the slope of the line, and b is the intercept on the Y axis). Using the equation derived, find the wetness value corresponding to a count rate of 30,000 cpm.

We first obtain the slope m:

$$m = (Y_2 - Y_1)/(X_2 - X_1) = (44{,}000 - 24{,}000)/(40 - 15)$$
$$= 800 \text{ cpm per } 1\% \text{ wetness}$$

We next obtain the Y intercept b:

$Y = 800X + b, \quad b = Y - 800X = 44{,}000 - 800 \times 40 = 12{,}000$ cpm
(or $24{,}000 - 800 \times 15 = 12{,}000$ cpm)

The complete equation is therefore

$$Y = 800X + 12{,}000$$

Now, to find the wetness value corresponding to 30,000 cpm, we set

$$30{,}000 = 800X + 12{,}000, \qquad X = (30{,}000 - 12{,}000)/800 = 22.5\%$$

3. The accompanying data were obtained with tension plate and pressure plate extraction devices from two soils of unknown texture.

Suction head		Volumetric wetness (%)	
(bar)	(cm)	Soil A	Soil B
0	0	44	52
0.01	10	44	52
0.02	20	43.9	52
0.05	50	38	51
0.1	100	22.5	48
0.3	300	12.5	32
1	1000	7	20
10	10,000	5.2	13.5
20	20,000	5.1	13
100	100,000	4.9	12.8

Plot the two soil-moisture characteristic curves on a semilog scale (logarithm of matric suction versus wetness). Estimate the bulk density, assuming that the soils do not swell or shrink. Estimate the volume and mass wetness values at suctions of $\frac{1}{3}$ bar and at 15 bar. How much water in depth units (mm) can each soil release per one meter depth of profile in transition from $\frac{1}{3}$ bar to to 15 bar of suction? Plot the volumetric water capacity against the logarithm of matric suction. What are the likely soil textures?

The curves shown in Fig. 7.16 are obtained. *Bulk density* can be estimated from the volumetric wetness at saturation (zero suction) assuming it to be equal to the porosity f (i.e., no occluded air):

$$f = 1 - \rho_b/\rho_s, \qquad \rho_b = \rho_s(1 - f)$$

For Soil A: $\rho_b = 2.65(1 - 0.44) = 1.48 \text{ gm/cm}^3$.
For Soil B: $\rho_b = 2.65(1 - 0.52) = 1.27 \text{ gm/cm}^3$.

Volume wetness values for different suctions can be read directly from the soil moisture characteristic curves:

For Soil A: 12% at $\frac{1}{3}$ bar; 5% at 15 bar.
For Soil B: 31% at $\frac{1}{3}$ bar; 13% at 15 bar.

Sample Problems

Fig. 7.16.

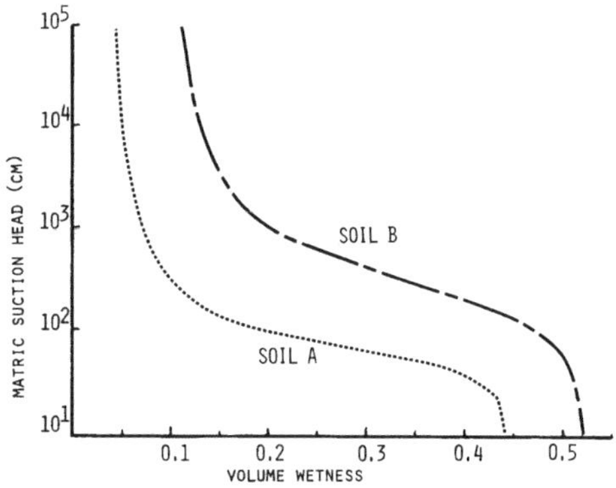

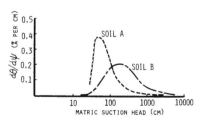

Mass wetness w values can be calculated from the volumetric wetness θ values and the bulk density ρ_b values by the equation

$$\theta = w(\rho_b/\rho_w) \quad \text{or} \quad w = \theta(\rho_w/\rho_b) \quad \text{where } \rho_w \text{ is the density of water)}$$

For Soil A at $\frac{1}{3}$ bar: $w = 12\% \times (1/1.48) = 8.1\%$; at 15 bar: $w = 5\% \times (1/1.48) = 3.4\%$.

For Soil B at $\frac{1}{3}$ bar: $w = 31\% \times (1/1.27) = 24.4\%$; at 15 bar: $w = 13\% \times (1/1.27) = 10.2\%$.

Water released per 1 meter depth in transition from $\frac{1}{3}$ to 15 bar:

For Soil A: $(12 - 5)\% \times 1000$ mm $= 70$ mm.
For Soil B: $(31 - 13)\% \times 1000$ mm $= 180$ mm.

Volumetric water capacity $(-d\theta/d\psi)$ can be obtained from the soil moisture characteristic curves by graphic differentiation (i.e., by measuring the slope of wetness versus suction curves at different values of suction). Curves of the type shown in Fig. 7.17 are obtained. *Note:* Soil A releases most of its water before a suction of 0.1 bar (100 cm) is obtained and nearly all of its water by the time a suction of 0.5 bar is reached. Soil B, on the other hand, still retains most of its water at a suction of 0.1 bar (100 cm), and can still supply substantial amounts of water at suctions exceeding 1 bar. From all indications, Soil A is a sand or sandy loam, and Soil B is probably a clay loam.

Fig. 7.17.

> I can fortell the way of celestial bodies,
> but can say nothing about the movement
> of a small drop of water.
> Galileo Galilei
> 1564–1642

8 Flow of Water in Saturated Soil

A. Laminar Flow in Narrow Tubes

Before we enter into a discussion of flow in so complex a medium as soil, it might be helpful to consider some basic physical phenomena associated with fluid flow in narrow tubes.

Early theories of fluid dynamics were based on the hypothetical concepts of a "perfect" fluid, i.e., one that is both frictionless and incompressible. In the flow of a perfect fluid, contacting layers can exhibit no tangential forces (shearing stresses), only normal forces (pressures). Such fluids do not in fact exist. In the flow of real fluids, adjacent layers do transmit tangential stresses (drag), and the existence of intermolecular attractions causes the fluid molecules in contact with a solid wall to adhere to it rather than slip over it. The flow of a real fluid is associated with the property of viscosity, defined in Chapter 3.

We can visualize the nature of viscosity by considering the motion of a fluid between two parallel plates, one at rest, the other moving at a constant velocity (Fig. 8.1). Experience shows that the fluid adheres to both walls, so that its velocity at the lower plate is zero, and that at the upper plate is equal to the velocity of the plate. Furthermore, the velocity distribution in the fluid between the plates is linear, so that the fluid velocity is proportional to the distance y from the lower plate.

To maintain the relative motion of the plates at a constant velocity, it is necessary to apply a tangential force, that force having to overcome the frictional resistance in the fluid. This resistance, per unit area of the plate, is proportional to the velocity of the upper plate U and inversely proportional to the distance h. The shearing stress at any point, is proportional to the

A. Laminar Flow in Narrow Tubes

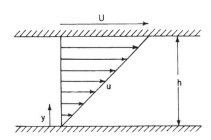

Fig. 8.1. Velocity distribution in a viscous fluid between two parallel flat plates, with the upper plate moving at a velocity U relative to the lower plate.

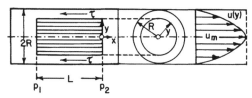

Fig. 8.2. Laminar flow through a cylindrical tube.

velocity gradient du/dy. The viscosity η is the proportionality factor between τ_s and du/dy[1]:

$$\tau_s = \eta \, du/dy \tag{8.1}$$

We can now apply these relationships to describe flow through a straight, cylindrical tube with a constant diameter $D = 2R$ (Fig. 8.2). The velocity is zero at the wall (because of adhesion), maximal on the axis, and constant on cylindrical surfaces which are concentric about the axis. Adjacent cylindrical *laminae*, moving at different velocities, slide over each other. A parallel motion of this kind is called *laminar*. Fluid movement in a horizontal tube is generally caused by a pressure gradient acting in the axial direction. A fluid "particle," therefore, is accelerated by the pressure gradient and retarded by the frictional resistance.

Now let us consider a coaxial fluid cylinder of length L and radius y. For flow velocity to be constant, the pressure force acting on the face of the cylinder $\Delta p \pi y^2$, where $\Delta p = p_1 - p_2$, must be equal to the frictional resistance due to the shear force $2\pi y L \tau_s$ acting on the circumferential area. Thus,

$$\tau_s = (\Delta p/L)(y/2)$$

Recalling Eq. (8.1)

$$\tau_s = -\eta \, du/dy$$

---

[1] Equation (8.1) bears an analogy to Hooke's law of elasticity. In an elastic solid, the shearing stress is proportional to the strain, whereas in a viscous fluid, the shearing stress is proportional to the time rate of the strain.

(the negative sign arises because in this case u decreases with increasing y), we obtain

$$du/dy = -(\Delta p/\eta L)(y/2)$$

which, upon integration, gives

$$u(y) = (\Delta p/\eta L)(c - y^2/4)$$

The constant of integration c is evaluated from the boundary condition of no slip at the wall; that is, $u = 0$ at $y = R$, so that $c = R^2/4$.

Therefore,

$$u(y) = (\Delta p/4\eta L)(R^2 - y^2) \tag{8.2}$$

Equation (8.2) indicates that the velocity is distributed parabolically over the radius, with the maximum velocity u_{max} being on the axis ($y = 0$):

$$u_{max} = \Delta p\, R^2/4\eta L$$

The discharge Q, being the volume flowing through a section of length L per unit time, can now be evaluated. The volume of a paraboloid of revolution is $\frac{1}{2}$(base × height), hence

$$Q = \tfrac{1}{2}\pi R^2 u_{max} = \pi R^4\, \Delta p/8\eta L \tag{8.3}$$

This equation, known as *Poiseuille's law*, indicates that the volume flow rate is proportional to the pressure drop per unit distance ($\Delta p/L$) and the fourth power of the radius of the tube.

The mean velocity over the cross section is

$$\bar{u} = \Delta p\, R^2/8\eta L = (R^2/a\eta)\, \nabla p \tag{8.4}$$

where ∇p is the pressure gradient. Parameter a, equal to 8 in a circular tube, varies with the shape of the conducting passage.

Laminar flow prevails only at relatively low flow velocities and in narrow tubes. As the radius of the tube and the flow velocity are increased, the point is reached at which the mean flow velocity is no longer proportional to the pressure drop, and the parallel *laminar flow* changes into a *turbulent flow* with fluctuating eddies. Conveniently, however, laminar flow is the rule rather than the exception in most water flow processes taking place in soils, because of the narrowness of soil pores (see the discussion of Reynolds number, Section H).

B. Darcy's Law

Were the soil merely a bundle of straight and smooth tubes, each uniform in radius, we could assume the overall flow rate to be equal to the sum of

B. Darcy's Law

the separate flow rates through the individual tubes. Knowledge of the size distribution of the tube radii could then enable us to calculate the total flow through a bundle caused by a known pressure difference, using Poiseuille's equation.

Unfortunately from the standpoint of physical simplicity, however, soil pores do not resemble uniform, smooth tubes, but are highly irregualr, tortuous, and intricate. Flow through soil pores is limited by numerous constrictions, or "necks," and occasional "dead end" spaces. Hence, the actual geometry and flow pattern of a typical soil specimen is too complicated to be described in microscopic detail, as the fluid velocity varies drastically from point to point, even along the same passage. For this reason, flow through complex porous media is generally described in terms of *a macroscopic flow velocity vector*, which is the overall average of the microscopic velocities over the total volume of the soil. The detailed flow pattern is thus ignored, and the conducting body is treated as though it were a uniform medium, with the flow spread out over the entire cross section, solid and pore space alike.[2]

Let us now examine the flow of water in a macroscopically uniform, saturated soil body, and attempt to describe the quantitative relations connecting the rate of flow, the dimensions of the body, and the hydraulic conditions at the inflow and outflow boundaries.

Figure 8.3 shows a horizontal column of soil, through which a steady flow of water is occurring from left to right, from an upper reservoir to a lower one, in each of which the water level is maintained constant.

Experience shows that the discharge rate Q, being the volume V flowing through the column per unit time, is directly proportional to the cross-sectional area and to the hydraulic head drop ΔH, and inversely proportional to the length of the column L:

$$Q = V/t \propto A \, \Delta H/L$$

The usual way to determine the hydraulic head drop across the system is to measure the head at the inflow boundary H_i and at the outflow boundary H_o, relative to some reference level. ΔH is the difference between these two heads:

$$\Delta H = H_i - H_o$$

Obviously, no flow occurs in the absence of a hydraulic head difference, i.e., when $\Delta H = 0$.

---

[2] An implicit assumption here is that the soil volume taken is sufficiently large relative to the pore sizes and microscopic heterogeneities to permit the averaging of velocity and potential over the cross section.

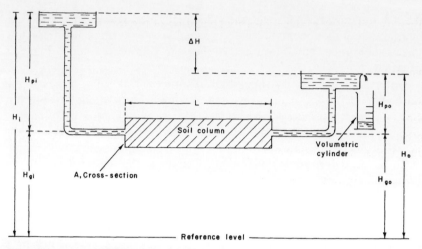

Fig. 8.3. Flow in a horizontal saturated column.

The head drop per unit distance in the direction of flow ($\Delta H/L$) is the *hydraulic gradient*, which is, in fact, the driving force. The specific discharge rate Q/A (i.e., the volume of water flowing through a unit cross-sectional area per unit time t) is called the *flux density* (or simply the *flux*) and is indicated by q. Thus, the flux is proportional to the hydraulic gradient:

$$q = Q/A = V/At \propto \Delta H/L$$

The proportionality factor K is generally designated as the *hydraulic conductivity*:

$$q = K \Delta H/L \tag{8.5}$$

This equation is known as *Darcy's law*, after Henri Darcy, the French engineer who discovered it over a century ago in the course of his classic investigation of seepage rates through sand filters in the city of Dijon (Darcy, 1856; Hubbert, 1956).

Where flow is unsteady (i.e., the flux changing with time) or the soil nonuniform, the hydraulic head may not decrease linearly along the direction of flow. Where the hydraulic head gradient or the conductivity is variable, we must consider the localized gradient, flux, and conductivity values rather than overall values for the soil system as a whole. A more exact and generalized expression of the Darcy law is, therefore, in differential form. Slichter (1899) generalized Darcy's law for saturated porous media into a three-

B. Darcy's Law

dimensional macroscopic differential equation of the form[3]

$$\mathbf{q} = -K \nabla H \tag{8.6}$$

where ∇H is the gradient of the hydraulic head in three-dimensional space.

Stated verbally, this law indicates that the flow of a liquid through a porous medium is in the direction of, and at a rate proportional to, the *driving force* acting on the liquid (i.e., the *hydraulic gradient*) and also proportional to the property of the conducting medium to transmit the liquid (namely, the *conductivity*).[4]

In a one-dimensional system, Eq. (8.6) takes the form

$$q = -K \, dH/dx \tag{8.7}$$

Mathematically, Darcy's law is similar to the linear transport equations of classical physics, including *Ohm's law* (which states that the current, or flow rate of electricity, is proportional to the electrical potential gradient),

---

[3] $K\nabla H$ is the product of a scalar K and a vector ∇H, and hence the flux $\mathbf{q}$ is a vector, the direction of which is determined by ∇H. This direction in an isotropic medium is orthogonal to surfaces of equal hydraulic potential H.

[4] Ultimately, both Poiseuille's and Darcy's laws rest upon the more general Navier–Stokes law, which describes the flow of viscous fluids and forms the basis of the science of fluid mechanics. For an incompressible fluid in isothermal flow,

$$\partial \mathbf{u}/\partial t + (\mathbf{u} \cdot \nabla)\mathbf{u} = -\nabla \phi - v_k \nabla^2 \mathbf{u}$$

where $\mathbf{u}$ is the vector flow velocity, t time, v_k the kinematic viscosity, and ϕ the potential including both the pressure term and the potential due to external or body forces (e.g., gravity).

This can be written as a set of three simultaneous differential equations of the type

$$\rho \underbrace{\left(\frac{\partial \mathbf{u}}{\partial t} + u \frac{\partial \mathbf{u}}{\partial x} + v \frac{\partial \mathbf{u}}{\partial y} + w \frac{\partial \mathbf{u}}{\partial z} \right)}_{(inertial\ terms)} = v_k \underbrace{\left(\frac{\partial^2 \mathbf{u}}{\partial x^2} + \frac{\partial^2 \mathbf{u}}{\partial y^2} + \frac{\partial^2 \mathbf{u}}{\partial z^2} \right)}_{(viscosity\ terms)} - \underbrace{\nabla p}_{\substack{(pressure \\ gradient)}} + \underbrace{X}_{\substack{(force \\ term)}}$$

where u, v, and w are component velocities along axes x, y, and z, respectively. Here, ρ is density p pressure, and X represents the x component of external forces such as gravity. Similar relations exist for v and w.

An analysis of how Poiseuille's and Darcy's laws relate to the Navier–Stokes law was given by Philip (1969). He showed that where the inertia terms are negligible [$(\mathbf{u} \cdot \nabla)\mathbf{u} \simeq 0$], and for steady flow ($\partial \mathbf{u}/\partial t = 0$),

$$v_k \nabla^2 \mathbf{u} = \nabla \phi$$

Philip showed that this equation is obeyed by both Poiseuille's and Darcy's laws provided, again, the inertia terms are negligible in relation to the viscous terms (i.e., where the capillaries or pores are sufficiently small and the flow velocity sufficiently slow).

Fourier's law (the rate of heat conduction is proportional to the temperature gradient), and *Fick's law* (the rate of diffusion is proportional to the concentration gradient).

C. Gravitational, Pressure, and Total Hydraulic Heads

The water entering the column of Fig. 8.3 is under a pressure P_i, which is the sum of the hydrostatic pressure P_s and the atmospheric pressure P_a acting on the surface of the water in the reservoir. Since the atmospheric pressure is the same at both ends of the system, we can disregard it and consider only the hydrostatic pressure. Accordingly, the water pressure at the inflow boundary is $\rho_w g H_{pi}$. Since ρ_w and g are both nearly constant, we can express this pressure in terms of the pressure head H_{pi}.

Water flow in a horizontal column occurs in response to a pressure head gradient. Flow in a vertical column can be caused by gravitation as well as pressure.[5] The *gravitational head* H_g at any point is determined by the height of the point relative to some reference plane, while the *pressure head* is determined by the height of the water column resting on that point.

The total hydraulic head H is composed of the sum of these two heads:

$$H = H_p + H_g \tag{8.8}$$

To apply Darcy's law to vertical flow, we must consider the total hydraulic head at the inflow and at the outflow boundaries (H_i and H_o, respectively):

$$H_i = H_{pi} + H_{gi}, \qquad H_o = H_{po} + H_{go}$$

Darcy's law thus becomes

$$q = K[(H_{pi} + H_{gi}) - (H_{po} + H_{go})]/L$$

The gravitational head is often designated as z, which is the vertical distance in the rectangular coordinate system x, y, z. It is convenient to set

[5] In classical hydraulics, the fluid potential Φ (the mechanical energy per unit mass) is generally stated in terms of the *Bernoulli equation*

$$\Phi = \int_{P_0}^{P} \frac{dP}{\rho} + gz + \frac{v^2}{2}$$

wherein P is pressure (P_0 being the pressure at the standard state), ρ is density of the fluid, g is gravitational acceleration, z is elevation above a reference level, and v is velocity. The three terms thus represent the pressure, gravity, and velocity potentials, respectively. Since flow is a porous medium is generally extremely slow, the third term can almost always be neglected (Freeze and Cherry, 1979). For an incompressible liquid (with ρ independent of P), moreover, the first term can be written as $(P - P_0)/\rho$. Finally, if P_0 is assumed equal to zero, we get

$$\Phi = P/\rho + gz$$

C. Gravitational, Pressure, and Total Hydraulic Heads

the reference level $z = 0$ at the bottom of a vertical column, or at the center of a horizontal column. However, the exact elevation of this hypothetical level is unimportant, since the absolute values of the hydraulic heads determined in reference to it are immaterial and only their differences from one point in the soil to another affect flow.

The pressure and gravity heads can be represented graphically in a simple way. To illustrate this, we shall immerse and equilibrate a vertical soil column in a water reservoir, so that the upper surface of the column will be level with the water surface, as shown in Fig. 8.4.

The coordinates of Fig. 8.4 are arranged so that the height above the bottom of the column is indicated by the vertical axis z; and the pressure, gravity, and hydraulic heads are indicated on the horizontal axis. The gravity head is determined with respect to the reference level $z = 0$, and increases with height at the ratio of 1:1. The pressure head is determined with reference to the free-water surface, at which the hydrostatic pressure is zero. Accordingly, the hydrostatic pressure head at the top of the column is zero, and at the bottom of the column it is equal to L, the column length. Just as the gravity head decreases from top to bottom, so the pressure head increases; thus, their sum, which is the hydraulic head, remains constant all along the column. This is a state of equilibrium in which no flow occurs.

This statement should be further elaborated. The water pressure is not equal along the column, being greater at the bottom than at the top of the column. Why, then, will the water not flow from a zone of higher to one of lower pressure? If the pressure gradient were the only force causing flow (as it is, in fact, in a horizontal column), the water would tend to flow upward. However, opposing the pressure gradient is a gravitational gradient of equal magnitude, resulting from the fact that the water at the top is at a higher

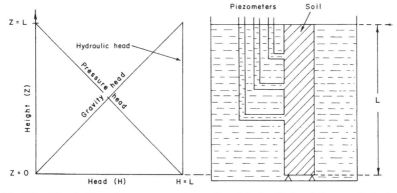

Fig. 8.4. Distribution of pressure, gravity, and total hydraulic heads in a vertical column immersed in water, at equilibrium.

gravitational potential than at the bottom. Since these two opposing gradients in effect cancel each other, the total hydraulic head is constant, as indicated by the standpipes (piezometer) connected to the column at the left.

As we have already point out, the reference level is generally set at the bottom of the column, so that the gravitational potential can always be positive. On the other hand, the pressure head of water is positive under a free-water surface (i.e., a water table) and negative above it. A "negative" hydraulic head signifies a pressure smaller than atmospheric, and it occurs whenever the soil becomes unsaturated. Flow under these conditions will be dealt with in the next chapter.

D. Flow in a Vertical Column

Figure 8.5 shows a uniform, saturated vertical column, the upper surface of which is ponded under a constant head of water H_1, and the bottom surface of which is set in a lower, constant-level reservoir. Flow is thus taking place from the higher to the lower reservoir through a column of length L.[6]

In order to calculate the flux according to Darcy's law, we must know the hydraulic head gradient, which is the ratio of the hydraulic head drop (between the inflow and outflow boundaries) to the column length as shown in the accompanying tabulation.

		Pressure head	Gravity head
Hydraulic head at inflow boundary	$H_i =$	H_1 +	L
Hydraulic head at outflow boundary	$H_o =$	0 +	0
Hydraulic head difference $\Delta H = H_i - H_o =$		H_1 +	L

The Darcy equation for this case is

$$q = K \Delta H/L = K(H_1 + L)/L, \qquad q = K H_1/L + K \qquad (8.9)$$

Comparison of this case with the horizontal one shows that the rate of downward flow of water in a vertical column is greater than in a horizontal column by the magnitude of the hydraulic conductivity. It is also apparent that, if the ponding depth H_1 is negligible, the flux is equal to the hydraulic conductivity. This is due to the fact that, in the absence of a pressure gradient,

---

[6] This is the same system that Darcy considered in his classic filter bed analysis.

D. Flow in a Vertical Column

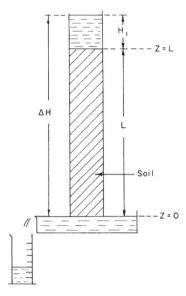

Fig. 8.5. Downward flow of water in a vertical saturated column.

the only driving force is the gravitational head gradient, which, in a vertical column, has the value of unity (since this head varies with height at the ratio of 1:1).

We shall now examine the case of upward flow in a vertical column, as shown in Fig. 8.6.

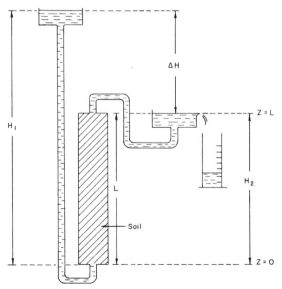

Fig. 8.6. Steady upward flow in a saturated vertical column.

In this case, the direction of flow is opposite to the direction of the gravitational gradient, and the hydraulic gradient becomes

		Pressure head	Gravity head
Hydraulic head at inflow boundary	$H_i =$	$H_1\ +$	0
Hydraulic head at outflow boundary	$H_o =$	$0\ +$	L
Hydraulic head difference $\Delta H = H_i - H_o =$		$H_1\ -$	L

The Darcy equation is thus

$$q = K(H_1 - L)/L = KH_1/L - K, \qquad q = K\,\Delta H/L$$

E. Flow in a Composite Column

We shall consider briefly a nonuniform soil column consisting of two distinct layers, each homogeneous within itself and characterized by its own thickness and hydraulic conductivity. Assume that layer 1 is at the inlet and layer 2 at the outlet side of the flowing column, and that H_1 is the hydraulic head at the inlet surface, H_2 at the interlayer boundary, and H_3 at the outlet end. At steady flow, the flux through both layers must be equal:

$$q = K_1(H_1 - H_2)/L_1 = K_2(H_2 - H_3)/L_2 \qquad (8.10)$$

where q is the flux, K_1 and L_1 are the conductivity and thickness (respectively) of the first layer, and K_2 and L_2 the same for the second layer. Here we have disregarded any possible contact resistance between the layers. Thus,

$$H_2 = H_1 - qL_1/K_1 \quad \text{and} \quad qL_2/K_2 = H_2 - H_3$$

Therefore,

$$qL_2/K_2 = H_1 - qL_1/K_1 - H_3, \qquad q = (H_1 - H_3)/(L_2/K_2 + L_1/K_1) \quad (8.11)$$

The reciprocal of the conductivity has been called the *hydraulic resistivity*, and the ratio of the thickness to the conductivity ($R_s = L/K$) has been called the *hydraulic resistance* per unit area. Hence,

$$q = \Delta H/(R_{s1} + R_{s2}) \qquad (8.12)$$

where H is the total *hydraulic head drop* across the entire system and R_{s1} and R_{s2} are the resistances of layers 1 and 2. Equation (8.12) is in complete analogy to Ohm's law for constant resistances in series.

F. Flux, Flow Velocity, and Tortuosity

As stated above, the *flux density* (hereafter, simply flux) is the volume of water passing through a unit cross-sectional area (perpendicular to the flow direction) per unit time. The dimensions of the flux are:

$$q = V/At = L^3/L^2T = LT^{-1}$$

i.e., length per time (in cgs units, centimeters per second). These are the dimensions of velocity, yet we prefer the term flux to flow velocity, the latter being an ambiguous term. Since soil pores vary in shape, width, and direction, the actual flow velocity in the soil is highly variable (e.g., wider pores conduct water more rapidly, and the liquid in the center of each pore moves faster than the liquid in close proximity to the particles). Strictly speaking, therefore, one cannot refer to a single velocity of liquid flow, but at best to an average velocity.

Yet, even the average velocity of the flowing liquid differs from the flux, as we have defined it. Flow does not in fact take place through the entire cross-sectional area A, since part of this area is plugged by particles and only the porosity fraction is open to flow. Since the real area through which flow takes place is smaller than A, the actual average velocity of the liquid must be greater than the flux q. Furthermore, the actual length of the path traversed by an average parcel of liquid is greater than the soil column length L, owing to the labyrinthine, or tortuous, nature of the pore passages, as shown in Fig. 8.7.

Tortuosity can be defined as the average ratio of the actual roundabout path to the apparent, or straight, flow path; i.e., it is the ratio of the average length of the pore passages (as if they were stretched out in the manner one can stretch out a coiled or tangled telephone wire) to the length of the soil specimen. Tortuosity is thus a dimensionless geometric parameter of porous media which, though difficult to measure precisely, is always greater than 1 and may exceed 2. The *tortuosity factor* is sometimes defined as the inverse of what we defined as the tortuosity.

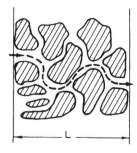

Fig. 8.7. Flow path tortuosity in the soil.

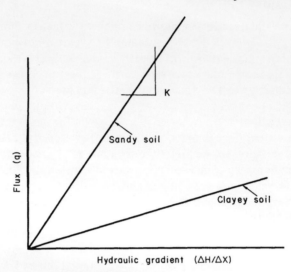

Fig. 8.8. The linear dependence of flux upon hydraulic gradient, the hydraulic conductivity being the slope (i.e., the flux per unit gradient).

G. Hydraulic Conductivity, Permeability, and Fluidity

The hydraulic conductivity, again, is the ratio of the flux to the hydraulic gradient, or the slope of the flux versus gradient curve (Fig. 8.8).

With the dimensions of flux being LT^{-1}, those of hydraulic conductivity depend on the dimensions assigned to the driving force (the potential gradient). In the last chapter, we showed that the simplest way to express the potential gradient is by use of length, or head, units. The hydraulic head gradient H/L, being the ratio of a length to a length, is dimensionless.[7] Accordingly, the dimensions of hydraulic conductivity are the same as the dimensions of flux, namely LT^{-1}. If, on the other hand, the hydraulic gradient is expressed in terms of the variation of pressure with length, then the hydraulic conductivity assumes the dimensions of $M^{-1}L^3T$. Since the latter is cumbersome, the use of head units is generally preferred.

In a saturated soil of stable structure, as well as in a rigid porous medium such as sandstone, for instance, the hydraulic conductivity is characteristically constant. Its order of magnitude is about 10^{-2}–10^{-3} cm/sec in a sandy soil and 10^{-4}–10^{-7} cm/sec in a clayey soil.

---

[7] Though, strictly speaking, H is not a true length, but a pressure equivalent in terms of a water column height; $H = P/\rho g$, and its gradient should be assigned the units of cm_{H_2O}/cm.

G. Hydraulic Conductivity, Permeability, and Fluidity

To appreciate the practical significance of these values in more familiar terms, consider the hypothetical case of an unlined (earth-bottom) reservoir or pond in which one wishes to retain water against losses caused by downward seepage. If the seepage into and through the underlying soil is by gravity alone (i.e., no pressure or suction gradients in the soil), we can assume it will take place at a rate approximately equal to the hydraulic conductivity. A coarse sandy soil might have a K value of, say, 10^{-2} cm/sec and would therefore lose water at the enormous rate of nearly 10 m/day. A fine loam soil with a K value of 10^{-4} cm/sec would lose "only" about 10 cm/day. Finally, and in contrast, a bed of clay with a conductivity of 10^{-6} cm/sec would allow the seepage of no more than 1 mm/day, much less than the expectable rate of evaporation. So, the retention of water in earthen dams and reservoirs and the prevention of seepage from unlined canals can be greatly aided by a bed of clay, particularly if the clay is dispersed to further reduce its hydraulic conductivity.

The hydraulic conductivity is obviously affected by structure as well as by texture, being greater if the soil is highly porous, fractured, or aggregated than if it is tightly compacted and dense. Hydraulic conductivity depends not only on total porosity, but also, and primarily, on the sizes of the conducting pores. For example, a gravelly or sandy soil with large pores can have a conductivity much greater than a clay soil with narrow pores though the total porosity of a clay is generally greater than that of a sandy soil. Cracks, worm holes, and decayed root channels are present in the field and may affect flow in different ways, depending on the direction and condition of the flow process. If the pressure head is positive, these passages will run full of water and contribute to the observed flux and measured conductivity. If the pressure head in the water is negative, that is, if soil water is under suction, large cavities will generally be drained and fail to transmit water.

In many soils, the hydraulic conductivity does not in fact remain constant. Because of various chemical, physical, and biological processes, the hydraulic conductivity may change as water permeates and flows in a soil. Changes occurring in the composition of the exchangeable-ion complex, as when the water entering the soil has a different composition or concentration of solutes than the original soil solution, can greatly change the hydraulic conductivity. In general, the conductivity decreases with decreasing concentration of electrolytic solutes, due to swelling and dispersion phenomena, which are also affected by the species of cations present. Detachment and migration of clay particles during prolonged flow may result in the clogging of pores. The interactions of solutes with the soil matrix, and their effect on hydraulic conductivity, will be discussed in Chapter 10.

In practice, it is extremely difficult to saturate a soil with water without trapping some air. Entrapped air bubbles may block pore passages, as shown

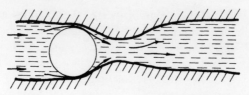

Fig. 8.9. An entrapped air bubble plugging flow.

in Fig. 8.9. Temperature changes may cause the flowing water to dissolve or to release gas, and will also cause a change in the volume of the gas phase, thus affecting conductivity.

The hydraulic conductivity K is not an exclusive property of the soil alone, since it depends upon the attributes of the soil and of the fluid together. The soil characteristics which affect K are the total porosity, the distribution of pore sizes, and tortuosity—in short, the pore geometry of the soil. The fluid attributes which affect conductivity are fluid density and viscosity.

It is possible in theory, and sometimes in practice, to separate K into two factors: *intrinsic permeability* of the soil k and *fluidity* of the liquid or gas f:

$$K = kf \tag{8.13}$$

When K is expressed in terms of cm/sec (LT^{-1}), k is expressed in cm² (L^2) and f in $1/(cm\ sec)(L^{-1}T^{-1})$.

Fluidity is inversely proportional to viscosity:

$$f = \rho g/\eta \tag{8.14}$$

hence,

$$k = K\eta/\rho g \tag{8.15}$$

where η is the viscosity in poise units (dyn sec/cm²), ρ is the fluid density (gm/cm³), and g is the gravitational acceleration (cm/sec²).

In an ordinary liquid, the density is nearly constant (though it varies somewhat with temperature and solute concentration), and changes in fluidity are likely to result primarily from changes in viscosity. In compressible fluids such as gases, on the other hand, changes in density due to pressure and temperature variation can also be considerable.

The use of the term permeability has in the past been a source of some confusion, as it has often been applied synonymously with hydraulic conductivity. Permeability has also been used in a loosely qualitative sense to describe the readiness with which a porous medium transmits water or various other fluids. For this reason, the use of *permeability* in a strict, quantitative sense with the dimensions of length squared as previously defined, in Eq. (8.15) may require the use of some such qualifying adjective as "intrinsic" permeability (Richards, 1954) or "specific" permeability (Scheidegger, 1957). For convenience, however, we shall henceforth refer to k simply as permeability.

It should be clear from the foregoing that, while fluidity varies with composition of the fluid and with temperature, the permeability is ideally an exclusive property of the porous medium and its pore geometry alone — provided the fluid and the solid matrix do not interact in such a way as to change the properties of either. In a completely stable porous body, the same permeability will be obtained with different fluids, e.g., with water, air, or oil.[8] In many soils, however, matrix–water interactions are such that conductivity cannot be resolved into separate and exclusive properties of water and of soil, and Eq. (8.13) is impractical to apply.

H. Limitations of Darcy's Law

Darcy's law is not universally valid for all conditions of liquid flow in porous media. It has long been recognized that the linearity of the flux versus hydraulic gradient relationship fails at high flow velocities, where inertial forces are no longer negligible compared to viscous forces (Hubbert, 1956). Darcy's law applies only as long as flow is laminar (i.e., nonturbulent slippage of parallel layers of the fluid one atop another), and where soil–water interaction does not result in a change of fluidity or of permeability with a change in gradient. Laminar flow prevails in silts and finer materials for any commonly occurring hydraulic gradients found in nature (Klute, 1965a). In coarse sands and gravels, however, hydraulic gradients much in excess of unity may result in nonlaminar flow conditions, and Darcy's law may not always be applicable.

The quantitative criterion for the onset of turbulent flow is the *Reynolds number* N_{Re}:

$$N_{Re} = d\bar{u}\rho/\eta \tag{8.16}$$

---

[8] Actually, the permeability of porous media to gases (at atmospheric pressure) is often a little greater than the value for liquids. This is because the gases do not behave exactly as a continuum in porous solids. The effect is caused by the fact that, for gases, the velocity of the fluid at the solid boundary may not actually be zero. It seems to depart from zero by a greater amount the smaller the pressure of the gas. Consequently, if the permeability to gas is determined at increasingly greater pressures, the permeability will be increasingly small. If a plot of k versus $1/P$ is made, the result is a straight line which, when extrapolated to $1/P = 0$, may intersect the k ordinate at a value corresponding to that which would be obtained for an adhering liquid. This phenomenon is known at the *Klinkenberg* effect after the man who discovered it. The effect tends to be greater the smaller the average pore size and the greater the *specific surface*.

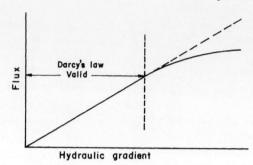

Fig. 8.10. The deviation from Darcy's law at high flux, where flow becomes turbulent.

where $\bar{u}$ is the mean flow velocity, d the effective pore diameter, p the liquid density, and η its viscosity. In straight tubes, the critical value of N_{Re} beyond which turbulence sets in has variously been reported to be of the order of 1000–2200 (Scheidegger, 1957; Childs, 1969). However, the critical Reynolds number at which water flowing in a tube becomes turbulent is apparently reduced greatly when the tube is curved and its diameter is variable. For porous media, therefore, it is safe to assume that flux remains linear with hydraulic gradient only as long as N_{Re} is smaller than unity. As flow velocity increases, especially in systems of large pores, the occurrence of turbulent eddies or nonlinear laminar flow results in waste of effective energy; i.e., the hydraulic potential gradient becomes less effective in inducing flow. This is illustrated in Fig. 8.10.

Deviations from Darcy's law may also occur at the opposite end of the flow-velocity range, namely at low gradients and in small pores. Some investigators (Swartzendruber, 1962; Miller and Low, 1963; Nerpin et al., 1966) have claimed that, in clayey soils, low hydraulic gradients may cause no flow or only low flow rates that are less than proportional to the gradient, while others have disputed some of these findings (Olsen, 1965). A possible reason for this anomaly is that the water in close proximity to the particles and subject to their adsorptive force fields may be more rigid than ordinary water, and exhibit the properties of a *Bingham liquid* (having a yield value) rather than a *Newtonian liquid*. (See Chapter 13.) The adsorbed, or bound, water may have a quasi-crystalline structure similar to that of ice. Some soils may exhibit an apparent threshold gradient, below which the flux is either zero (the water remaining apparently immobile), or at least lower than predicted by the Darcy relation,[9] and only at gradients exceeding the

---

[9] Another possible cause for apparent flow anomalies in clay soils is their compressibility. Darcy's law applies to flow relative to the particle, and it may seem to fail when the particles themselves are moving relative to a fixed frame of reference.

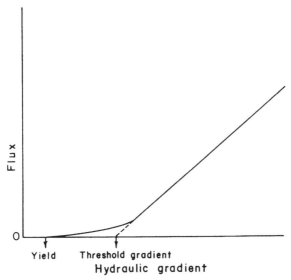

Fig. 8.11. Possible deviations from Darcy's law at low gradients.

threshold value does the flux become proportional to the gradient (Fig. 8.11). These phenomena and their possible explanations, though highly interesting, are generally of little or no importance in practice, and Darcy's law can be employed in the vast majority of cases pertaining to the flow of water in soil.

I. Relation of Conductivity and Permeability to Pore Geometry

Since permeability is a characteristic physical property of a porous medium, it would seem only reasonable to assume that it relates in some functional way to certain measurable properties of the soil pore geometry, e.g., porosity, pore-size distribution, internal surface area, etc. However, numerous attempts to discover a functional relation of universal applicability have so far met with disappointing results.

Perhaps the simplest approach is to seek a correlation between permeability and porosity (e.g., Jacob, 1946; Franzini, 1951). The reader should have concluded by now, however, that this is, in general, a futile approach (except for the comparison of otherwise identical media), owing to the strong dependence of flow rate upon the width, continuity, shape, and tortuosity of the conducting channels. Thus, a medium composed of numerous small pores with a high total porosity is likely to exhibit a lower saturated conductivity than a medium of lesser porosity but larger (though fewer) individual pores.

Attempts have also been made to find correlations between permeability and grain-size distribution. Such correlations may indeed be found for similar materials, but hardly for materials of different grain shapes and aggregations (e.g., sand versus clay). Efforts to refine this approach by introducing empirical grain shape and packing parameters have not won general acceptance.

Numerous theoretical models have been introduced to represent porous media by a set of relationships that are amenable to mathematical treatment. Some of these idealized models are highly elegant, yet their worth must depend on experiment, which alone can show how well, or how poorly, they portray the behavior of real porous media. Scheidegger (1957) gave a comprehensive review of such models, including the straight capillaric, parallel, serial, and branching models. He pointed out that, in general, natural porous media are extremely *disordered*, so that it seems a rather poor procedure to represent them by something which is intrinsically *ordered*.[10] He therefore suggested that the preferred model of a porous medium should be based upon statistical concepts.

One of the most widely accepted theories on the relation of permeability to the geometric properties of porous media is the Kozeny theory and particularly its modification by Carman (1939). This theory is based on the concept of a *hydraulic radius*, i.e., a characteristic length parameter presumed to be linked with the hypothetical channels to which the porous medium is thought to be equivalent. The measure of hydraulic radius is the ratio of the volume to the surface of the pore space, or the average ratio of the cross-sectional area of the pores to their circumferences. The following is known as the Kozeny–Carman equation:

$$k = f^3/ca^2(1-f)^2 \qquad (8.17)$$

where f is the porosity, a the specific surface exposed to the fluid, and c a constant representing a particle shape factor. For a critique of this theory, see Scheidegger (1957). A criticism particularly apt in relation to soils is that the hydraulic radius theory fails to describe structured bodies, such as, for example, fissured clays, where the structural fissures contribute negligibly both to porosity and specific surface, and yet they may dominate the permeability.

A promising approach to the prediction of permeability from basic physical properties of the porous medium is to seek a connection between permeability and pore-size distribution. Since, however, there is no direct or

---

[10] An *ordered* medium consists of a sequence of internally identical units having some consistent geometric pattern.

simple way to obtain or characterize this distribution per se, it is only possible to work with parameters which are based indirectly upon the pore-size distribution, namely parameters based on the suction or capillary pressure versus sorption or desorption. Since flow through an irregular pore is limited by the narrow "necks" along the flow paths, one needs also to consider or estimate the number and size of necks and the interconnections of pores of different widths. Work along these lines has been published by Childs and Collis-George (1950), Marshall (1958),[11] and Millington and Quirk (1959). The results of these theories, while more generally applicable than those based on earlier models, still appear to be valid mostly for coarse materials in which capillary phenomena predominate.

J. Homogeneity and Isotropy

The hydraulic conductivity (or permeability) may be uniform throughout the soil, or may vary from point to point, in which case the soil is said to be hydraulically *inhomogeneous*. If the conductivity is the same in all directions, the soil is *isotropic*. However, the conductivity at each point may differ for different directions (e.g., the horizontal conductivity may be greater, or smaller, than the vertical), a condition known as *anisotropy*. A soil may be homogeneous and nevertheless anisotropic, or it may be inhomogeneous (e.g., layered) and yet isotropic at each point. Some soils exhibit both inhomogeneity and anisotropy. In certain cases, K may also be *asymmetrical* (or directional), that is to say, indicate a different value depending on the direction of flow along a given line. Measurement of the directional permeability of soils was discussed by Maasland and Kirkham (1955). A comprehensive review of anisotropy and layering is given in the book by Bear *et al.* (1968). Anisotropy is generally due to the structure of the soil, which may be laminar, or platy, or columnar, etc., thus exhibiting a pattern of micropores or macropores with a distinctly directional bias.

To illustrate the meaning of the concepts of homogeneity and isotropy, let us consider the hypothetical situation of two continguous bodies of soil,

---

[11] Marshall's approach is based on dividing the soil moisture characteristic curve into equally spaced segments, each characterized by a value of matric suction sufficient to empty it (assuming the capillary equation $P = -2\gamma/r$). His equation is

$$K = (\varepsilon^2/8n^2)[r_1^2 + 3r_2^2 + 5r_3^2 + \cdots + (2n - 1)r_n^2]$$

where ε is porosity and r_1, r_2, and r_n represent the mean radii of the pores (in decreasing order of size) in each of the n equal fractions of the total pore space. For a more detailed description of this theory and its application to flow in unsaturated soils, see Chapter 9.

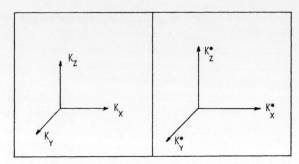

Fig. 8.12. Schematic illustration of homogeneity and isotropy for two adjacent realms in the soil. Explanation is given in the text.

or better yet, two realms within the same body of soil, as shown in Fig. 8.12. The hydraulic conductivities for the three principal axes of the x, y, z coordinate system are designated K_x, K_y, K_z, respectively, for the soil body on the left, and the corresponding conductivities are designated K_x^*, K_y^*, K_z^* for the body on the right. Now let us list four possible cases:

Case I: The soil is homogeneous and isotropic throughout both realms.

Homogeneous: $K_x = K_x^*$, $K_y = K_y^*$, $K_z = K_z^*$ ⎱ single value
Isotropic: $K_x = K_y = K_z$, $K_x^* = K_y^* = K_z^*$ ⎰ of K

Case II: The soil is homogeneous but anisotropic.

Homogeneous: $K_x = K_x^*$, $K_y = K_y^*$, $K_z = K_z^*$ ⎱ 3 values
Anisotropic: $K_x \neq K_y \neq K_z$, $K_x^* \neq K_y^* \neq K_z^*$ ⎰ of K

Case III: The soil is inhomogeneous but isotropic.

Inhomogeneous: $K_x \neq K_x^*$, $K_y \neq K_y^*$, $K_z \neq K_z^*$ ⎱ 2 values
Isotropic: $K_x = K_y = \mathbf{K}_z$, $K_x^* = K_y^* = K_z^*$ ⎰ of K

Case IV: The soil is both inhomogeneous and anisotropic.

Inhomogeneous: $K_x \neq K_x^*$, $K_y \neq K_y^*$, $K_z \neq K_z^*$ ⎱ 6 values
Anisotropic: $K_x \neq K_y \neq K_z$, $K_x^* \neq K_y^* \neq K_z^*$ ⎰ of K

Analysis of anisotropic flow systems is also complicated by the fact that the flow direction may not be orthogonal to the equipotential lines or planes; that is to say, the flow direction is not necessarily the same as the direction of the steepest potential gradient. (Analogous is the tendency of a round ball to run straight down a smooth slope whereas the same ball will tend to roll sideways while running downslope if the surface has slanted furrows or grooves).

K. Measurement of Hydraulic Conductivity of Saturated Soils

Methods for measuring hydraulic conductivity in the laboratory were reviewed by Klute (1965a), and for measurement in the field by Talsma (1960) and by Boersma (1965a,b). The use of permeameters for laboratory determinations is illustrated in Figs. 8.13 and 8.14. Such determinations can

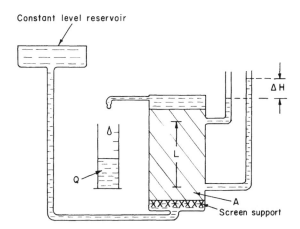

Fig. 8.13. The measurement of saturated hydraulic conductivity with a constant head permeameter; $K = VL/At \, \Delta H$.

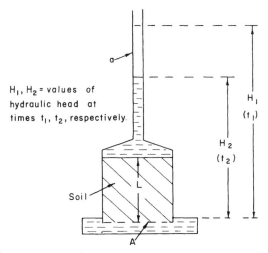

Fig. 8.14. The measurement of saturated hydraulic conductivity with a falling head permeameter; $K = [2.3aL/A(t_2 - t_1)](\log H_1 - \log H_2)$. H_1 and H_2 are the values of hydraulic head at times t_1 and t_2, respectively.

be made with dried and fragmented specimens, which then must be packed into the flow cells in a standard manner, or, preferably, with undisturbed core samples taken directly from the field. In either case, provision must be made to avoid boundary flow along the walls of the container. Field measurements can be made most conveniently below the water table, as by the *augerhole method* (Luthin, 1957) or by the *piezometer method* (Johnson *et al.*, 1952). Techniques have also been proposed for measurements above the water table, as by the double-tube method (Bouwer, 1961, 1962), the shallow-well pump-in method, and the field-permeameter method (Winger, 1960). A more recent review of methods for measuring the hydraulic conductivity of saturated soils was given by van Schilfgaarde (1974).

L. Equations of Saturated Flow

Darcy's law, by itself, is sufficient only to describe *steady*, or *stationary*, flow processes, in which the flux remains constant and equal along the conducting system (and hence the potential and gradient at each point remain constant with time). *Unsteady*, or *transient*, flow processes, in which the magnitude and possibly even the direction of the flux and potential gradient vary with time, require the introduction of an additional law, namely, the law of conservation of matter. To understand how this law applies to flow phenomena, consider a small volume element (say, a cube) of soil, into and out of which flow takes place at possibly differing rates. The mass conservation law, expressed in the *equation of continuity*, states that if the rate of inflow into the volume element is greater than the rate of outflow, then the volume element must be storing the excess and increasing its water content (and, conversely, if outflow exceeds inflow, storage must decrease).

Consider first the simplest case of one-dimensional flow, with q_x being the flux in the x direction. The rate of increase of q_x with x must equal the rate of decrease of volumetric water content θ with time t:

$$\partial\theta/\partial t = -\partial q_x/\partial x \tag{8.18}$$

which, in multidimensional systems, becomes

$$\partial\theta/\partial t = -\nabla \cdot \mathbf{q} \tag{8.19}$$

We recall Darcy's law,

$$\mathbf{q} = -K\nabla H \tag{8.20}$$

which in one dimension is

$$q_x = -K\, dH/dx \tag{8.21}$$

L. Equations of Saturated Flow

(where H is the hydraulic head and K the hydraulic conductivity). Now, we combine (8.20) with the continuity equation (8.19) to obtain the *general flow equation*:

$$\partial \theta / \partial t = \nabla \cdot K \nabla H \qquad (8.22)$$

In applying this equation, the assumptions are usually made that inertial forces are negligible in comparison with viscous forces, that the water is continuously connected throughout the flow region, that isothermal conditions prevail, and that no chemical or biological phenomena change the fluid or the porous medium.

In one dimension, Eq. (8.22) becomes

$$\frac{\partial \theta}{\partial t} = \frac{\partial}{\partial x}\left(K \frac{\partial H}{\partial x}\right) \qquad (8.23)$$

Since the hydraulic head can be resolved into a pressure head H_p and a gravitational head (an elevation above some reference datum, z), we can rewrite (8.22)

$$\frac{\partial \theta}{\partial t} = \nabla \cdot [K(\nabla H_p + \nabla z)] \qquad (8.24)$$

In horizontal flow, $\nabla z = 0$, so for this case,

$$\frac{\partial \theta}{\partial t} = \frac{\partial}{\partial x}\left[K\left(\frac{\partial H_p}{\partial x}\right)\right] \qquad (8.25)$$

while in vertical flow, $\partial z/\partial z = 1$, and therefore, for this case,

$$\frac{\partial \theta}{\partial t} = \frac{\partial}{\partial z}\left[K\left(\frac{\partial H_p}{\partial z} + 1\right)\right] \qquad (8.26)$$

In a saturated soil with an incompressible matrix, $\partial \theta/\partial t = 0$, the conductivity is usually assumed to remain constant, hence Eq. (8.23) becomes

$$K_s \, \partial^2 H / \partial x^2 = 0 \qquad (8.27)$$

where K_s is the hydraulic conductivity of the saturated soil (the saturated conductivity). For three-dimensional flow conditions, and allowing for anisotropy, the equation is

$$K_x \frac{\partial^2 H}{\partial x^2} + K_y \frac{\partial^2 H}{\partial y^2} + K_z \frac{\partial^2 H}{\partial z^2} = 0 \qquad (8.28)$$

where K_x, K_y, and K_z represent the hydraulic conductivity values in the three principal directions x, y, z.

In an isotropic soil (where $K_x = K_y = K_z$ at each point) that is also homogeneous (K equal at all points), we obtain the well-known *Laplace equation*:

$$\partial^2 H/\partial x^2 + \partial^2 H/\partial y^2 + \partial^2 H/\partial z^2 = 0 \tag{8.29}$$

This is a second-order partial differential equation of the elliptical type, and it can be solved in certain cases to obtain a quantitative description of water flow in various systems.

In general, a differential equation can have an infinite number of solutions. To determine the specific solution in any given case, it is necessary to specify the boundary conditions, and, in the case of unsteady flow, of the initial conditions as well. Various types of boundary conditions can exist (e.g., impervious boundaries, free water surfaces, boundaries of known pressure, or known inflow or outflow rates, etc.), but in each case the flux and pressure head must be continuous throughout the system. In layered soils, the hydraulic conductivity and water content may be discontinuous across interlayer boundaries (that is, they may exhibit abrupt changes). Flow equations for inhomogeneous, anisotropic, and compressible systems were given by Bear *et al.* (1968).

Philip (1969a) analyzed flow in swelling (compressible) media. In unsteady flow, the solid matrix of a swelling soil undergoes motion, so that Darcy's law applies to water movement relative to the particles, rather than relative to physical space. Experimental work with such soils was carried out by Smiles and Rosenthal (1968).

Sample Problems

1. Water in an irrigation hose is kept at an hydrostatic pressure of 1 bar. Five drip-irrigation emitters are inserted into the wall of the hose. Calculate the drip rate (liter/hr) from the emitters if each contained a coiled capillary tube 1 m long and the capillary diameters are 0.2, 0.4, 0.6, 0.8, and 1.0 mm. (Assume laminar flow.) What percentage of the total discharge is due to the single largest emitter?

Use Poiseuille's law to calculate the discharge Q:

$$Q = \pi R^4 \, \Delta P / 8 \eta L$$

Substituting the values for π (3.14), the pressure differential ΔP (10^6 dyn/cm^2), the viscosity η [10^{-2} gm/cm sec at 20°C], the capillary tube length (100 cm), and the appropriate tube radii (0.01, 0.02, 0.03, 0.04, and

0.05 cm), we obtain

Emitter #1: $Q_1 = \dfrac{3.14 \times (10^{-2})^4 \times 10^6}{8 \times 10^{-2} \times 10^2} = 3.91 \times 10^{-3}$ cm^3/sec
$= 0.014$ liter/hr

Emitter #2: $Q_2 = \dfrac{3.14 \times (2 - 10^{-2})^4 \times 10^6}{8 \times 10^{-2} \times 10^2} = 6.28 \times 10^{-2}$ cm^3/sec
$= 0.226$ liter/hr

Emitter #3: $Q_3 = \dfrac{3.14 \times (3 \times 10^{-2})^4 \times 10^6}{8 \times 10^{-2} \times 10^2} = 3.18 \times 10^{-1}$ cm^3/sec
$= 1.14$ liter/hr

Emitter #4: $Q_4 = \dfrac{3.14 \times (4 \times 10^{-2})^4 \times 10^6}{8 \times 10^{-2} \times 10^2} = 1$ cm^3/sec
$= 3.6$ liter/hr

Emitter #5: $Q_5 = \dfrac{3.14 \times (5 \times 10^{-2})^4 \times 10^6}{8 \times 10^{-2} \times 10^2} = 2.45$ cm^3/sec
$= 8.83$ liter/hr

Total discharge from all five emitters:

$$Q_{tot} = 0.014 + 0.226 + 1.14 + 3.6 + 8.83 = 13.81 \text{ liter/hr}$$

Percentage contribution of the largest emitter:

$$\frac{Q_5}{Q_{tot}} \times 100 = \frac{8.83}{13.81} \times 100 = 63.9\%$$

The single largest emitter thus accounts for nearly two-thirds of the total discharge, while the smallest emitter accounts for only 0.1% (though its diameter is only $\frac{1}{5}$ that of the largest emitter).

Note: Modern drip-irrigation emitters generally depend on partly turbulent (rather than completely laminar) flow, to reduce sensitivity to pressure fluctuations and vulnerability to clogging by particles.

2. Over a century ago, the population of the French mustard making town of Dijon was, perhaps, 10,000. Since they drank mostly wine, their daily water requirements were, say, no more than 20 liter per person. Then the denizens of the town began to notice that their water supply had somehow become polluted. Since they were unable at the time to find any bona fide soil physicists, they invited an engineer named Darcy to design a filtration system. He must have looked for a textbook or a handbook on the topic but found none, so he had to experiment from scratch. Darcy ended up promul-

gating a new law and achieving immortal, if posthumous, fame. Now supposing we were given the same task today (with the benefits of hindsight) and we knew that a column thickness of 30 cm was needed for adequate filtration and that the hydraulic conductivity of the available sand was 2×10^{-3} cm/sec. Could we calculate the area of filter bed needed under an hydrostatic pressure (ponding) head of 0.7 m? Consider the flow to be vertically downward to a fixed drainage surface. We begin by calculating the discharge Q needed:

$$Q = \frac{10^4 \text{ persons} \times 20 \text{ liter/person day} \times 10^3 \text{ cm}^3/\text{liter}}{8.64 \times 10^4 \text{ sec/day}}$$
$$= 2.31 \times 10^3 \text{ cm}^3/\text{sec}$$

We recall Darcy's law:

$$Q = AK \, \Delta H/L$$

Hence, the area A needed is

$$A = QL/(K \, \Delta H)$$

The hydraulic head drop ΔH equals the sum of the pressure head and gravitational head drops. Hence

$$\Delta H = 70 + 30 = 100 \text{ cm}$$

We now substitute the appropriate values for L (30 cm), ΔH (100 cm), and K (2×10^{-3} cm/sec), to obtain

$$A = \frac{2.31 \times 10^3 \text{ cm}^3/\text{sec} \times 30 \text{ cm}}{2 \times 10^{-3} \text{ cm/sec} \times 10^2 \text{ cm}} = 3.5 \times 10^5 \text{ cm}^2 = 35 \text{ m}^2$$

Note: Since populations and per capita water use tend to increase whereas filter beds tend to clog, it might behoove use to apply a factor of safety to our calculations and increase the filtration capacity severalfold (particularly to accommodate peak demand periods). Per capita water use in the U.S., incidentally, ranges from 100 to 400 liter/day.

3. Consider two cases of steady downward percolation through a two-layer soil profile, the top of which is submerged under a 1 m head of water and the bottom of which is defined by a water table. Each of the two layers is 50 cm thick. In the one case, the conductivity of the top layer is 10^{-4} cm/sec and that of the sublayer is 10^{-5} cm/sec. In the second case, the same layers are reversed (i.e., the less conductive soil overlies the more conductive). See Fig. 8.15.

Calculate the flux, and the hydraulic and pressure heads at the interface

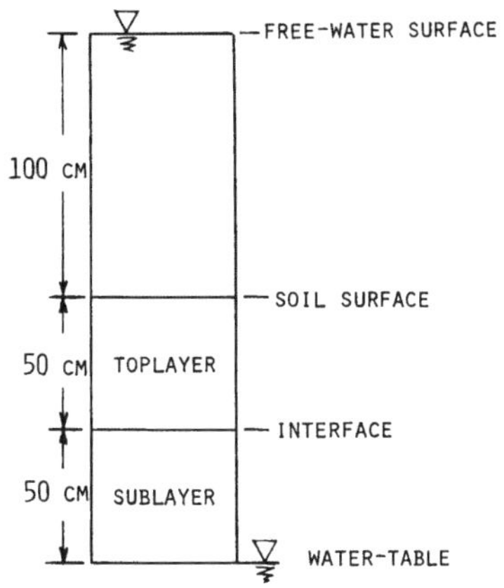

Sample Problems

Fig. 8.15. Steady percolation through a two-layer profile.

between the layers, for each of the two cases. Using the Ohm's law analogy for steady flow through two resistors in series, we can write:

$$q = \Delta H/(R_1 + R_2)$$

where q is the flux; ΔH the total hydraulic head drop across the profile; and R_1, R_2 are the hydraulic resistances of the top layer and sublayer, respectively. Each resistance is proportional directly to the layer's thickness and inversely to its conductivity (i.e., the resistance is equal to the ratio of the layer's thickness to its conductivity). The pressure head at the soil surface is 100 cm and the gravity head (with reference to the soil's bottom) is also 100 cm. Both the pressure and gravity heads at the bottom boundary are zero. Hence,

$$q = \frac{100 \text{ cm} + 100 \text{ cm}}{(50 \text{ cm}/10^{-4} \text{ cm/sec}) + (50 \text{ cm}/10^{-5} \text{ cm/sec})}$$
$$= \frac{200}{5 \times 10^5 + 5 \times 10^6} \text{ cm/sec} = 3.64 \times 10^{-5} \text{ cm/sec}$$

We can now apply Darcy's equation to the top layer alone to obtain the hydraulic head at the interlayer interface:

$$q = K_1 \Delta H_1/L_1 = K_1(H_{\text{surface}} - H_{\text{interface}})/L_1$$

Therefore

$$H_{\text{interface}} = H_{\text{surface}} - qL_1/K_1 = 200 - (3.64 \times 10^{-5})(5 \times 10^1)/10^{-4}$$

$$H_{\text{interface}} = 181.8 \text{ cm}$$

Since the gravity head at the interface is 50 cm (above our reference datum at the bottom of the profile), the pressure head H_p is

$$H_p = H - H_g = 171.8 - 50 = 121.8 \text{ cm}$$

We now reverse the order of the layers, so that the less conductive overlies the more conductive. The total head drop remains the same, as does the total resistance. Therefore the flux remains the same (assuming that both layers are still saturated and the conductivity of each does not change).
Applying Darcy's equation to the top layer, we have, as previously,

$$H_{\text{interface}} = 200 - \frac{(3.64 \times 10^{-5})(5 \times 10^1)}{10^{-5}} = 18 \text{ cm}$$

In this case the pressure head at the interface is

$$H_p = H - H_g = 18 - 50 = -32 \text{ cm}$$

Note: The comparison between the interface pressures of the two cases reveals and illustrates an important principle regarding flow in layered profiles. With the more conductive layer on top, flow is impeded at the interface and there is a pressure build-up, which in our case caused an increase of the pressure head from 100 cm at the soil surface to 121.8 cm at the interface. The opposite occurs whenever the upper layer is less conductive. In this case the pressure is dissipated through the top layer, often to the extent that negative pressures develop at the interface. If this negative pressure exceeds the sublayer's air-entry value, the sublayer will become (or remain, depending on its antecedent state) *unsaturated*, a condition which is the subject of our next chapter.

> Remember when discoursing about water
> to adduce first experience and then reason.
> Leonardo da Vinci
> 1452–1519

9 Flow of Water in Unsaturated Soil

A. Introduction

Most of the processes involving soil–water interactions in the field, and particularly the flow of water in the rooting zone of most crop plants, occur while the soil is in an unsaturated condition. Unsaturated flow processes are in general complicated and difficult to describe quantitatively, since they often entail changes in the state and content of soil water during flow. Such changes involve complex relations among the variable soil wetness, suction, and conductivity, whose interrelations may be further complicated by hysteresis. The formulation and solution of unsaturated flow problems very often require the use of indirect methods of analysis, based on approximations or numerical techniques. For this reason the development of rigorous theoretical and experimental methods for treating these problems was rather late in coming. In recent decades, however, unsaturated flow has become one of the most important and active topics of research in soil physics, and this research has resulted in significant theoretical and practical advances.

B. Comparison of Flow in Unsaturated versus Saturated Soil

In the previous chapter, we stated that soil-water flow is caused by a driving force resulting from a potential gradient, that flow takes place in the direction of decreasing potential, and that the rate of flow (flux) is proportional to the potential gradient and is affected by the geometric properties of the pore channels through which flow takes place. These principles apply in unsaturated, as well as saturated, soils.

The moving force in a saturated soil is the gradient of a positive pressure potential.[1] On the other hand, water in an unsaturated soil is subject to a subatmospheric pressure, or suction, which is equivalent to a negative pressure potential. The gradient of this potential likewise constitutes a moving force. Matric suction is due, as we have pointed out, to the physical affinity of water to the soil-particle surfaces and capillary pores. Water tends to be drawn from a zone where the hydration envelopes surrounding the particles are thicker to where they are thinner, and from a zone where the capillary menisci are less curved to where they are more highly curved.[2] In other words, water flows spontaneously from where matric suction is lower to where it is higher. When suction is uniform all along a horizontal column of soil the column is at equilibrium and there is no moving force. Not so when a suction *gradient* exists. In that case, water will flow in the pores which are water filled at the existing suction and will creep along the hydration films over the particle surfaces, in a tendency to equilibrate the potential. Vapor transfer is an additional mechanism of water movement in unsaturated soils. In the absence of temperature gradients, it is likely to be much slower than liquid flow as long as the soil is fairly moist. In the surface zone, however, where the soil becomes desiccated and strong temperature gradients occur, vapor transfer can become the dominant mechanism of water movement.

The moving force is greatest at the *wetting front* zone, where water invades and advances into an originally dry soil (see Fig. 9.3). In this zone, the suction gradient can amount to many bars per centimeter of soil. Such a gradient constitutes a moving force thousands of times greater than the gravitational force. As we shall see later on, such strong forces are sometimes required for water movement to take place in the face of the extremely low hydraulic conductivity which a relatively dry soil often exhibits.

Perhaps the most important difference between unsaturated and saturated flow is in the hydraulic conductivity. When the soil is saturated, all of the pores are water filled and conducting, so that continuity and hence con-

---

[1] We shall disregard, for the moment, the gravitational force, which is completely unaffected by the saturation or unsaturation of the soil.

[2] The question of how water-to-air interfaces behave in a conducting porous medium that is unsaturated is imperfectly understood. It is generally assumed, at least implicitly, that these interfaces, or menisci, are anchored rigidly to the solid matrix so that, as far as the flowing water is concerned, air-filled pores are like solid particles. The presence of organic surfactants which adsorb to these surfaces is considered to increase their rigidty or viscosity. Even if the air–water interfaces are not entirely stationary, however, the drag, or momentum transfer, between flowing water and air appears to be very small.

B. Comparison of Flow in Unsaturated versus Saturated Soil

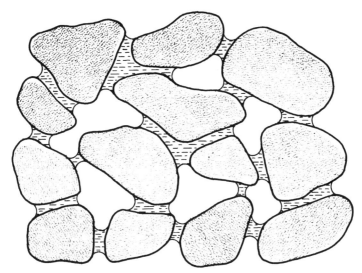

Fig. 9.1. Water in an unsaturated coarse-textured soil.

ductivity are maximal. When the soil desaturates, some of the pores become air filled and the conductive portion of the soil's cross-sectional area decreases correspondingly. Furthermore, as suction develops, the first pores to empty are the largest ones, which are the most conductive,[3] thus leaving water to flow only in the smaller pores. The empty pores must be circumvented, so that, with desaturation, tortuosity increases. In coarse-textured soils, water sometimes remains almost entirely in capillary wedges at the contact points of the particles, thus forming separate and discontinuous pockets of water (see Fig. 9.1). In aggregated soils, too, the large interaggregate spaces which confer high conductivity at saturation become (when emptied) barriers to liquid flow from one aggregate to its neighbors.

For these reasons, the transition from saturation to unsaturation generally entails a steep drop in hydraulic conductivity, which may decrease by several orders of magnitude (sometimes down to 1/100,000 of its value at saturation) as suction increases from 0 to 1 bar. At still higher suctions, or lower wetness values, the conductivity may be so low[4] that very steep suction gradients, or very long times, are required for any appreciable flow to occur.

---

[3] By Poiseuille's law, the total flow rate of water through a capillary tube is proportional to the fourth power of the radius, while the flow rate per unit cross-sectional area of the tube is proportional to the square of the radius. A 1-mm-radius pore will thus conduct as 10,000 pores of radius 0.1 mm.

[4] As very high suctions develop, there may (in addition to the increase in tortuosity and the decrease in number and sizes of the conducting pores) also be a change in the viscosity of the (mainly adsorbed) water, tending to further reduce the conductivity. As yet, however, there is no conclusive evidence of this.

At saturation, the most conductive soils are those in which large and continuous pores constitute most of the overall pore volume, while the least conductive are the soils in which the pore volume consists of numerous micropores. Thus, as is well known, a saturated sandy soil conducts water more rapidly than a clayey soil. However, the very opposite may be true when the soils are unsaturated. In a soil with large pores, these pores quickly empty and become nonconductive as suction develops, thus steeply decreasing the initially high conductivity. In a soil with small pores, on the other hand, many of the pores retain and conduct water even at appreciable suction, so that the hydraulic conductivity does not decrease as steeply and may actually be greater than that of a soil with large pores subjected to the same suction.

Since in the field the soil is unsaturated much, and perhaps most, of the time, it often happens that flow is more appreciable and persists longer in clayey than in sandy soils. For this reason, the occurrence of a layer of sand in a fine-textured profile, far from enhancing flow, may actually impede insaturated water movement until water accumulates above the sand and suction decreases sufficiently for water to enter the large pores of the sand. This simple principle is all too often misunderstood.

C. Relation of Conductivity to Suction and Wetness

Let us consider an unsaturated soil in which water is flowing under suction. Such flow is illustrated schematically in the model of Fig. 9.2. In this model, the potential difference between the inflow and outflow ends is maintained not by different heads of positive hydrostatic pressure, but by different imposed suctions. In general, as the suction varies along the sample, so will the wetness and conductivity. If the suction head at each end of the sample is maintained constant, the flow process will be steady but the suction gradient will vary along the sample's axis. Since the product of gradient and conductivity is constant for steady flow, the gradient will increase as the conductivity decreases with the increase in suction along the length of the sample. This phenomenon is illustrated in Fig. 9.3.

Since the gradient along the column is not constant, as it is in uniform saturated systems, it is not possible, strictly speaking, to divide the flux by the overall ratio of the head drop to the distance ($\Delta H/\Delta x$) to obtain the conductivity. Rather, it is necessary to divide the flux by the exact gradient at each point to evaluate the exact conductivity and its variation with suction. In the following treatment, however, we shall assume that the column of Fig. 9.2 is sufficiently short to allow us to evaluate at least an average conductivity for the sample as a whole (i.e., $K = q\,\Delta x/\Delta H$).

C. Relation to Conductivity of Suction and Wetness

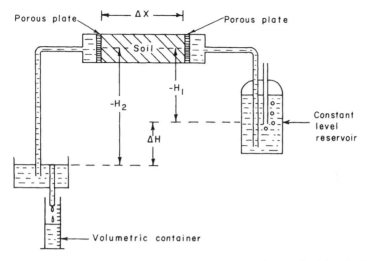

Fig. 9.2. A model illustrating unsaturated flow (under a suction gradient) in a horizontal column.

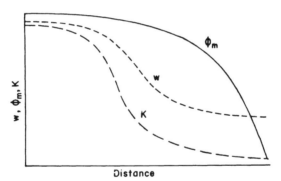

Fig. 9.3. The variation of wetness w, matric potential ϕ_m, and conductivity K along a hypothetical column of unsaturated soil conducting a steady flow of water.

The average negative head, or suction, acting in the column is

$$-\bar{H} = \bar{\psi} = -\tfrac{1}{2}(H_1 + H_2)$$

We shall further assume that the suction everywhere exceeds the air-entry value so that the soil is unsaturated throughout.

Let us now make successive and systematic measurements of flux versus suction gradient for different values of average suction. The results of such a series of measurements are shown schematically in Fig. 9.4. As in the case of saturated flow, we find that the flux is proportional to the gradient. However, the slope of the flux versus gradient line, being the hydraulic conductivity,

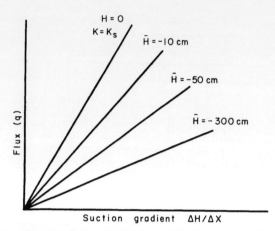

Fig. 9.4. Hydraulic conductivity, being the slope of the flux versus gradient relation, depends upon the average suction in an unsaturated soil.

varies with the average suction. In a saturated soil, by way of contrast, the hydraulic conductivity is generally independent of the magnitude of the water potential, or pressure.

Figure 9.5 shows the general trend of the dependence of conductivity on suction[5] in soils of different texture. It is seen that, although the saturated conductivity of the sandy soil K_{s1} is typically greater than that of the clayey soil K_{s2}, the unsaturated conductivity of the former decreases more steeply with increasing suction and eventually becomes lower.

Although attempts have been made to develop theoretically based equations for the relation of conductivity to suction or to wetness, existing knowledge still does not allow reliable a priori prediction of the unsaturated conductivity function from basic soil properties such as texture. Various empirical equations have been proposed, however, including (Gardner, 1960)

$$K(\psi) = a/\psi^n \qquad (9.1\text{a})$$

$$K(\psi) = a/(b + \psi^n) \qquad (9.1\text{b})$$

$$K(\psi) = K_s/[1 + (\psi/\psi_c)^m] \qquad (9.1\text{c})$$

$$K(\theta) = a\theta^m \qquad (9.1\text{d})$$

$$K(\theta) = K_s s^m = K_s(\theta/f)^m \qquad (9.1\text{e})$$

[5] K versus suction curves are usually drawn on a log–log scale, as both K and ψ vary over several orders of magnitude within the suction range of general interest (say, 0–10,000 cm of suction head).

C. Relation to Conductivity of Suction and Wetness

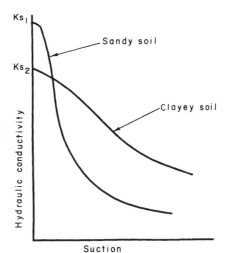

Fig. 9.5. Dependence of conductivity on suction in soils of different texture (log–log scale).

where K is the hydraulic conductivity at any degree of saturation (or unsaturation): K_s is the saturated conductivity of the same soil; a, b, and m are empirical constants (different in each equation); ψ is matric suction head; θ is volumetric wetness; s is the degree of saturation; and ψ_c is the suction head at which $K = \frac{1}{2} K_s$. Note that $s = \theta/f$, where f is porosity.

Of these various equations, the most commonly employed are the first two (of which the first is the simplest, but cannot be used in the suction range approaching zero). In all of the equations, the most important parameter is the exponential constant, since it controls the steepness with which conductivity decreases with increasing suction or with decreasing water content. The m value of the first two equations is about 2 or less for clayey soils and may be 4 or more for sandy soils. For each soil, the equation of best fit, and the values of the parameters, must be determined experimentally.

The relation of conductivity to suction depends upon hysteresis, and is thus different in a wetting than in a drying soil. At a given suction, a drying soil contains more water than a wetting one. The relation of conductivity to water content, however, appears to be affected by hysteresis to a much lesser degree.

Some investigators have used a different designation, usually "capillary conductivity," to distinguish the hydraulic conductivity of a soil at unsaturation from that at saturation. This, however, is generally unnecessary and the adjective capillary can be misleading, since unsaturated flow may not conform to the capillary model any more than does saturated flow.

D. General Equation of Unsaturated Flow

1. THE CONTINUITY EQUATION

Students often feel that the *equation of continuity* is intuitively understandable or logically self-evident so that it requires no formal proof. For those who like formalism, however, we offer the following. Consider a volume element of soil in the shape of a rectangular parallelepiped inside a space defined by a set of the rectangular coordinates x, y, z, as shown in Fig. 9.6. Assume that the sides of the volume element are Δx, Δy, and Δz and hence its volume is $\Delta x \, \Delta y \, \Delta z$. Now consider, for a moment, only the net flow in the x direction. If the flux emerging from the right-hand face exceeds the flux entering the left-hand face by the amount $(\partial q/\partial x) \, \Delta x$, then the difference in discharge (volume of water per unit time) flowing through the two faces must equal (discharge being equal to flux multiplied by area $\Delta y \, \Delta z$):

change in volume discharge = $\quad q \, \Delta y \, \Delta z \quad -[q + (\partial q/\partial x) \, \Delta x]\Delta y \, \Delta z$
(net inflow) $\qquad\quad$ (volume inflowing $\quad$ (volume outflowing $\quad$ (9.2)
$\qquad\qquad\qquad\qquad\quad$ per unit time) $\qquad$ per unit time)

The net inflow must equal the rate of gain of water by the volume element of soil per unit time, which is given by

$$\text{net inflow} = -(\partial q/\partial x) \, \Delta x \, \Delta y \, \Delta z \qquad (9.3)$$

The rate of gain of water by the volume element of soil can also be expressed in terms of the time rate of change of the volume concentration of water, θ, multiplied by the volume of the element, thus

$$\text{rate of gain} = (\partial \theta/\partial t) \, \Delta x \, \Delta y \, \Delta z$$

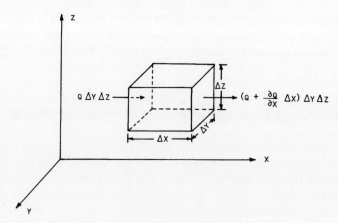

Fig. 9.6. The continuity principle: A volume element of soil gaining or losing water in accordance with the divergence of the flux.

D. General Equation of Unsaturated Flow

Now setting the two alternative expressions equal to each other, we obtain

$$(\partial\theta/\partial t)\,\Delta x\,\Delta y\,\Delta z = -(\partial q/\partial x)\,\Delta x\,\Delta y\,\Delta z$$

or
$$\partial\theta/\partial t = -\partial q/\partial x \qquad (9.4)$$

If we also consider the fluxes in the y and z directions, we obtain the three-dimensional form of the continuity equation, namely,

$$\frac{\partial\theta}{\partial t}\,\Delta x\,\Delta y\,\Delta z = -\left(\frac{\partial q_x}{\partial x} + \frac{\partial q_y}{\partial y} + \frac{\partial q_x}{\partial z}\right)\Delta x\,\Delta y\,\Delta z \qquad (9.5)$$

$$\frac{\partial\theta}{\partial t} = -\left(\frac{\partial q_x}{\partial x} + \frac{\partial q_y}{\partial y} + \frac{\partial q_z}{\partial z}\right) \qquad (9.6)$$

where q_x, q_y, q_z are the fluxes in the x, y, and z directions, respectively.

In shorthand mathematical notation, the last equation is written as

$$\partial\theta/\partial t = -\nabla\cdot\mathbf{q} \qquad (9.7)$$

where the symbol ∇(del) is the vector differential operator, representing the three-dimensional gradient in space (in our case it is the spatial gradient of the flux $\mathbf{q}$). The scalar product of the del operator and a vector function is called the divergence and designated div. An alternative expression of Eq. (9.7) is therefore

$$\frac{\partial\theta}{\partial t} = -\operatorname{div}\mathbf{q} \qquad (9.8)$$

2. The Combined Flow Equation

Darcy's law, though originally conceived for saturated flow only, was extended by Richards (1931) to unsaturated flow, with the provision that the conductivity is now a function of the matric suction head (i.e., $K = K(\psi)$]:

$$\mathbf{q} = -K(\psi)\,\nabla H \qquad (9.9)$$

where ∇H is the hydraulic head gradient, which may include both suction and gravitational components.

As pointed out by Miller and Miller (1956), this formulation fails to take into account the hysteresis of soil-water characteristics. In practice, the hysteresis problem can sometimes be evaded by limiting the use of Eq. (9.1) to cases in which the suction (or wetness) change is monotonic—that is, either increasing or decreasing continuously. In processes involving both wetting and drying phases, Eq. (9.1) is difficult to apply, as the $K(\psi)$ function may be highly hysteretic. As mentioned in the previous section, however, the relation of conductivity to volumetric wetness $K(\theta)$ or to degree of saturation $K(s)$ is affected by hysteresis to a much lesser degree than is the $K(\theta)$ function. Thus, Darcy's law for unsaturated soil can also be written

$$\mathbf{q} = -K(\theta)\,\nabla H \tag{9.10}$$

which, however, still leaves us with the problem of dealing with the hysteresis between ψ and θ.

To obtain the general flow equation and account for transient, as well as steady, flow processes, we must introduce the continuity principle [Eq. (9.7)]:

$$\partial\theta/\partial t = -\nabla\cdot\mathbf{q}$$

Thus,

$$\partial\theta/\partial t = \nabla\cdot[K(\psi)\,\nabla H] \tag{9.11}$$

Remembering that the hydraulic head is, in general, the sum of the pressure head (or its negative, the suction head ψ) and the gravitational head (or elevation) z, we can write

$$\partial\theta/\partial t = -\nabla\cdot[K(\psi)\,\nabla(\psi - z)] \tag{9.12}$$

Since ∇z is zero for horizontal flow and unity for vertical flow, we can rewrite (9.6) as follows:

$$\frac{\partial\theta}{\partial t} = -\nabla\cdot(K(\psi)\,\nabla\psi) + \frac{\partial K}{\partial z} \tag{9.12a}$$

or

$$\frac{\partial\theta}{\partial t} = -\frac{\partial}{\partial x}\left(K\frac{\partial\psi}{\partial x}\right) - \frac{\partial}{\partial y}\left(K\frac{\partial\psi}{\partial y}\right) - \frac{\partial}{\partial z}\left(K\frac{\partial\psi}{\partial z}\right) + \frac{\partial K}{\partial z} \tag{9.13}$$

Processes may also occur in which ∇z (the gravity gradient) is negligible compared to the strong matric suction gradient $\nabla\psi$. In such cases,

$$\partial\theta/\partial t = \nabla\cdot[K(\psi)\,\nabla\psi] \tag{9.14}$$

or, in a one-dimensional horizontal system,

$$\frac{\partial\theta}{\partial t} = \frac{\partial}{\partial x}\left[K(\psi)\frac{\partial\psi}{\partial x}\right] \tag{9.15}$$

E. Hydraulic Diffusivity

To simplify the mathematical and experimental treatment of unsaturated flow processes, it is often advantageous to change the flow equations into a

E. Hydraulic Diffusivity

form analogous to the equations of diffusion and heat conduction, for which ready solutions are available (e.g., Carslaw and Jaeger, 1959; Crank, 1956) in some cases involving boundary conditions applicable to soil-water flow processes. To transform the flow equation, it is sometimes possible to relate the flux to the water content (wetness) gradient rather than to the suction gradient.

The matric suction gradient $\partial\psi/\partial x$ can be expanded by the chain rule as follows

$$\frac{\partial\psi}{\partial x} = \frac{d\psi}{d\theta}\frac{\partial\theta}{\partial x} \tag{9.16}$$

where $\partial\theta/\partial x$ is the wetness gradient and $d\psi/d\theta$ is the reciprocal of the specific water capacity $c(\theta)$:

$$c(\theta) = d\theta/d\psi \tag{9.17}$$

which is the slope of the soil-moisture characteristic curve at any particular value of wetness θ.

We can now rewrite the Darcy equation as follows:

$$q = K(\theta)\frac{\partial\psi}{\partial x} = -\frac{K(\theta)}{c(\theta)}\frac{\partial\theta}{\partial x} \tag{9.18}$$

To cast this equation into a form analogous to Fick's law of diffusion, a function was introduced (Childs and Collis-George, 1950), originally called the *diffusivity* D, where

$$D(\theta) = K(\theta)/c(\theta) = K(\theta)\, d\psi/d\theta \tag{9.19}$$

D is thus defined as the ratio of the hydraulic conductivity K to the specific water capacity c, and since both of these are functions of soil wetness, D must also be so. To avoid any possibility of confusion between the classical concept of diffusivity pertaining to the *diffusive transfer* of components in the gaseous and liquid phases (see, for example, our chapters on solute movement and on gas exchange in the soil) and this borrowed application of the same term to describe *convective flow*, we propose to qualify it with the adjective hydraulic. In this book, therefore, we shall henceforth employ the term *hydraulic diffusivity* when referring to D of Eq. (9.19). Now we can rewrite Eq. (9.19)

$$\mathbf{q} = -D(\theta)\,\nabla\theta \tag{9.20}$$

or, in one dimension,

$$q = -D(\theta)\,\partial\theta/\partial x \tag{9.21}$$

which is mathematically identical with Fick's first equation of diffusion.

Hydraulic diffusivity can thus be viewed as the ratio of the flux to the soil-water content (wetness) gradient. As such, D has dimensions of length squared per unit time (L^2T^{-1}), since K has the dimensions of volume per unit area per time (LT^{-1}) and the specific water capacity c has dimensions of volume of water per unit volume of soil per unit change in matric suction head (L^{-1}). In the use of Eq. (9.21), the gradient of wetness is taken to represent, implicitly, a gradient of matric potential, which is the true driving force.

Introducing the hydraulic diffusivity into Eq. (9.15), for one-dimensional flow in the absence of gravity, we obtain[6]

$$\frac{\partial \theta}{\partial t} = \frac{\partial}{\partial x}\left[D(\theta)\frac{\partial \theta}{\partial x}\right] \qquad (9.22)$$

which has only one dependent variable (θ) rather than the two (θ and ψ) of Eq. (9.15).

In the special case that the hydraulic diffusivity remains constant (though it is not generally safe to assume this except for a very small range of wetness), Eq. (9.22) can be written in the form of Fick's second diffusion equation:

$$\partial \theta / \partial t = D\, \partial^2 \theta / \partial x^2 \qquad (9.23)$$

A word of caution is now in order. In employing the diffusivity concept, and all relationships derived from it, we must remember that the process of liquid water movement in the soil is not one of diffusion but of mass flow, or convection. As we have already suggested, the borrowed term diffusivity, if taken literally, can be misleading. Furthermore, the diffusivity equations become awkward whenever the hysteresis effect is appreciable or where the soil is layered, or in the presence of thermal gradients, since under such conditions flow bears no simple or consistent relation to the decreasing water-content gradient and may actually be in the opposite direction to it. On the other hand, an advantage in using the hydraulic diffusivity is in the fact that its range of variation is smaller than that of hydraulic conductivity.[7]

To take account of gravity [as in Eq. (9.6)], a diffusivity equation can be written in the form

$$\frac{\partial \theta}{\partial t} = \mathbf{\nabla}\cdot[D(\theta)\,\mathbf{\nabla}\theta] + \frac{\partial K(\theta)}{\partial z} = \mathbf{\nabla}\cdot[D(\theta)\,\mathbf{\nabla}\theta] + \frac{dK}{d\theta}\frac{\partial \theta}{\partial z} \qquad (9.24)$$

---

[6] In two dimensions, $\partial\theta/\partial t = \partial(D\,\partial\theta/\partial x)/\partial x + \partial(D\,\partial\theta/\partial y)/\partial y$; in three dimensions, $\partial\theta/\partial t = \partial(D\,\partial\theta/\partial x)/\partial x + \partial(D\,\partial\theta/\partial y)/\partial y + \partial(D\,\partial\theta/\partial z)/\partial z$; in two dimensions, cylindrical geometry, $\partial\theta/\partial t = (1/r)\,\partial(Dr\,\partial\theta/\partial r)/\partial r$; in three dimensions, spherical geometry, $\partial\theta/\partial t = (1/r^2)\,\partial(Dr^2\,\partial\theta/\partial r)/\partial r$.

[7] The maximum value of D found in practice is of the order of 10^4 cm^2/day. D generally decreases to about 1–10 cm^2/day at the lower limit of wetness normally encountered in the root zone. It thus varies about a thousandfold rather than about a millionfold, as does the hydraulic conductivity in the same wetness range.

F. The Boltzmann Transformation

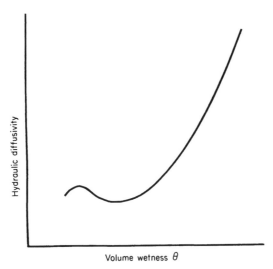

Fig. 9.7. Relation of diffusivity to soil wetness.

The relation of hydraulic diffusivity to wetness is shown in Fig. 9.7. This relation is sometimes expressed in the empirical equation (Gardner and Mayhugh, 1958)

$$D(\theta) = ae^{b\theta} \qquad (9.25)$$

This equation applies only to the right-hand section of the curve showing a rise in diffusivity with wetness. In the very dry range, the diffusivity often indicates an opposite trend—namely, a rise with decreasing soil wetness. This is apparently due to the contribution of vapor movement (Philip, 1955). In the very wet range, as the soil approaches complete saturation, the diffusivity becomes indeterminate as it tends to infinity [since $c(\theta)$ tends to zero].

F. The Boltzmann Transformation

Because of its nonlinearity, the flow equation (9.16) for water in unsaturated soil is much more difficult to solve than are the classical linear equations for the flow of heat or electricity.

The ability of modern high speed computers to solve nonlinear differential equations by successive numerical approximation is rapidly opening the door to practical success in broad areas of soil physics. There are also a number of simple analytical techniques that facilitate the application of the unsaturated-flow equation to particular problems.

When gravity can be neglected and flow is monotonic (no hysteresis effects), the simple form $\partial\theta/\partial t = \mathbf{\nabla}\cdot(D\,\mathbf{\nabla}\theta)$ can be put into a form that is amenable to solution by the separation-of-variables technique so familiar in the solution of linear partial differential equations. By this means, families of one-dimensional flow patterns can be computed for Cartesian, cylindrical, and spherical coordinates. The Cartesian pattern conforms to an important and widely studied boundary condition, namely that of a long ("semi-infinite") horizontal column of soil, initially at some uniform wetness and suction, which is suddenly subjected at one end to a different suction (either higher *or* lower). To convert the diffusivity equation from its original form as a partial differential equation with two independent variables (space and time, x and t), namely,

$$\frac{\partial \theta}{\partial t} = \frac{\partial}{\partial x}\left(D \frac{\partial \theta}{\partial x}\right) \tag{9.26}$$

to the form of an ordinary differential equation (i.e., with only one independent variable), we can employ the so-called Boltzmann transformation as follows:

First we assume a composite variable[8] $B(\theta)$ such that

$$B(\theta) = xt^{-1/2} \tag{9.27}$$

We now introduce this variable into Eq. (9.26) by means of the chain rule to obtain

$$\frac{\partial \theta}{\partial B}\frac{\partial B}{\partial t} = \frac{\partial}{\partial B}\left(D \frac{\partial \theta}{\partial B}\frac{\partial B}{\partial x}\right)\frac{\partial B}{\partial x} \tag{9.28}$$

Differentiating B with respect to x gives

$$\partial B/\partial x = t^{-1/2} \tag{9.29}$$

and with respect to t:

$$\partial B/\partial t = -\tfrac{1}{2}xt^{-3/2} \tag{9.30}$$

Substituting these expressions in Eq. (9.28), we get

$$-\frac{1}{2}xt^{-3/2}\frac{d\theta}{dB} = \frac{d}{dB}\left(D \frac{d\theta}{dB}t^{-1/2}\right)t^{-1/2} \tag{9.31}$$

---

[8] One can perhaps perceive intuitively why the composite variable $x/\sqrt{t}$ should apply to Eq. (9.16), since this equation is first order with respect to t and second order with respect to x.

A common error is to regard this transformation—i.e., that θ is a simple function of the combined variable $(x/\sqrt{t})$—as a testable *assumption* pertaining to the behavior of soils. It is not. It is simply a *mathematical consequence* of the form of the differential equation. When an actual experiment fails to conform accurately to the $x/\sqrt{t}$ relation, the discrepancy can only be attributed to an imperfect description of the behavior of the soil system by the assumed differential equation and/or its assumed boundary conditions, or to errors of the experiment.

Rearranging, we get

$$\frac{xt}{2t^{3/2}}\frac{d\theta}{dB} = \frac{d}{dB}\left(D\frac{d\theta}{dB}\right)$$

or

$$\frac{B}{2}\frac{d\theta}{dB} = \frac{d}{dB}\left[D(\theta)\frac{d\theta}{dB}\right] \tag{9.32}$$

which is indeed an ordinary differential equation with B as the independent variable.

This technique has been used by several investigators to obtain solutions for soil-water flow problems that involve horizontal, semi-infinite, homogeneous media of uniform initial wetness. See Chapter 2 in Hillel (1980) for an application of the Boltzmann transformation to the solution of the flow equation for horizontal infiltration.

G. Theoretical Calculation of the Hydraulic Conductivity Function

We have already stated earlier in this chapter that as yet no fundamentally based method of universal validity is available for a priori prediction of hydraulic conductivity as a function of suction or wetness. There have been, however, some interesting approaches to this important problem, particularly with regard to the relation between the hydraulic conductivity of an unsaturated soil and the effective pore-size distribution as determined from the soil moisture characteristic.

In the preceding chapter, we introduced the concept underlying the relation of K_s, the hydraulic conductivity of a soil in a state of water saturation, to its pore-size distribution. The same approach, based on the capillary-tube analogy, has been used to calculate the hydraulic conductivity of unsaturated soil on the basis of the soil moisture characteristic. To begin, we refer to the work of Childs and Collis-George (1950), Marshall (1958), and Millington and Quirk (1959). Conceptually, their approach rests upon the assumptions that a soil contains distinct pores of various radii which are randomly distributed in space and that when adjacent planes or sections of the soil are brought into contact the overall hydraulic conductance across the plane depends statistically upon the number of pairs of superposed (interconnected) pores and geometrically upon their configurations. The conductance of each pair of interconnected pores is determined by the narrower of the two, i.e., by the constricting passage or "neck." Uncontacted or "bypassing" pore sequences are neglected. The conceptual model is illustrated in Fig. 9.8.

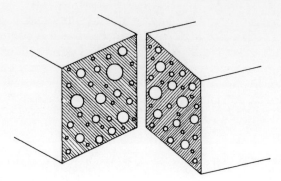

Fig. 9.8. Schematic representation of a sectioned soil. Each of the adjacent planes to be brought into contact exhibits a random distribution of various pore sizes. The pores are represented as circular in cross section.

For a given volume of pores, the number of pore connections N across any plane is inversely related to the cross-sectional area of the pores (since smaller pores are more numerous and hence make more connections), i.e., $N \propto 1/r^2$, wherein r is the pore radius. According to Poiseuille's law, the discharge Q through each pore is proportional to the fourth power of the radius, i.e., $Q \propto r^4$. The overall conductance of the plane under consideration due to each class of pores should be proportional to the product of the two: $K \propto NQ \propto r^2$. When the two planes are brought into contact, furthermore, the probability of any two pores overlapping is equal to the products of their partial areas. If we divide the overall porosity of a soil into distinct pore-size classes and can somehow measure the volume of each such class, we can then assess the probability of pores of various radii contacting pores of larger or smaller radii and thus obtain a value of conductivity for different degrees of saturation. These are some of the basic ideas underlying the theory.

In practice, the porosity is divided into pore-size classes and corresponding partial volumes by means of the soil moisture characteristic. That function relates the partial volume of water-filled pores (as a fraction of total soil volume) θ to suction ψ. Since suction ("negative" pressure) can be related to pore radius by the equation of capillarity ($r = -2\gamma/\psi$ if ψ is in pressure units, or $r = -2\gamma/\psi\rho g$ if ψ is in head units), we can divide the soil moisture characteristic into pore radii increments and corresponding water volume increments.

The original equation by Childs and Collis-George (1950) was

$$K = F\frac{\rho g}{\eta} \sum_{\Gamma=0}^{\Gamma=R} \sum_{\delta=0}^{\delta=R} \delta^2 f(\Gamma)\, dr\, f(\delta)\, dr \qquad (9.33)$$

G. Theoretical Calculation of Hydraulic Conductivity Function

where K is hydraulic conductivity, ρ density of water, g gravitational acceleration, η viscosity of water, $f(\Gamma)\,dr$ the partial area occupied by pores of radii Γ to $\Gamma\,dr$, and $f(\delta)\,dr$ the partial area of pores with radii $\delta\,dr$. The calculation is carried out successively for different wetness values, and in each case the summation is ended at the pore size R associated with the largest pore radius which remains water filled at the specified soil wetness value. F is a matching factor, by which the calculated function is pegged to an experimentally determined point of K for a known value of θ. Childs and Collis-George showed that their calculated $K(\theta)$ function gave an adequate representation of the actual conductivity function as measured for sands and slate dust. However, their method of calculation was cumbersome.

The method of calculation was subsequently improved by Marshall (1958) and by Millington and Quirk (1959), whose equations differed only by an arbitrary factor. Methods of computation based on the same theory were further simplified and codified by Kunze et al. (1968), Green and Corey (1971), and Jackson (1972). Although theoretically the calculation based on Poiseuille's law does not require a matching factor, in actual practice a matching factor is needed to adjust the computed and measured conductivities at saturation.

Using the ratio of the measured (K_s) to the calculated (K_0) saturated conductivity, K_s/K_0, as the matching factor, Jackson's formulation is as follows:

$$K_i = K_s(\theta_i/\theta_s)^c \sum_{j=i}^{m}[(2j + 1 - 2i)\psi_j^{-2}]/\sum_{j=1}^{m}[(2j - 1)\psi_j^{-2}] \quad (9.34)$$

wherein K_i is the hydraulic conductivity at a wetness value θ_i, m is the number of increments of θ (equal intervals from dryness to saturation, i.e., from $\theta = 0$ to $\theta = \theta_s$), ψ_i is the suction head at the midpoint of each θ increment, and j and i are summation indices. Finally, c is an arbitrary constant assigned values of 0 to $\frac{4}{3}$ by various workers [a value of unity was found to be satisfactory by Kunze et al. (1968) and by Jackson (1972)]. A numerical example of the use of this equation will be given at the end of this chapter.

Because it is based on the capillary hypothesis, we can expect the theory described to apply more to coarse-grained than to fine-grained soils, as the latter might exhibit phenomena such as film flow and ionic effects unaccounted for in the simple theory. Another complication arises where the soil is strongly aggregated and two types of flow occur *within* aggregates and *between* aggregates. When the interaggregate cavities are emptied, the water remaining inside the aggregates may remain relatively immobile. Brooks and Corey (1964) showed that in such cases the theory should be modified so as to replace θ, the total wetness, by $\theta - \theta_r$, where θ_r is the residual immobile wetness.

Still another problem is inherent in the use of the soil moisture characteristic obtained by static equilibrium techniques to characterize the relation of wetness θ to suction ψ in dynamic flow systems. As shown by Hillel and Mottes (1962), Davidson *et al.* (1966), and Topp *et al.* (1967), the θ versus ψ relation, quite apart from its strong dependence on hysteresis (i.e., the direction of θ change preceding the measurement), is markedly affected by the rate and duration of the change, as some of the factors involving this relation are time dependent. Included among these factors are air entrapment and dissolution, as well as diffusion, dissolution, and exchange of organic and mineral components.

Recently, the theory of hydraulic conductivity prediction has been expanded and generalized by Mualem (1976).

H. Measurement of Unsaturated Hydraulic Conductivity and Diffusivity in the Laboratory

Knowledge of the unsaturated hydraulic conductivity and diffusivity at different suction and wetness values is generally required before any of the mathematical theories of water flow can be applied in practice. Since there is as yet no universally proven way to predict these values from more basic or more easily obtainable soil properties, K and D must be measured experimentally. In principle, K and D can be obtained from either *steady-state* or *transient-state* flow systems. In *steady flow systems*, flux, gradient, and water content are constant in time, while in *transient flow systems*, they vary. In general, therefore, measurements based on steady flow are more convenient to carry out and often more accurate. The difficulty, however, lies in setting up the flow system, which may take a very long time to stabilize.

Techniques for measurement of conductivity and diffusivity of soil samples or models in the laboratory were described by Klute (1965b). The conductivity is usually measured by applying a constant hydraulic head difference across the sample and measuring the resulting steady flux of water. Soil samples can be desaturated either by tension-plate devices or in a pressure chamber. Measurements are made at successive levels of suction and wetness, so as to obtain the functions $K(\psi)$, $K(\theta)$, and $D(\theta)$. The $K(\psi)$ relationship is hysteretic, and therefore, to completely describe it, measurements should be made both in desorption and in sorption, as well, perhaps, as in intermediate scanning. This is difficult, however, and requires specialized apparatus (Tanner and Elrick, 1958), so that all too often only the desorption curve is measured (starting at saturation and proceeding to increase the suction in increments).

Such laboratory techniques can also be applied to the measurement of undisturbed soil cores taken from the field. This is certainly preferable to measurements taken on fragmented and artificially packed samples, though it should be understood that no field sampling technique yet available provides truly undisturbed samples. Moreover, any attempt to represent a field soil by means of extracted samples incurs the problem of field soil heterogeneity as well as the associated problem of determining the appropriate scale (i.e., the representative volume) for realistic measurement of parameters.

A widely used transient flow method for measurement of conductivity and diffusivity in the laboratory is the *outflow method* (Gardner, 1956). It is based on measuring the falling rate of outflow from a sample in a pressure cell when the pressure is increased by a certain increment. One problem encountered in the application of this method is that of the hydraulic resistance (also called impedance) of the porous plate or membrane upon which the sample is placed, and of the soil-to-plate contact zone. Techniques to account for this resistance were proposed by Miller and Elrick (1958), Rijtema (1959), and Kunze and Kirkham (1962).

Laboratory measurements of conductivity and diffusivity can also be made on long columns of soil, not only on small samples contained in cells. In such columns, steady-state flow can be induced by evaporation (e.g., Moore, 1939) or by infiltration (Youngs, 1964). If the column is long enough to allow the measurement of suction gradients (e.g., by a series of tensiometers) and of wetness gradients (by sectioning, or, preferably, by some nondestructive technique such as gamma-ray scanning), the $K(\theta)$ and $K(\psi)$ relationship can be obtained for a range of θ with a single column or with a series of similarly packed columns.

Measurements in columns under transient flow conditions have also been made (e.g., the horizontal infiltration technique of Bruce and Klute, 1956). If periodic suction and wetness profiles are measured, the flux values at different time and space intervals can be evaluated by graphic integration between successive moisture profiles. This procedure has been called the *instantaneous profile* technique (Watson, 1966) and it can be applied in the field as well (Rose *et al.*, 1965).

I. Measurement of Unsaturated Hydraulic Conductivity of Soil Profiles in Situ

In recent years research in soil physics has resulted in the development of mathematical theories and models describing the state and movement of water in both saturated and unsaturated soil bodies. Moreover, experimental work has resulted in the development of more precise and reliable techniques for the measurement of flow phenomena and of pertinent soil parameters. All too often, however, our theories and models have been validated, if at all, only in highly artificial sets of laboratory produced conditions. Yet too rare are the instances in which theories and models have been applied and found to be valid and useful under realistic conditions, in actual soil management practice. In fact, there is still a great dichotomy in this area between fundamental knowledge and practical application.

Application of the theories of soil physics to the description or prediction of actual processes in the field (e.g., processes involved in irrigation, drainage, water conservation, groundwater recharge and pollution, as well as infiltration and runoff control) depends upon knowledge of the pertinent hydraulic characteristics of the soil, including the functional relation of hydraulic conductivity and of matric suction to soil wetness as well as the spatial and temporal variation of these in the often heterogeneous field situation.

The problem of field soil heterogeneity relates to a fundamental theoretical question which is too often ignored, namely the characteristic scale of the system. Obviously such soil properties as conductivity, porosity, and pore-size distribution are scale dependent and their magnitudes should be considered in relation to some specified or implied size of sample. All soils are inherently inhomogeneous in that their primary and secondary particles and pore spaces differ from point to point and their geometry is too complicated to characterize in microscopic detail. For this reason, the soil is generally characterized in macroscopic terms based on the gross averaging of microscopic and sometimes of mesoscopic heterogeneities. An implicit assumption is that the physical properties are measured on a volume of soil sufficiently large relative to the microscopic heterogeneities to permit such an averaging. Yet how large must the measured volume be? This is generally left unspecified. For instance, in the case of a uniform sand soil, a measurement made on a cubic decimeter may be sufficient to characterize the entire soil. On the other hand, in some cases the soil may be layered, aggregated, and fissured, with relatively large cracks present, so that the hydraulic properties of the medium as a whole can be represented only by a volume as great perhaps as several cubic meters.

From the considerations given, it seems basically unrealistic to try to measure the unsaturated hydraulic conductivity of field soil by making laboratory determinations on discrete and small samples removed from their natural continuum, particularly when such samples are fragmented or otherwise disturbed. Hence it is necessary to devise and test practical methods for measuring soil hydraulic conductivity on a macroscale in situ.

We note that in the excellent book on "Methods of Soil Analysis," published by the American Society of Agronomy as recently as 1965, there is specification of methods for measuring saturated hydraulic conductivity both below and above the water table (Boersma, 1965a,b) and specification of methods for measuring saturated and unsaturated hydraulic conductivity in the laboratory (Klute, 1965a,b), but no mention at all of methods for measuring the unsaturated hydraulic conductivity function of soil profiles in the field. It is therefore extremely important to disseminate information on the methods which have since become available for measuring this vital property of natural soils.

We shall now proceed to give a brief description of several of these methods.

1. METHOD I: SPRINKLING INFILTRATION

This method has been described in principle by Youngs (1964), who, however, tested it only under uniform laboratory conditions. The principle of the method is that the continued supply of water to the soil, as under sprinkling, at a constant rate lower than the effective hydraulic conductivity of the soil, eventually results in the establishment of a steady moisture distribution in the conducting profile. Once steady-state conditions are established, a constant flux exists. In a uniform soil the suction gradients will tend to zero, and with only a unit gravitational gradient in effect the hydraulic conductivity becomes essentially equal to the flux. If this test is carried out on an initially dry soil with a series of successively increasing application rates (sprinkling intensities), it becomes possible to obtain different values of hydraulic conductivity corresponding to different values of soil wetness. The theoretical relationship between rain intensity and the soil moisture profile during infiltration was described by Rubin (1966).

The difficulty of the steady sprinkling infiltration test in the field is that it requires rather elaborate equipment which must be maintained in continuous operation for considerable periods of time. The requirement of maintaining continuous operation becomes increasingly important, and difficult, as one attempts to extend the test toward the greater suction range by reducing the application rate below 1 mm/hr. Another difficulty is to avoid the raindrop impact effect which can cause the exposed surface soil to disperse and seal, thus reducing infiltrability. Variable intensity sprinkling infiltrometers adaptable to field use have been described by Steinhardt and Hillel (1966), Morin *et al.* (1967), and more recently by Amerman *et al.* (1970) and Rawitz *et al.* (1972). The field plot should normally be at least 1 m^2 in size surrounded by a buffer area under the same sprinkling so as to minimize the lateral flow component in the test plot. The assumption of one-dimensional (vertical) flow becomes questionable and perhaps invalid in the case of a soil profile with distinct impeding layers, over which temporary perched water table conditions can develop, leading to subsurface lateral flows beyond the boundaries of the test plot. A comparison between the steady sprinkling method and the instantaneous-profile unsteady drainage method for determination of $K(\theta)$ was presented by Hillel and Benyamini (1974).

2. METHOD II: INFILTRATION THROUGH AN IMPEDING LAYER

This method was suggested by Hillel and Gardner (1970), who showed that an impeding layer at the surface of the soil can be used to achieve the desired boundary conditions to allow measurements of the unsaturated hydraulic conductivity and diffusivity as a function of soil wetness. The effect of an impeding layer present over the top of the profile during infiltration is to decrease the hydraulic potential in the profile under the impeding layer. Thus, the soil wetness, and correspondingly the conductivity and diffusivity values of the infiltrating profile, are reduced. These authors held that their proposed test can be applied during the transient stage of infiltrations, but it becomes most reliable if the test is continued long enough to attain steady state. Qualitatively, when the surface is covered by an impeding layer (crust) with the saturated conductivity smaller than that of the soil, the steady flow conditions which develop are such that the head gradient through the impeding layer is necessarily greater than unity. If the ponding depth is negligible, such an impeding layer thus induces the development of suction in the subsoil, the magnitude of which will increase with increasing hydraulic resistance of the crust. When steady infiltration is achieved, the flux and the conductivity of the subcrust soil become equal.

The advantage of this procedure over those using constant application rates is the simplicity of the experimental system. It may require a relatively long time at high values of soil moisture suction. However, there is no a priori theoretical limit to the suction or conductivity range measurable, which depends only on the range of soil moisture suction one can find in the field at the outset of the test. In principle, one could use a nearly impervious plate and continue the experiment almost indefinitely, provided the flux can be measured accurately while evaporation is prevented. The test procedure obviously involves a series of infiltration trials through capping plates (or crusts) of different hydraulic resistance values. The surface crust can be made artificially by dispersing the surface of the soil itself or by applying a puddled slurry of dispersed soil or clay of different hydraulic resistance. The use of a series of crusts of progressively lower hydraulic resistance can give progressively high K values corresponding to higher water content up to saturation. Such a series of tests can be carried out if the soil is initially fairly dry, either successively in the same location or concurrently on adjacent locations.

This method was applied to field use by Bouma *et al.* (1971). They used ring infiltrometers and puddled the surface soil to obtain the necessary boundary condition.[9] They also made tensiometric measurements in the infiltrating profile to monitor the vertical and horizontal gradients. Each infiltration run yielded a point on the K versus suction or K versus θ curves. A serious problem which has come to light recently and which may invalidate this test in some cases is related to the so-called unstable flow phenomenon. It has been observed that in transition from a fine-textured zone to a coarse-textured zone during infiltration, the advance of the water may not be even, but that sudden breakthrough flows may occur in specific locations where fingerlike intrusions take place. Despite several theoretical studies (e.g., Raats, 1974; Philip, 1974; Parlange, 1975) this phenomenon is still insufficiently understood.

3. METHOD III: INTERNAL DRAINAGE

a. *General Description*

This method is based upon monitoring the transient state internal drainage of a profile (Fig. 9.9) as suggested originally by Richards and Weeks (1953) and Ogata and Richards (1957), and later by Rose *et al.* (1965) and Watson (1966). This method, when carried out in the field, can help to eliminate the possible alteration of soil hydraulics due to disturbance of structure as well as the doubtful procedure of applying steady-state methods to transient state processes. The use of strain gauge pressure transducers in tensiometry make possible the rapid and automatic acquisition of soil moisture suction data while the soil wetness data required for the method are obtainable with a neutron moisture gauge.

Work along these lines has been carried out by Rose and Stern (1967), Van Bavel *et al.* (1968a,b), Davidson *et al.* (1969), Gardner (1970), and Giesel *et al.* (1970). These workers have proven the feasibility of determining the unsaturated hydraulic conductivity function of soils in the field. More recently, Hillel *et al.* (1972) gave a detailed description of a simplified procedure for determining the intrinsic hydraulic properties of a layered soil profile in situ.

The method requires frequent and simultaneous measurements of the soil wetness and matric suction profiles under conditions of drainage alone (evapotranspiration prevented). From these measurements it is possible to obtain instantaneous values of the potential gradients and fluxes operating within the profile and hence also of hydraulic conductivity values. Once the

---

[9] Attempts have also been made to form a crust by using gypsum (calcium sulfate). However, the addition of electrolytes into the soil solution may affect and modify soil properties and therefore seems undesirable.

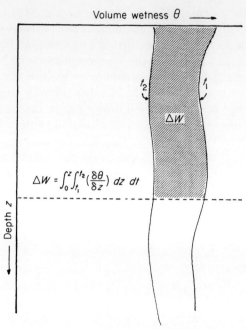

Fig. 9.9. Change in profile water content ΔW in the depth interval from $Z = 0$ (soil surface) to a given depth Z_1, in the time interval from t_1 to t_2, during internal drainage in the absence of either evaporation or lateral flow (schematic).

hydraulic conductivity at each elevation within the profile is known in relation to wetness, the data can be applied to the analysis of drainage and evapotranspiration in a vegetated field. To apply this method in the field, one must choose a characteristic fallow plot that is large enough (say, 10 × 10 m, or at least 5 × 5 m) so that processes at its center are unaffected by boundaries. Within this plot at least one neutron access tube is installed as deeply as possible through and below the root zone. The desirable depth will sometimes exceed 2 m. A series of tensiometers is installed at various depths near the access tube so as to represent profile horizons. The depth intervals between succeeding tensiometers should not exceed 30 cm. Water is then ponded on the surface and the plot is irrigated long enough so that the entire profile becomes as wet as it can be. The soil is then covered by a sheet of plastic so as to prevent any water flux across the surface. As internal drainage proceeds, periodic measurements are made of water content distribution and tension throughout the profile. The handling of the data and a sample calculation of the hydraulic conductivity values of a soil were described by Hillel *et al.* (1972).

This method appears to be the most practical one at the present time for obtaining the unsaturated hydraulic conductivity function of soils in situ. The only instruments required are a neutron moisture meter (or some other nondestructive method for repeated determination of the soil moisture profile) and a set of tensiometers. As such it is easier to carry out than alternative methods based on controlled infiltration. Measurements can be made over several succeeding irrigation-drainage cycles in the same location. Moreover, the method does not assume previous knowledge of the soil moisture characteristic (matric suction versus water content) and in fact can yield information of this function in situ for the range of change occurring during the process of soil moisture redistribution (internal drainage) in the field. The method as described is not applicable where lateral movement of soil moisture is appreciable. This movement is not normally significant when the soil profile is unsaturated, but lateral movement can become significant wherever an impeding layer occurs upon which saturated conditions might prevail for some time. In practice, the moisture range for which conductivity can be measured by the internal drainage method is generally limited to suctions not exceeding about 0.5 bar, as the drainage process often slows down within a few days or weeks to become practically imperceptible.

b. Theoretical Basis

The method we shall describe is based upon the *instantaneous profile method* developed by Watson (1966). However, whereas this method originally was developed and tested in connection with uniform laboratory columns draining into a water table located at a constant depth, the method described herein applies similar principles to field situations where a water table may be absent (or too deep to affect soil moisture flow) and where the soil profile might be heterogeneous (e.g., layered).

The general equation describing the flow of water in a vertical soil profile is

$$\frac{\partial \theta}{\partial t} = \frac{\partial}{\partial z}\left[K(\theta) \frac{\partial H}{\partial z} \right] \tag{9.35}$$

where θ is volumetric wetness (measurable by means of the neutron meter), t time, z the vertical depth coordinate here taken as positive downward, K the hydraulic conductivity, which is a function of soil wetness, and H the hydraulic head (the sum of gravitational and matric suction heads, the latter measurable by means of tensiometers).

Integrating, we obtain

$$\int_0^z \frac{\partial \theta}{\partial t} dz = \left[K \frac{\partial H}{\partial z} \right]_z \tag{9.36}$$

Here Z is the soil depth to which the measurement applies. If the soil surface is covered to prevent evaporation and only internal drainage is allowed, the total water content change per unit time (obtainable by integrating between successive soil moisture profiles down to the depth Z) is thus

$$[dW/dt]_Z = K[\partial H/\partial z]_Z \tag{9.37}$$

Here W is the total water content of the profile to depth Z, i.e.,

$$W = \int_0^Z \theta \, dz \tag{9.38}$$

During the internal drainage of a deeply wetted profile $(\partial H/\partial z)_Z$ (the hydraulic gradient at the depth Z) is often found to be near unity; that is, the suction gradient is nil and only the gravitational gradient operates. If so, then $K = (dW/dt)_Z$.[10] Otherwise, the suction gradient must be taken into account and the hydraulic conductivity is obtained from the ratio of flux to the total hydraulic head gradient (gravitational plus matric). This can be done successively at gradually diminishing water content during drainage to obtain a series of K versus θ values and thus to establish the functional dependence of hydraulic conductivity upon soil wetness for each layer in the profile.

Once the relation of hydraulic conductivity to soil wetness is known for the profile, it is possible to interpret water content and tensiometry data from a vegetated field so as to compute the drainage component of the field water balance. The actual rate of evapotranspiration E_t can then be obtained from the relation:

$$\frac{dE_t}{dt} = \left(\frac{dW}{dt}\right)_{Z_r} - \left(K \frac{\partial H}{\partial z}\right)_{Z_r} \tag{9.39}$$

The first term on the right-hand side of Eq. (9.39) is the rate of total diminu-

---

[10] Gardner (1970) showed how the internal drainage method can serve to measure hydraulic diffusivity as well as conductivity. If the hydraulic head gradient is near unity, the drainage rate approximates the conductivity. If the profile drains fairly uniformly, θ can be assumed to be a function of time but not of depth, and Eq. (9.36) reduces to $L \, \partial \bar{\theta}/\partial t = -K(\partial H/\partial z)|_L$, where $\bar{\theta}$ is the average wetness above depth L, and the conductivity and hydraulic gradient are evaluated at the depth L. If we assume a unique relation between $\bar{\theta}$ and matric suction ψ, we can write $L(d\theta/d\psi)(\partial \psi/\partial t) = -K(\partial H/\partial z)|_L$. Rearranging terms and remembering that by definition the diffusivity $D = K(d\psi/d\theta)$, we have $D = L(d\psi/dt)/(dH/dz)$.

Thus, D can be determined from the time rate of change of the matric suction, and the hydraulic gradient. In cases where the hydraulic gradient is nearly unity, only the time rate of change of the matric suction is needed. The only instrumentation required is one tensionmeter or, preferably, several tensionmeters at various depths in the profile. If the soil is not draining uniformly, the diffusivity calculated will be an average over the entire profile above the depth L.

J. Vapor Movement

tion of soil-water content in the root zone (to depth Z_r) per unit time, and the second term is the flow rate (flux) across the bottom of the root zone (that is, the product of the hydraulic conductivity and the hydraulic gradient operating across the Z_r plane). Since in practice these computations will be made for finite time periods, average rather than instantaneous values of gradient and conductivity will be used.

When the hydraulic gradient below the root zone is unity, which happens frequently but should not be taken for granted, Eq. (9.39) reduces to

$$\frac{dE_t}{dt} = \left(\frac{dW}{dt}\right)_{Z_r} - K_{zr} \tag{9.40}$$

where K_{Z_r} is the hydraulic conductivity prevailing at the depth Z_r, a function of the wetness θ at that depth.

J. Vapor Movement

We have already stated that liquid water moves in the soil by *mass flow*, a process by which the entire body of a fluid flows in response to differences in hydraulic potential. In certain special circumstances, water vapor movement can also occur as mass flow; for instance, when wind gusts induce bulk movement of air and vapor mixing in the surface zone of the soil. In general, however, vapor movement through most of the soil profile occurs by *diffusion*, a process in which different components of a mixed fluid move independently, and at times in opposite directions, in response to differences in concentration (or partial pressure) from one location to another. Water vapor is always present in the gaseous phase of an unsaturated soil, and vapor diffusion occurs whenever differences in vapor pressure develop within the soil.

The diffusion equation for water vapor is

$$q_v = -D_v \, \partial \rho_v / \partial x \tag{9.41}$$

wherein ρ_v is the vapor density (or concentration) in the gaseous phase and D_v the diffusion coefficient for water vapor. D_v in the soil is lower than in open air because of the restricted volume and the tortuosity of air-filled pores (Currie, 1961).

By considering that the liquid water serves as a source and sink for water vapor and assuming that changes in liquid water content with time are much greater than vapor density changes with time, Jackson (1964) derived the equation

$$\frac{\partial \theta}{\partial t} = \frac{\partial}{\partial x}\left[D_v \frac{\partial \rho_v}{\partial \theta} \frac{\partial \theta}{\partial x}\right] \tag{9.42}$$

which describes nonsteady vapor transfer in terms of the liquid water content θ (the soil's volumetric wetness). For the simultaneous transfer of both liquid and vapor, the following equation applies:

$$\frac{\partial \theta}{\partial t} = \frac{\partial}{\partial x}\left[\left(D_v \frac{\partial \rho_v}{\partial \theta} + D_\theta\right)\frac{\partial \theta}{\partial x}\right] \quad (9.43)$$

in which D_θ is the hydraulic diffusivity for liquid water described in Section E.

The foregoing equations consider water vapor diffusion as an isothermal process, assuming that both viscous flow in the liquid phase and diffusion of vapor are impelled by the force fields of soil particle surface and capillarity. No explicit account has been taken of osmotic or solute effects on vapor pressure, though they can obviously be significant. More importantly, the foregoing discussion disregards the simultaneous and interactive transport of both water and heat in nonisothermal situations, which will be described in a subsequent chapter.

At constant temperature, the vapor-pressure differences which may develop in a nonsaline soil are likely to be very small. For example, a change in matric suction between 0 and 100 bar is accompanied by a vapor pressure change of only 17.54–16.34 torr,[11] a difference of only 1.6 mbar. For this reason, it is generally assumed that under normal field conditions soil air is nearly vapor saturated at almost all times. Vapor-pressure gradients can be caused by differences in the concentration of dissolved salts, but this effect is probably appreciable only in soils which contain zones of high salt concentration.

When temperature differences occur, however, they might cause considerable differences in vapor pressure. For example, a change in water temperature from 19 to 20°C results in an increase in vapor pressure of 1.1 torr. In other words, a change in temperature of 1°C has nearly the same effect upon vapor pressure as a change in suction of 100 bar!

In the range of temperatures prevailing in the field, the variation of saturated vapor pressure (that is, the vapor pressure in equilibrium with pure, free water) is as follows:

Temperature (°C)	0	20	30	40
Vapor pressure (torr)	4.58	17.5	38.0	55.8

Vapor movement tends to take place from warm to cold parts of the soil. Since during the daytime the soil surface is warmer, and during the night colder, than the deeper layers, vapor movement tends to be downward during the day and upward during the night. Temperature gradients can also induce liquid flow.

Since liquid movement includes the solutes, while vapor flow does not,

[11] Torr is a unit of pressure equal to 1.316×10^{-3} atm (i.e., 1 mm of mercury under standard conditions).

Sample Problems

there have been attempts to separate the two mechanisms by monitoring salt movement in the soil (Gurr *et al.*, 1952; Deryaguin and Melnikova, 1958). It has been observed that the rate of vapor movement often exceeds the rate which could be predicted on the basis of diffusion alone (Cary and Taylor, 1962). It appears to be impossible to separate absolutely the liquid from the vapor movement, as overall flow can consist of a complex sequential process of evaporation, short-range liquid flow, reevaporation, etc. (Philip and de Vries, 1957). The two phases apparently move simultaneously and interdependently as a consequence of the suction and vapor-pressure gradients in the soil. It is commonly assumed, however, that liquid flow is the dominant mode in moist, nearly isothermal soils (Miller and Klute, 1967) and hence that the contribution of vapor diffusion to overall water movement is negligible in the main part of the root zone, where diurnal temperature fluctuations are slight.

Sample Problems

1. Use the theory described in Section G of this chapter, as formulated by Eq. (9.34), to calculate the hydraulic conductivity function for a sand with the

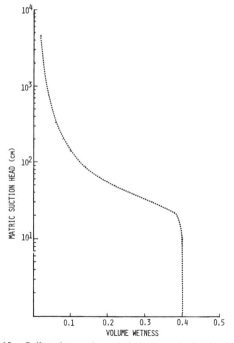

Fig. 9.10. Soil-moisture characteristic curve for Problem 1.

soil-moisture characteristic function shown in Fig. 9.10. Assume the saturated hydraulic conductivity value $K_s = 5.0 \times 10^{-3}$ cm/sec.

Solution. *Step 1:* Divide the total wetness range, 0–0.40, into 20 equal intervals, and tabulate the wetness and suction values. Experience shows that 20 is a sufficient number of intervals, though a greater number may be preferable in some cases. Calculate the indices for the denominator of Eq. (9.34) according to the accompanying table.

Volumetric wetness θ_i	Matric suction head (cm) (ψ_i at θ-interval midpoint)	Pore class increment number j	Denominator index $2j - 1$
0.40	10	1	1
0.38	22	2	3
0.36	24	3	5
0.34	27	4	7
0.32	30	5	9
0.30	32.5	6	11
0.28	36	7	13
0.26	40	8	15
0.24	44	9	17
0.22	49	10	19
0.20	56	11	21
0.18	62	12	23
0.16	72	13	25
0.14	88	14	27
0.12	110	15	29
0.10	140	16	31
0.08	200	17	33
0.06	330	18	35
0.04	750	19	37
0.02	4500	20	39

Step 2: Calculate successively the conductivity values K_i for each of the θ values, using the summation equation

$$K_i = K_s \left(\frac{\theta_i}{\theta_s}\right)^1 \sum_{j=i}^{m} [(2j + 1 - 2i)\psi_j^{-2}] \bigg/ \sum_{j=1}^{m} [(2j - 1)\psi_j^{-2}]$$

(1) $\theta_1 = 0.40$ (the saturation value). Obviously, in this case $\theta_i = \theta_s$, and the summations in the numerator and the denominator are equal to each other; hence $K_i = K_s = 5.0 \times 10^{-3}$ cm/sec.

(2) $\theta_i = 0.38$. For this second pore class $i = 2$, hence $2j + 1 - 2i = 1, 3, 5$, etc.

$$K_2 = 5 \times 10^{-3} \left(\frac{0.38}{0.40}\right) \frac{1/22^2 + 3/24^2 + 5/27^2 + \cdots + 37/4500^2}{1/10^2 + 3/22^2 + 5/24^2 + \cdots + 39/4500^2}$$

Sample Problems

Note: The denominator need only be calculated once, as it is common to all succeeding $K_i(\theta_i)$ calculations. In this example its value is 0.118952.

(3) $\theta_i = 0.36$. Here $i = 3$; hence

$$K_3 = 5 \times 10^{-3} \left(\frac{0.36}{0.40}\right) \frac{1/24^2 + 3/27^2 + \cdots + 35/4500^2}{1/10^2 + 3/22^2 + 5/24^2 + \cdots + 39/4500^2}$$

(4) $\theta_i = 0.34$, $i = 4$; hence

$$K_4 = 5 \times 10^{-3} \left(\frac{0.34}{0.40}\right) \frac{1/27^2 + 3/30^2 + \cdots + 33/4500^2}{1/10^2 + 3/22^2 + 5/24^2 + \cdots + 39/4500^2}$$

This series of calculations continues through $\theta_i = 0.02$, for which

$$K_{20} = 5 \times 10^{-3} \left(\frac{0.02}{0.40}\right) \frac{1/4500^2}{1/10^2 + 3/22^2 + 5/24^2 + \cdots + 39/4500^2}$$

Step 3: Tabulate the K_i versus θ_i values, as calculated.

θ	K (cm/sec)	θ	K	θ	K
0.40	5×10^{-3}	0.26	4.28×10^{-4}	0.12	5.65×10^{-6}
0.38	3.48×10^{-3}	0.24	2.77×10^{-4}	0.10	1.97×10^{-6}
0.36	2.56×10^{-3}	0.22	1.72×10^{-4}	0.08	5.27×10^{-7}
0.34	1.86×10^{-3}	0.20	1.02×10^{-4}	0.06	9.44×10^{-8}
0.32	1.33×10^{-3}	0.18	5.65×10^{-5}	0.04	8.22×10^{-9}
0.30	9.35×10^{-4}	0.16	2.90×10^{-5}	0.02	1.05×10^{-10}
0.28	6.42×10^{-4}	0.14	1.36×10^{-5}		

Step 4: Plot K_i versus θ_i on a semilog scale as shown in Fig. 9.11.

2. The hydraulic conductivity versus matric suction functions of two hypothetical soils, a sandy and a clayey soil, conform to the empirically based equation

$$K = a/[b + (\psi - \psi_a)^n] \quad \text{for} \quad \psi \geq \psi_a$$

where K is hydraulic conductivity (cm/sec); ψ is suction head (cm H_2O) and ψ_a is air-entry suction; and a, b, n are constants, with a/b representing the saturated soil's hydraulic conductivity K_s. The exponential parameter n characterizes the steepness with which K decreases with increasing ψ. Assume that in the sandy soil $K_s = 10^{-3}$ cm/sec, $a = 1$, $b = 10^3$, $\psi_a = 10$ cm, and $n = 3$; whereas in the clayey soil $K_s = 2 \times 10^{-5}$, $a = 0.2$, $b = 10^4$, $\psi_a = 20$, and $n = 2$.

Plot the K versus ψ curves. Note that the curves cross, and the relative conductivities are reversed: the sandy soil with the higher K_s exhibits a steeper decrease of K and falls below the clayey soil beyond a certain suction

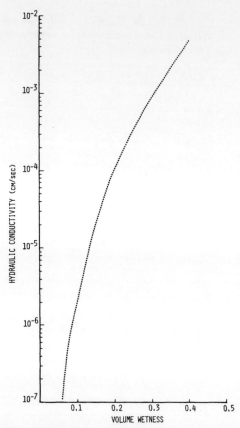

Fig. 9.11. Hydraulic conductivity as a function of wetness, calculated from the data of Problem 1.

value ψ_c. Calculate the values of ψ (designated $\psi_{1/2}$) at which each K equals $\frac{1}{2}K_s$, and estimate the common value of ψ_c, at which the two curves intersect.

For the sandy soil:

$$K = [10^3 + (\psi - 10)^3]^{-1}$$

The suction $\psi_{1/2}$ at which $K = 0.5\ K_s = a/2b$ can be obtained by substituting $\tilde{\psi}$ for $(\psi - \psi_a)$ and setting $\tilde{\psi}^3 = 10^3$ (thus doubling the denominator). Therefore,

$$\tilde{\psi} = 10 \quad \text{and} \quad \psi_{1/2} = \tilde{\psi} + \psi_a = 10 + 10 = 20 \text{ cm}$$

For the clayey soil:

$$K = 0.2/[10^4 + (\psi - 20)^2]$$

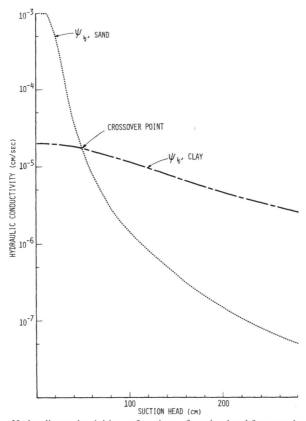

Fig. 9.12. Hydraulic conductivities as functions of suction head for a sand and a clay.

Following the previous procedure, and setting $\bar{\psi}^2 = 10^4$, we obtain

$$\bar{\psi} = 10^2 \quad \text{and} \quad \psi_{1/2} = \bar{\psi} + \psi_a = 100 + 20 = 120 \text{ cm}$$

"*Crossover suction value* (ψ_c): We can attempt to obtain this value algebraically (by setting the two expressions for K equal to each other and solving for the common ψ value) or graphically (Fig. 9.12) by reading the ψ value at which the two $K(\psi)$ curves intersect. The latter procedure is easier in this case, and it shows ψ_c to be about 48 cm.

Note: The soils depicted are completely hypothetical and are not to be taken as typical of real sandy and clayey soils.

3. Using the "instantaneous profile" internal drainage method (Section I.3.a of this chapter) compute the hydraulic conductivity K as a function of volume wetness θ for the set of data shown in the following two graphs: (1) Figure 9.13 shows a plot of volumetric wetness variation with time for each

9. Flow of Water in Unsaturated Soil

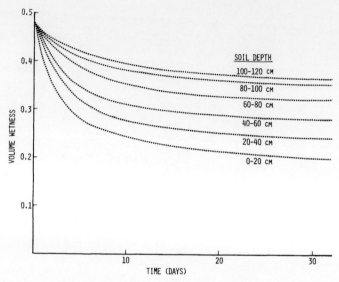

Fig. 9.13. Volumetric wetness as a function of time for different depth layers in a draining profile.

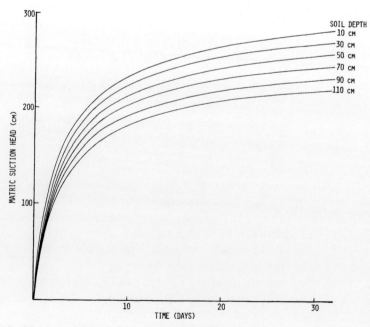

Fig. 9.14. Matric suction variation with time for different depth layers in a draining profile.]

Sample Problems

of five depth layers in a 1-m deep profile (neutron meter data); (2) Fig. 9.14 shows a plot of matric suction variation with time for each depth (tensiometric data).

From the first graph, we calculate the soil moisture flux through the bottom of each depth increment (or layer) by integrating moisture–time curves with respect to depth. First, the slopes of the wetness curves ($-\partial\theta/\partial t$) are measured at days 1, 2, 4, 8, 16, and 32. Second, these slopes are multiplied

Table 9.1

CALCULATION OF SOIL MOISTURE FLUX

t (day)	z (cm)	$-\partial\theta/\partial t$ (day^{-1})	$-dz(\partial\theta/\partial t)$ (cm/day)	$q = dz(\partial\theta/\partial t)$ (cm/day)
	0–20	0.085	1.7	1.7
	20–40	0.065	1.3	3.0
1	40–60	0.045	0.9	3.9
	60–80	0.033	0.66	4.56
	80–100	0.030	0.60	5.16
	0–20	0.05	1.0	1.0
	20–40	0.04	0.8	1.8
2	40–60	0.035	0.7	2.5
	60–80	0.02	0.4	2.9
	80–100	0.016	0.32	3.22
	0–20	0.02	0.4	0.4
	20–40	0.02	0.4	0.8
4	40–60	0.02	0.4	1.2
	60–80	0.016	0.32	1.52
	80–100	0.012	0.24	1.76
	0–20	0.01	0.2	0.2
	20–40	0.01	0.2	0.4
8	40–60	0.008	0.16	0.56
	60–80	0.008	0.16	0.72
	80–100	0.006	0.12	0.84
	0–20	0.006	0.12	0.12
	20–40	0.005	0.10	0.22
16	40–60	0.005	0.10	0.32
	60–80	0.003	0.06	0.38
	80–100	0.003	0.06	0.44
	0–20	0.004	0.08	0.08
	20–40	0.003	0.06	0.14
32	40–60	0.003	0.06	0.20
	60–80	0.0025	0.05	0.25
	80–100	0.0025	0.05	0.30

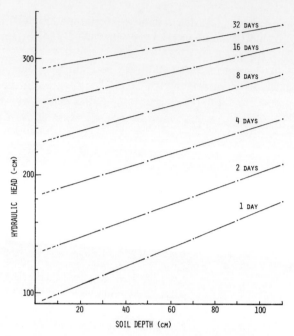

Fig. 9.15. Hydraulic head variation with time and depth during drainage.

by their respective depth increments to obtain the per layer rate of water content change $dZ(\partial\theta/\partial t)$. Then the flux q through the bottom of each depth is obtained by accumulating the water content increments of all layers overlying that depth, i.e., $q = dz(\partial\theta/\partial t)$. The whole procedure is presented in tabular form in Table 9.1. We next use Fig. 9.14 (the time variation of matric suction at each depth) to obtain the hydraulic head profiles existing at the times for which the flux values were calculated (Table 9.1). We do this by adding the depth of each tensiometer to its matric suction values. The resulting hydraulic head profiles are plotted in Fig. 9.15. We can now divide each of the fluxes listed in Table 9.1 by its corresponding hydraulic gradient (namely, the slopes of the lines plotted in Fig. 9.15). We also note the volumetric wetness values which prevailed at each depth and time for which the hydraulic conductivity has been computed. This procedure is shown in Table 9.2. Our last step is to plot the hydraulic conductivity values against volumetric wetness and draw a best-fit curve through the plotted points (Fig. 9.16). In the example given, all the points seem to lie, more or less, on a straight line when plotted on a semilog coordinate system (log K versus $\bar\theta$).

Table 9.2
CALCULATION OF HYDRAULIC CONDUCTIVITY

z (cm)	q^a (cm/day)	$\partial H/\partial z^b$ (cm/cm)	K^c (cm/day)	$\bar{\theta}$ (%)
20	1.7	0.8	2.25	41.0
	1.0	0.68	1.47	38.2
	0.4	0.6	0.67	31.5
	0.2	0.52	0.38	27.0
	0.12	0.44	0.27	24.0
	0.08	0.36	0.22	22.0
40	3.0	0.8	3.75	44.5
	1.8	0.68	2.65	42.0
	0.8	0.6	1.33	37.0
	0.4	0.52	0.77	31.0
	0.22	0.44	0.50	28.0
	0.14	0.36	0.39	26.0
60	3.9	0.8	4.88	46.2
	2.5	0.68	3.68	42.5
	1.2	0.6	2.00	39.0
	0.56	0.52	1.08	35.0
	0.32	0.44	0.73	31.0
	0.20	0.36	0.56	30.0
80	4.56	0.8	5.70	46.6
	2.9	0.68	4.26	44.5
	1.52	0.6	2.53	41.5
	0.72	0.52	1.38	38.0
	0.38	0.44	0.86	34.5
	0.25	0.36	0.69	33.0
100	5.16	0.8	6.45	47.0
	3.22	0.68	4.74	45.0
	1.76	0.6	2.93	42.0
	0.84	0.52	1.62	39.0
	0.44	0.44	1.00	37.0
	0.30	0.36	0.83	35.0

[a] Copied from Table 9.1
[b] Obtained by measuring the slopes of the curves in Fig. 9.15 at the appropriate depths. In reality, these slopes are not necessarily constant.
[c] Obtained by dividing each flux q by its corresponding hydraulic gradient $\partial H/\partial z$.

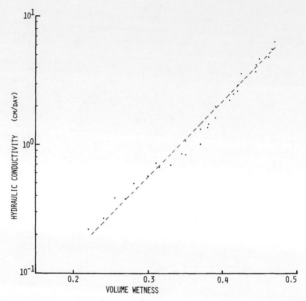

Fig. 9.16. Hydraulic conductivity as a function of volume wetness, calculated from the data of Problem 3.

This is not merely fortuitous: it is, admittedly, contrived. We chose a simple and well-behaved set of data for the sake of illustration. The real world is seldom so convenient. A more realistic example, pertaining to a texturally layered soil, was presented (in similarly excruciating detail) by Hillel *et al.* (1972).

> Ye are the salt of the earth;
> but if the salt shall have lost its savour
> ... it is henceforth good for nothing
> but to be cast out and trodden underfoot.
> Matthew 5:13

10 Movement of Solutes and Soil Salinity

A. Introduction

In the preceding chapters, we discussed soil water with only occasional reference to solutes. In actual fact, water present in the soil and constituting its liquid phase is never chemically pure. To begin with, the water entering the soil as rain or irrigation is itself a solution. Rainwater is of course distilled and essentially pure when it first condenses to form clouds, but as it descends through the atmosphere it generally dissolves such atmospheric gases as carbon dioxide and oxygen, often together with such products of our industrial civilization as oxides of sulfur and nitrogen, as well as—along the coast—generous amounts of salt which enter the air as sea spray. Irrigation water, generally obtained from surface or subterranean reservoirs, frequently contains very significant quantities of dissolved salts. Finally, during its residence in the soil, the infiltrated water tends to dissolve additional solutes, mostly electrolytic salts. The range of salt concentrations encountered in soil-water relations begins at 5–10 mg/liter (also designated parts per million, or ppm) in rainwater and as much as 1000 mg/liter in irrigation water, and goes as high as 10,000 mg/liter in drainage from saline soil.

As it moves through the profile, soil water carries its solute load in its convective stream, leaving some of it behind to the extent that the component salts are adsorbed, taken up by plants, or precipitated whenever their concentration exceeds their solubility (e.g., at the soil surface during evaporation). Solutes move not only with soil water, but also within it, in response to concentration gradients. At the same time, solutes react among themselves and interact with the solid matrix of the soil in a continuous cyclic succession of interrelated physical and chemical processes. These interactions involve and are strongly influenced by such variable factors as acidity, temperature, oxidation–reduction potential, composition, and concentration of the soil solution.

In former times, processes involving solutes might have been considered to belong in the exclusive realm of soil chemistry and outside the scope of a treatise in soil physics. Nowadays, however, we have come to recognize that artificial and arbitrary separations among traditionally distinct, but in fact complementary, disciplines should not be allowed to hinder our quest for an ever more comprehensive understanding of interactive phenomena in the environoment.

A better understanding of the simultaneous movement and interactions of water and solutes in the soil is essential to the improvement of soil fertility through control of nutrients in the root zone, as well as to the prevention of soil salinity and alkalinity. Such an understanding has also become crucial in the area of environmental management whenever soil-borne solutes migrate to, and threaten the quality of, groundwater or surface water resources.

B. Convective Transport of Solutes

We begin our discussion of the mechanisms of solute movement by considering convection alone. The convection (or mass flow) of soil water, sometimes called the Darcian flow, carries with it a convective flux of solutes J_c proportional to their concentration c:

$$J_c = qc = -c(K\, dH/dx) \qquad (10.1)$$

where $q = -K\, dH/dx$ is Darcy's law, discussed in the preceding two chapters. Since q is usually expressed as volume of liquid (solution) flowing through a unit area (perpendicular to the flow direction) per unit time, and c as mass of solute per unit volume of solution, J_c is given in terms of mass of solute passing through a unit cross-sectional area of a soil body per unit time.

To estimate the distance of travel of a solute per unit time, we consider the average apparent velocity $\bar{v}$ of the flowing solution:

$$\bar{v} = q/\theta \qquad (10.2)$$

wherein θ is volumetric wetness and $\bar{v}$ is taken as the straight line length of path traversed within the soil in unit time. In this formulation we disregard the roundabout labyrinthine path caused by the geometric tortuosity of soil pores. Since actual velocities vary over several orders of magnitude within

B. Convective Transport of Solutes

pores and between pores, the concept of average velocity is obviously a gross approximation (much like lumping insects and elephants in computing the "average" size animal). However, assuming $\bar{v}$ to be a working approximation, we get

$$J_c = \bar{v}\theta c \qquad (10.3)$$

It is sometimes useful to have an estimate of the average residence time t_r of a solute within a layer of soil of thickness L (especially if we are concerned with some time-dependent interactive process involving the solute under consideration). Accordingly,

$$t_r = L/\bar{v} \qquad (10.4)$$

If the flow is impelled by gravity alone, that is to say if there are no pressure gradients, then the downward flux of liquid is equal to the hydraulic conductivity K of the medium, which in turn depends on its wetness θ:

$$q = K(\theta)$$

Using (10.2), we thus obtain for (10.4)

$$t_r = L\theta/K(\theta) \qquad (10.5)$$

The foregoing equations allow us to estimate, for instance, the distance of travel of a soluble pollutant from the bottom of, say, a septic tank or sanitary landfill to the water table, through the so-called unsaturated zone, as follows:

$$L_t = tK(\theta)/\theta \qquad (10.6)$$

where L_t is the average distance of convective transport in time t. If the values of wetness and hydraulic conductivity vary within the soil profile, as is typically the case, the foregoing calculations must be carried out layer by layer to determine the space-variable flux, average distance of travel, and residence time per layer.

The serious flaw in this entire approach is that one can seldom, if ever, assume that the transport of solutes occurs by convection alone. In fact, we know that solutes do not merely flow with the water, as sedentary passengers in a train, but actually move within the flowing water in response to concentration gradients in the twin processes of diffusion and hydrodynamic dispersion. Furthermore, solutes are not always as inert as we have thus far assumed, since they tend to interact with the biological system within the soil (e.g., be taken up or released by microbes and the roots of higher plants) and with the physiocochemical system (e.g., be adsorbed or exchanged by the soil's exchange complex) and since they enter into chemical reactions or are precipitated or volatilized. In other words, solutes resemble a group of rather rowdy passengers who constantly move from car to car and occasionally jump off the train entirely while others join in their stead.

We shall now proceed to describe several of these processes in turn.

C. Diffusion of Solutes

Diffusion processes commonly occur within gaseous and liquid phases, in consequence of the random thermal motion (often called Browian motion) and repeated collisions and deflections of molecules in the fluid. The net effect is a tendency to equalize the spatial distribution of diffusible components in any mixed or multicomponent fluid. As such, diffusion processes are extremely important in the soil. As we shall attempt to show in our chapter on soil aeration, diffusion in the air phase of such gases as oxygen, carbon dioxide, nitrogen (in the elemental, oxide, and ammonia forms), and of course water vapor can have a decisive influence on the soil's chemical and biological processes. Equally important are diffusion processes involving solutes in the soil's liquid phase, including nutrients as well as potentially harmful salts and toxic compounds.

If solutes do not happen to be distributed uniformly throughout a solution, concentration gradients will exist and solutes will tend to diffuse from where their concentration is higher to where it is lower. In bulk water at rest, the rate of diffusion J_d is related by *Fick's first law* to the gradient of the concentration c:

$$J_d = -D_0 \, dc/dx \tag{10.7}$$

in which D_0 is the diffusion coefficient for the solute diffusing in bulk water and dc/dx is the concentration gradient.

For diffusion in the soil's liquid phase, the effective diffusion coefficient is generally less than the diffusion coefficient in bulk water D_0 for several reasons. In the first place, the liquid phase occupies only a fraction of the soil volume; at most, in a state of saturation, its volume fraction equals the soil's porosity. Second, the soil's pore passages are tortuous so that the actual path length of diffusion is significantly greater than the apparent straight line distance. In an unsaturated soil, as soil wetness is decreased, the fractional volume available for diffusion in the liquid phase decreases still further, while the tortuous length of path increases.

If fractional water volume θ and tortuosity ξ were the only factors affecting the diffusion coefficient in the soil D_s, we could write

$$D_s = D_0 \theta \xi \tag{10.8}$$

where ξ, the tortuosity factor, is an empirical parameter smaller than unity,

C. Diffusion of Solutes

expressing the ratio of the straight line length of a soil sample to the average roundabout path length through the water-filled pores for a diffusing molecule or ion. This parameter has been found to depend on both the fractional volume and geometric configuration of the water phase, and hence to decrease with decreasing θ. Thus, D_s is seen to be strongly dependent upon θ, both directly and through its dependence on ξ, which itself is a function of θ. To show this dependence, we can write $D_s(\theta)$.

Other factors, in addition to the geometric ones considered in Eq. (10.8), serve to further reduce the effective diffusion coefficient, particularly in unsaturated soils with an appreciable content of clay. As soil wetness is reduced and the water films coating the particles contract, the increasing density of the exchangeable cations adsorbed to the clay surfaces and the corresponding exclusion of anions, as well as the possible increase in viscosity of the liquid, might combine to further retard diffusion. Because these and other complicating factors are not mutually independent, it has seemed impossible thus far to formulate them separately. Hence, it is tempting to lump them all together into a single *complexity factor*, which we can designate α. Thus,

$$D_s = D_0 \theta \alpha \tag{10.9}$$

We can now rewrite Eq. (10.7) for diffusion in the liquid phase of an unsaturated soil:

$$J_d = -D_s(\theta) \, dc/dx \tag{10.10}$$

By itself, this equation can only describe steady-state diffusion processes. In order to proceed to a more generalized diffusion equation capable of describing transient-state processes (in which the rate and concentration vary in time), we must invoke the mass conservation law, as formulated in the continuity equation. Let us assume that there are no sources or sinks for the diffusing solute in the soil body in which diffusion is occurring. Now let us consider a rectangular volume element of soil which contains a liquid phase and which is bounded by two parallel square planes, of area A, separated by a distance Δx. The amount of solute diffusing through one of these planes into the volume element per unit time is AJ_d, and the amount diffusing out of the volume element through the second plane is $A[J_d + (\partial J_d/\partial x) \Delta x]$. The rate of accumulation of the solute in the volume element is $-A(\partial c/\partial t)\Delta x$, where $\partial c/\partial t$ is the time rate of change of concentration. The negative sign expresses the fact that any increase of diffusive flux along the direction of diffusion, from, say, the entry face to the outlet face of the volume element, necessarily depletes the concentration in the volume element and vice versa (the concentration increases if the out-diffusion rate is less than the in diffusion). Thus,

$$A(\partial c/\partial t)\,\Delta x = A[J_{\text{d}} + \partial J_{\text{d}}/\partial x\,\Delta x] - AJ_{\text{d}}$$

which reduces to

$$\partial c/\partial t = -\partial J_{\text{d}}/\partial x \qquad (10.11)$$

Combining this with Eq. (10.10), we obtain a second-order equation as follows:

$$\partial c/\partial t = \partial[D_{\text{s}}(\theta)\partial c/\partial x]/\partial x \qquad (10.12)$$

In the special case when the diffusion coefficient D_{s} is constant, Eq. (10.12) assumes a form analogous to the well-known *Fick's second law*:

$$\partial c/\partial t = D_{\text{s}}\,\partial^2 c/\partial x^2 \qquad (10.13)$$

However, D_{s} is generally not constant, and in fact might be not only wetness dependent but also concentration dependent, i.e., a function of both θ and c.

Attempts to apply the foregoing equations to describe the diffusion of solutes in the soil solution, particularly in unsaturated soil conditions, have encountered numerous complications. The soil varies both in space and time. Solutes interact with, and modify, the solid matrix—and hence affect the pore space. Different species of solutes in the solution phase interact with each other, as well as with those in the adsorbed phase. Evaporation and condensation of water in pores further modify the concentration gradients and pattern of diffusion. Finally, the convective flow of the solution affects the diffusion process by changing the distribution of solutes and by inducing a process called hydrodynamic dispersion, which is the topic of the next section.

D. Hydrodynamic Dispersion

Underlying the foregoing discussion of diffusion was the implicit assumption that, whenever concentration gradients occur in the soil's liquid phase, the inhomogeneously distributed solutes will diffuse according to Eq. (10.12) regardless of whether or not the liquid phase itself is in motion. However, the motion of any inhomogeneous solution in a porous body brings about another process which differs from diffusion in its mechanism but which tends to produce an analogous or synergetic effect, which is to mix and eventually to even out the concentration or composition differences between different portions of the flowing solution. This process, which at times predominates over diffusion, is called *hydrodynamic dispersion*. It results from the microscopic nonuniformity of flow velocity in the soil's conducting pores.

D. Hydrodynamic Dispersion

Since water moves faster through wide than through narrow pores and faster in the center of each pore than along its walls, some parts of the flowing solution move ahead of other parts which lag behind.

Consider, for example, the laminar flow of a solution through a single capillary pore hypothetically shaped like a cylindrical tube. From our earlier derivation of Poiseuille's law (Chapter 8) we know that flow velocity v in such a tube is a decreasing function of radial distance r from the center:

$$v = 2\bar{v}(1 - r^2/R^2) \tag{10.14}$$

Here $\bar{v}$ is average velocity and R is the radius of the tube. Thus the velocity of a solute molecule carried by the convective stream depends on its position within the pore passage. At $r = R$, which is at the wall of the tube, the velocity is zero, and at $r = 0$, in the center of the tube, the velocity is maximal and equal to twice the average velocity.

Added to the nonuniformity of velocity within each pore is the fact that pores vary widely in radius, over several orders of magnitude, say from 1 μm to 1 mm (10^{-4}–10^{-1} cm). We recall Poiseuille's law [Eq. (8.3)],

$$Q = 4R^4 \, \Delta p/8\eta L$$

which states that the discharge Q (volume flow per unit time) is proportional to the fourth power of the radius R. Accordingly, a pore with an effective radius of 1 mm will conduct a volume of water $(10^3)^4 = 10^{12}$ (a million million) times greater than a pore having an effective radius of 1 μm. It should be clear by now that the microscopic-scale variation of pore water velocity in the soil is great indeed!

The fact that some portions of a flowing solution move ever so much faster than other portions causes an incoming solution to mix with or disperse within an antecedent solution. The degree of mixing depends on such factors as average flow velocity, pore-size distribution, degree of saturation, concentration gradients, etc. When the convective velocity is sufficiently high, the relative effect of hydrodynamic dispersion can greatly exceed that of molecular diffusion and the latter can be neglected in the analysis of solute movement. On the other hand, when the soil solution is at rest, the hydrodynamic dispersion effect does not come into play at all.

Mathematically, hydrodynamic dispersion is formulated in a manner analogous to the formulation of diffusion as given by Eq. (10.10) and (10.12), except that, instead of a diffusion coefficient a *dispersion coefficient* is introduced. This coefficient, which we will designate D_h, has been found (Bresler, 1972) to depend more or less linearly on the average flow velocity $\bar{v}$. Thus

$$D_h = a\bar{v} \tag{10.15}$$

with a an empirical parameter.

Because of the similarity in effect (though not in mechanism) between diffusion and dispersion, it is tempting to assume the two effects to be additive. Accordingly, the diffusion and dispersion coefficients are often combined into a single term, namely the *diffusion–dispersion coefficient* D_{sh}, which is a function of both the fractional water volume θ and the average velocity $\bar{v}$:

$$D_{sh}(\theta, \bar{v}) = D_s(\theta) + D_h(\bar{v}) \tag{10.16}$$

We are now ready for an attempt to formulate a set of equations for the overall, steady and unsteady, movement of solutes in an inhomogeneous solution flowing in a porous medium, including the mechanisms of convection, diffusion, and hydrodynamic dispersion, as well as the effects of sources and sinks. Before we do so, however, we wish to describe the important phenomenon of *miscible displacement*, as exemplified by experimentally determined *breakthrough curves*.

E. Miscible Displacement and Breakthrough Curves

When a liquid different in composition or concentration from the preexisting pore liquid is introduced into a column of soil, and the outflow from the end of the column is collected and analyzed, its composition is seen to change in time as the old liquid is displaced and replaced by the new one. If the two liquids are not mutually soluble (as is the case, for example, with oil and water), then the process is called *immiscible displacement*. If, on the other hand, the two liquids mix readily—as do many aqueous solutions—the process just described is referred to as *miscible displacement*, and plots of the outflowing solutions's solute content versus time or versus cumulative discharge are called *breakthrough curves*.

If the soil were saturated throughout the experiment, and if neither diffusion nor hydrodynamic dispersion were to take place at the boundary between the displacing and the displaced solutions, then that boundary would remain a sharp front moving along the length of the column at a velocity equal to the flux. If we were to monitor the composition at the column's outlet, we would notice an abrupt change in composition at the moment the last portion of the old solution was completely driven out and the new solution arrived. Such displacement without mixing can be called *piston flow*. It is seldom if ever encountered in practice. What normally happens at the boundary or front between the two solutions is a gradual mixing resulting from both diffusion and hydrodynamic dispersion so that the boundary becomes increasingly diffuse about the mean position of the advancing front as shown in Figs. 10.1 and 10.2.

E. Miscible Displacement and Breakthrough Curves

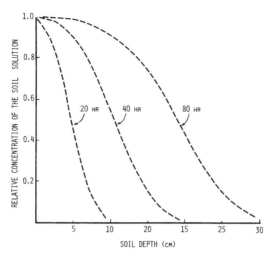

Fig. 10.1. Solute concentration versus soil column depth at various times, during the infiltration of a saline solution into a nonsaline soil. Note that the concentration "front" becomes increasingly diffuse about its mean position as it advances into the soil.

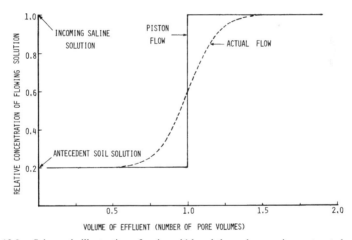

Fig. 10.2. Schematic illustration of a sigmoid breakthrough curve in a saturated sandy soil during the replacement of a dilute antecedent solution by an entering saline solution of fivefold greater concentration.

Ideally, breakthrough curves should be symmetrical about the front of the advancing solution, with the inflection representing 50% displacement at a cumulative flow of one pore volume if the soil is saturated. A typical curve of this sort is shown in Fig. 10.2 for a sand. Curves obtained with finer textured soils differ from the symmetrical ideal, particularly in the case of structured (aggregated) soils, owing to various interactions between solutes and the soil matrix and to the possible existence of pore spaces where water movement is sluggish and portions of the antecedent solution lag behind and are displaced only very slowly.

Miscible displacement phenomena and breakthrough curves are not merely of academic interest but are highly pertinent to the solution of such real world problems as the leaching of excess salts from saline soils, the distribution of nutrient solutions, and the possible pollution of groundwater by soil-borne solutes of various types, including radioactive waste products, toxic chemicals, and pesticide residues.

F. Combined Transport of Solutes

To take into account the three mechanisms of solute movement described thus far (namely, convention, molecular diffusion, and hydrodynamic dispersion), we can combine Eq. (10.3), (10.10), and (10.16) to obtain

$$J = \bar{v}\theta c - [D_s(\theta)\, dc/dx + D_h(\bar{v})\, dc/dx]$$

which can be stated in words:

[combined solute flux] = [flux due to convection]
  + [flux due to diffusion] + [flux due to hydrodynamic dispersion]

Since in practice the diffusion and dispersion phenomena cannot be separated, the foregoing equation is usually written in the form

$$J = \bar{v}\theta c - D_{sh}(\theta, \bar{v})\, dc/dx \qquad (10.17)$$

Here J is the total mass of a solute transported across a unit cross-sectional area of soil per unit time, D_{sh} is the lumped diffusion–dispersion coefficient (a function of volumetric wetness θ and average pore-water velocity $\bar{v}$), c is the solute concentration, and dc/dx is the solute gradient.

Equation (10.17) by itself can describe only steady state (time invariant) processes. Moreover, it is limited to "noninteracting" solutes, by which term we designate solutes not subject to adsorption by soil solids nor subject to chemical or biological reactions. Strictly speaking, truly noninteracting solutes scarcely exist. Moreover, the parameters D_s, D_h, $\bar{v}$, θ, c can only be defined in macroscopic terms as gross spatial averages. Hence Eq. (10.17) is only an approximation of the process it purports to depict.

Turning now to transient-state processes, in which fluxes and concentrations can vary both in time and in space, we once again invoke the con-

F. Combined Transport of Solutes

tinuity condition, which for combined convective–diffusive–dispersive transport can be written

$$\partial(c\theta)/\partial t = -\partial J/\partial x \tag{10.18}$$

For the rate of change of the solute mass present in a volume element of soil to equal the difference between the incoming and outgoing fluxes of the solute for that volume element, we must assume that there are no gains or losses of the solute by any mechanisms operating within the volume element itself.

Combining Eq. (10.8) and (10.17), we get

$$\frac{\partial(c\theta)}{\partial t} = -\frac{\partial(\bar{v}\theta c)}{\partial x} + \frac{\partial}{\partial x}\left(D_{\text{sh}} \frac{\partial c}{\partial x}\right) \tag{10.19}$$

For steady water flow (which, however, does not necessarily imply steady solute movement) θ, $\bar{v}$, and D_{sh} can be taken as constant, and Eq. (10.19) simplifies to

$$\frac{\partial c}{\partial t} = -\bar{v}\frac{\partial c}{\partial x} + \frac{D_{\text{sh}}}{\theta}\frac{\partial^2 c}{\partial x^2} \tag{10.20}$$

Certain solutes are generated within the soil. An example is the nitrate ion, which evolves under appropriate conditions from nitrification of organic matter. Under different conditions, sometimes within the same soil profile, certain solutes may disappear from the soil volume (e.g., as in the case of nitrates being taken up by plants or undergoing denitrification). To account for these *sources* and *sinks*, we can modify Eq. (10.18) to include a composite source–sink S expressing the rate of production or disappearance of the particular solute:

$$\partial(c\theta)/\partial t = -\partial J/\partial x + S \tag{10.21}$$

The term S represents the net sum of all n possible sources $\sum_{i=1}^{n} s_i$ and all m possible sinks $\sum_{j=1}^{m} s_j$. Accordingly,

$$S = \sum_{i=1}^{n} s_i - \sum_{j=1}^{m} s_j \tag{10.22}$$

The terms of this equation must be specified as quantities of the solute generated or dissipated per unit volume of soil and per unit length of time.

An additional possibility we might consider is the existence of a dynamic storage for the solute outside the soil's liquid phase, as, for instance, in precipitated form or in the soil's exchange complex. In this case, the left-hand side of Eq. (10.21) can be expanded to include the quantity of solute in storage (σ_s). Accordingly, the left-hand side of Eq. (10.21) becomes $\partial(\theta c + \sigma_s)/\partial t$. The time derivative of σ_s, namely, $\partial \sigma_s/\partial t$, expresses the rate of increase of storage outside the solution phase for the solute under consideration.

We can now write a comprehensive equation of transient state solute dynamics to include convective–dispersive–diffusive movement as well as sources, sinks, and storage changes. For a vertical soil profile, we substitute the axis z (representing depth below the soil surface) for x, to obtain

$$\frac{\partial(\theta c + \sigma_s)}{\partial t} = \frac{\partial}{\partial z}\left(D_{sh}\frac{\partial c}{\partial z}\right) - \frac{\partial(qc)}{\partial z} + S \tag{10.23}$$

wherein, as previously defined, the convective flux of the solution q is equal to the product of average velocity and volume fraction of the solution; i.e., $q = \bar{v}\theta$.

The adsorption of ions in the soil can be positive, as is the attachment of cations to clay surfaces, or negative, as is the repulsion or partial exclusion of anions from the electrostatic double layer of the same clay. This, incidentally, is why anions in a solution tend to travel somewhat faster than cations when an electrolytic solution is passed through a soil, as the cations undergo exchange phenomena within the soil's cation exchange complex and are retarded in the process. If the attainment of equilibrium between the ions in the solution phase and the soil's exchange complex is rapid enough to be considered instantaneous, then the amount adsorbed A can be taken as dependent only on the concentration c of the soil solution. Assuming that storage outside the solution phase is due entirely to adsorption (i.e., no other storage mechanisms, such as precipitation of a component of limited solubility), then the amount in storage σ_s is equal to A. Using these assumptions, Eq. (10.23) can be rewritten

$$[\theta + A(c)]\frac{\partial c}{\partial t} = \frac{\partial}{\partial z}\left(D_{sh}\frac{\partial c}{\partial z}\right) - \frac{\partial(qc)}{\partial z} + S \tag{10.24}$$

Here $A(c)$ is the slope of the *adsorption isotherm*, a function of concentration. Note that θ, c, s, and q are functions of depth and time, and D_{sh} is a function of θ and average velocity $\bar{v} = q/\theta$. Because of all these mutual dependencies, Eq. (10.24) is indeed complicated, and the only way to solve it at present is by means of numerical methods, based on the use of a computer (Bresler, 1973). Several investigators have used these and similar equations to describe solute movement in laboratory columns under controlled conditions, but to date the theory has not been adequately tested and validated sufficiently under actual field conditions.

G. Effects of Solutes on Water Movement

Thus far we have discussed the effect of soil-water movement on solute transport. Now we wish to consider the reciprocal effect of solutes on soil-water movement. Recalling Darcy's law, which states that the flux of soil water is determined by the product of the driving force acting on the water and the hydraulic conductivity of the porous medium, we ask first of all whether solutes can affect the nature or magnitude of the driving force for water movement.

1. Effect of Solutes on the Driving Force

Normally, the driving force for the movement of liquid water in the soil is composed of the sum of the gravitational and the pressure potential gradients. On the other hand, we know that osmotic potential gradients can induce water flow across *selective* (semipermeable) *membranes*, so termed because they restrict the passage of solute molecules or ions while allowing the transmission of water. The question arises as to whether, and to what extent, the soil matrix itself—particularly clay layers within the soil—can restrict the passage of various solutes, thus in effect acting as a selective membrane. A corollary question is whether, in the event that a soil layer does exhibit selective permeability, an osmotic potential gradient can be as effective as an hydraulic gradient in inducing convective water flow.

The evidence available on the latter question (e.g., Low, 1965; Kemper and Evans, 1963) suggests that when the solute is completely restricted by the porous medium, water flow in response to an osmotic potential difference is equivalent to that caused by an hydraulic pressure difference of equal magnitude, and that the two driving forces are therefore additive and act through the same transmission coefficient K. If so, the flux of water can be formulated as follows:

$$q = -K\left(\frac{dH}{dx} + \frac{1}{\rho g}\frac{d\Pi}{dx}\right) \tag{10.25}$$

wherein H is the hydraulic head, being the sum of the pressure and gravity potential heads, Π is osmotic pressure of the flowing solution, and, as in our previous notation, ρ is density of the solution, g gravitational acceleration, and x distance along the direction of flow. The osmotic pressure is related to salt concentration c by the expression $\Pi = ycRT$, where y is the osmotic coefficient of the solute, R the gas constant, and T absolute temperature.

When solutes are restricted in their movement relative to that of the water solvent, an effect known as *salt sieving* or *reverse osmosis* can develop (Kemper and Maasland, 1964; Bolt and Bruggenwert, 1976; Nielsen *et al.*,

1972). The exclusion of free electrolytes from the vicinity of negatively charged mineral surfaces is predicted by the Gouy-Chapman double layer theory (see Chapter 5). If soil water occurs as capillary "pockets" connected by thin films of water on and between adjacent soil particles, then one can conjecture that a tendency might exist for exclusion of the free electrolytes from the thin connecting films. When soil wetness decreases and the water films shrink to a thickness of the same order as the diffuse layer of adsorbed cations (say, 20–300 Å) an appreciable degree of solute restriction can be expected to take place. This restriction could well result in the soil exhibiting membranelike properties of selective permeability when a solution is driven through the soil by an hydraulic and/or osmotic gradient.

As water flows along the charged particle surfaces, a related phenomenon comes into play, namely, the development of a *streaming potential*, resulting from the tendency of the flowing "stream" to drag along some of the cations (especially the highly hydrated ones) of the diffuse double layer. The downstream accumulation of the positively charged ions creates a potential gradient which, in turn, acts in opposition to the hydraulic gradient and thus tends to retard flow. The occurrence of a streaming potential, incidentally, has been implicated in purported deviations from Darcy's law (i.e., in causing the convective flux to be smaller than predicted by Darcy's law at low hydraulic gradients and wetness values). However, this conjecture remains to be proven, as the effect described is normally very small and very difficult to measure. Another moot question is the fate of the solutes excluded or retarded from passage through a differentially permeable layer, and the possibility that, because of salt sieving, they might accumulate on the inflow side sufficiently to create an osmotic gradient counter to the flow direction. A related phenomenon has been noticed in connection with the extraction of the soil solution by means of the pressure plate and pressure membrane technique (Chapter 7). The first increments of the solution extracted from the soil often have a lower concentration of the solutes than the remaining solution. This, however, may also be due to an initial lack of osmotic equilibrium between water in large (interaggregate) and water in small (intraaggregate) pores.

Any assumption of complete restriction of solutes by a soil matrix is, in general, likely to be highly unrealistic. For partial restriction, Eq. (10.25) takes the form

$$q = -K\left(\frac{dH}{dx} + \frac{F_0}{\rho g}\frac{d\Pi}{dx}\right) \quad (10.26)$$

Here F_0 is an *osmotic efficiency factor*, or *reflection coefficient*, which represents a membrane's fractional degree of selectivity. A value of unity indicates total selectivity (i.e., transmission of solvent but "reflection" of solutes) whereas a value of zero indicates indiscriminant permeability to the solution—including molecules of the solvent and solute alike.

Experimental information on water movement as affected by osmotic gradients indicates that F_0 is close to zero in saturated and nearly saturated soil conditions, and becomes nonnegligible only at high suction values. Letey (1968) reported an F_0 value of about 0.03 in the intermediate suction range of 0.25–1 bar. It is tempting, therefore, to say that for most cases of practical interest the effect of salt concentration gradients on the driving force for water movement is small enough to be disregarded. Certainly it is so for nonclayey soils. One must therefore be wary of such oft-repeated but misleading blanket statements to the effect that soil water movement occurs in response to a gradient in the *total* soil water potential. Apart from the transfer of water across biological membranes, soil water flow is generally impelled by a gradient of an *hydraulic potential*, as described in the preceding two chapters.

2. Effect of Solutes on Hydraulic Conductivity

If the possible effect of solutes on the driving force for soil water movement is questionable, their effect on soil hydraulic conductivity is certain, and it can be very great indeed.

The first phenomenon to consider in this regard is the hydration of clay particles and consequent swelling of the soil. We recall that electrostatically charged clay particles attract cations, with which they form a diffuse double layer. This process of imbibition, causing swelling, is especially pronounced when the ambient solution (that is, the soil solution away from the particle surfaces) is the more dilute. Imbibition between clay platelets is constrained by interparticle bonds, and usually ceases at ambient solution concentrations above 200–400 meq/liter (McNeal, 1974). With more dilute solutions, continued swelling—to relieve the osmotic pressure differential between the clay domains and the ambient solution—weakens the interparticle bonds. A combination of this osmotic swelling with mechanical disturbance of the soil system can lead to a rupturing of interparticle bonds, so that adjacent particles separate and the clay fraction undergoes *dispersion*, which in turn alters the geometry of soil pores and results in a decrease of intrinsic permeability.

Combinations of low salt concentration and high *exchangeable sodium percentage* (ESP) are the conditions most likely to cause swelling, dispersion, and reduction of permeability. The collapse of aggregates resulting from dispersion of clay tends to plug the large interaggregate pores, particularly in the top layer, so that an "open" surface can become sealed. Moreover, dispersed particles can move with percolating water and migrate into the soil profile. Evidence of such migration can be seen in the occurrence of clay skins over aggregates deeper in the profile and in the natural deposition of clay in formation of distinct layers called *clay pans*.

Loss of permeability resulting from the adverse effect of salt concentration and exchangeable sodium percentage[1] can become a major problem in irrigated agriculture, as well as in the operation of municipal and industrial waste disposal systems such as septic tank drainage fields, where decreases in the hydraulic conductivity by one or more orders of magnitude have been reported.

In an effort to quantify the effects of various combinations of exchangeable cations and total concentration of the soil solution on hydraulic conductivity, Quirk and Schofield (1955) introduced the *threshold salt concentration* concept, defined as the level to which salt concentration of the soil solution must be decreased to produce a 10–15% reduction of hydraulic conductivity at various values of ESP (exchangeable sodium percentage). For their soil, threshold concentration values increased from 5 meq/liter at an ESP of 10 to 250 meq/liter at an ESP of 100 (respresenting complete saturation of the exchange complex with sodium ions). Similar work, reported by McNeal and Coleman (1966), is illustrated in Fig. 10.3.

A more theoretically based attempt to predict variation of hydraulic conductivity of soils equilibrated with different solutions was developed by Lagerwerf *et al.* (1969). They applied double-layer and clay swelling theory with a model of viscous flow through porous media based on the Kozeny–Carman theory (see Chapter 8) to obtain estimates of hydraulic conductivity decreases due to swelling of the clay matrix.

[1] The index used most often to characterize the soil solution with respect to its likely influence on the exchangeable sodium percentage is the *sodium adsorption ratio* (SAR), which has been used extensively to assess the quality of irrigation water. SAR is defined as follows:

$$\text{SAR} = [\text{Na}^+]/\{([\text{Ca}^{2+}] + [\text{Mg}^{2+}])/2\}^{1/2} \quad (10.27)$$

In words, it is the ratio of the sodium ion concentration to the square root of the average concentration of the divalent calcium and magnesium ions. In this context, all concentrations are expressed in milliequivalents per liter. SAR is thus an approximate expression for the relative activity of Na^+ ions in exchange reactions in soils. A high SAR, particularly at low concentrations of the soil solution, causes a high ESP and is also likely to cause a decrease of soil permeability.

The relationship between ESP and SAR of the soil solution was measured on numerous soil samples in the Western States by the U.S. Salinity Laboratory and reported (Richards, 1954) to be

$$\text{ESP} = 100[-a + b(\text{SAR})]/\{1 + [-a + b(\text{SAR})]\}$$

wherein $a = 0.0126$ and $b = 0.01475$.

G. Effects of Solutes on Water Movement

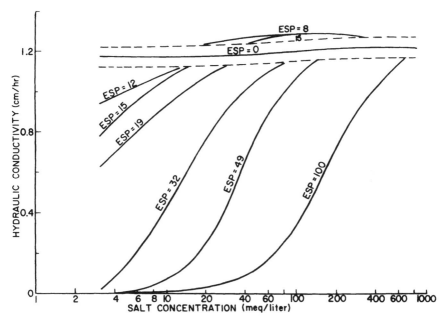

Fig. 10.3. Hydraulic conductivity of a sandy loam as related to total salt concentration of the soil solution and to the soil's exchangeable sodium percentage (ESP). (From McNeal and Coleman, 1966.)

The relation of hydraulic conductivity to composition and concentration of the soil solution and to the composition and capacity of the soil's exchange complex is influenced by the quantity and nature of the clay present (whether kaolinitic or montmorillonitic, for example), as well as by the content of such cementing agents as humus and sesquioxides.

So far we have referred only to the effect of solutes on the hydraulic conductivity of soils saturated with water. Little is known about the analogous effect on the hydraulic conductivity function of unsaturated soils. As pointed out by McNeal (1974), extension of known relationships from saturated to unsaturated soils requires an estimate of the inhibition of swelling for sodium-adsorbed clays at various matric potential values and depths within the profile and of the extent to which clay swelling affects different size sequences of soil pore. In principle, unsaturated conductivity can be expected to be less sensitive to total salt and sodium ion concentrations than saturated hydraulic conductivity. At high suction values, the conductivity of swelled soils may even be higher than that of nonswelled soils, owing to enhanced continuity of the fine-pore network and to greater retention of moisture at equivalent suctions by the matrix of the swelled soil.

H. Soil Salinity and Alkalinity

An excessive accumulation of salts in the soil profile causes a decline in productivity. *Soil salinity* is the term used to designate a condition in which the soluble salt content of the soil reaches a level harmful to crops. Soil salinity affects plants directly by reducing the osmotic potential of the soil solution and by the toxicity of specific ions such as boron, chloride, and sodium. If the salts are primarily sodic salts, as is frequently the case, their accumulation increases the concentration of sodium ions in the soil's exchange complex, which in turn affects soil properties and behavior. Thus, salinity can also have indirect effects on plant growth through deleterious modification of such soil properties as permeability and porosity.

Soluble salts may be present in the original soil material or be brought to the soil by invading surface water or groundwater. The ocean is a source of salt for low lying soils along the coast, where it may also contribute salt to rainfall by sea spray. Except along the seacoast, however, saline soils seldom occur in humid regions, thanks to the typically downward net seepage of water through the soil profile brought about by the excess of rainfall over evapotranspiration. In arid regions, on the other hand, there may be no net downward percolation of water and hence no effective leaching, so salts can—and often do—accumulate. Here the combination of a high evaporation rate, the occurrence of salt-bearing sediments, and the presence of shallow, brackish groundwater gives rise to a group of soils known in classical pedology as *solochaks*.

Less obvious than the appearance of naturally saline soils, but certainly more ominous, is the inadvertently induced salinization of originally productive soils by the injudicious practice of irrigation in certain arid regions. Typically located in the river valleys of the dry zone, irrigated soils are particularly vulnerable to the insidious, and for a time invisible, rise of the water table caused by failure to provide for adequate drainage of groundwater. Proper irrigation maintains a supply of good-quality moisture needed by plants to answer the climatically imposed transpirational demand, while at the same time ensuring a favorable salt balance and nutrient supply, as well as adequate aeration and temperature regimes throughout the root zone. Efficient irrigation, furthermore, avoids wasting water through runoff or excessive drainage, except insofar as some drainage is necessary to flush out potentially harmful salts which would otherwise accumulate in the root zone. Crop plants extract water from the soil while leaving most of the salt behind. Unless leached away (preferably continuously, but at least periodically) such salts will sooner or later begin to hinder crop growth.

The classical and still pertinent publication on soil salinity is the U. S. Department of Agriculture Handbook 60 (L. A. Richards, editor, 1954). It seems that a new and much revised issue is now under preparation, but until it is issued we have no choice but to refer to the definitions and concepts of the original publication, which have been adopted almost universally. The accepted classification of soil salinity (as well as of irrigation water quality) is based on total salt concentration of an equilibrated solution, as measured by its electrical conductivity, and on the relative concentration of sodium ions. *Soil alkalinity* (or *sodicity* is characterized by the *exchangeable sodium percentage* (ESP), being the content of exchangeable sodium ions (in milliequivalents per 100 grams of soil) as a fraction (percent) of the total cation exchange capacity. According to these criteria, a *saline soil* is one whose saturation extract[2] indicates an electrical conductivity exceeding 4 mmho/cm at 25°C but whose ESP is less than 15. Such soils usually have pH values less than 8.5 and are ordinarily well flocculated. Reclamation of saline soils requires removal of excess salts by leaching, with due care to prevent a rise in the exchangeable sodium percentage. *Saline–alkali soils* have been defined as having a saturation-extract conductivity greater than 4 mmho/cm and an exchangeable sodium percentage greater than 15. Such soils, when leached, become highly dispersed and exhibit higher pH values. To be reclaimed, these soils require, in addition to leaching, treatment with soluble calcium salts (e.g., gypsum) to replace the excess of exchangeable sodium. Finally, *nonsaline, alkali soils* are such that exhibit an ESP greater than 15 but a salinity less than the 4 mmho/cm level. Such soils, which often exhibit very high pH values (8.5–10) have traditionally been called *solonetz* by pedologists, and sometimes *black alkali* (in reference to the black color resulting from the highly dispersed organic matter at the soil surface).

These classifications are obviously somewhat arbitrary and may appear to be anachronistic. However, they seem to have stood the test of time and still to serve as useful diagnostic criteria. It should be obvious, though, that different kinds of plants can be expected to react differently to total salt concentration and composition of the soil solution under different soil water regimes and climatic conditions. What with new methods of irrigation and water management in general, the time may have come for reexamination of soil salinity criteria, which ought to be based on dynamic (rather than static) concepts related to water and solute movement in the soil.

I. Salt Balance of the Soil Profile

First introduced by Schofield (1940), the *salt balance* is a quantified summary of all salt inputs and outputs for a defined volume or depth of soil

---

[2] A saturation extract is obtained after adding an amount of water just sufficient to saturate a soil sample, which is then stirred to form a slurry or "paste" in a standard manner. The water is later sampled by filtration (using suction) for determination of salt concentration.

during a specified period of time. If salts are conserved (that is to say, if they are neither generated nor decomposed chemically in the soil) then the difference between the total input and output must equal the change in salt content of the soil zone monitored, and if input exceeds output then salt must be accumulating. The salt balance has been used as an indicator of salinity trends and the need for salinity control measures in large-scale irrigation projects as well as in individual irrigated fields. The following equation applies to the amount of salt in the liquid phase of the root zone per unit area of land:

$$[\rho_w(V_r c_r + V_i c_i + V_g c_g) + M_s + M_a] - (M_p + M_c + \rho_w V_h c_h) = \Delta M_{sw} \quad (10.28)$$

Here V_r is the volume of rain water entering the soil with a salt concentration c_r, V_i volume of irrigation water with a concentration c_i, V_g volume of groundwater with a concentration c_g entering the root zone by capillary rise, V_h volume of water drained from the soil with a concentration c_h, M_s and M_a are masses of salt dissolved from the soil and from agricultural inputs (fertilizers and soil amendments), respectively, M_p and M_c are mass of salt precipitated (or adsorbed) in situ and mass removed by the crop, respectively, ρ_w is the density of water, and, finally, ΔM_{sw} is the change in mass of salt in the soil's liquid phase during the period considered.

A simplified form of the salt balance equation was given by Buras (1974) for a complete annual period. Assuming that the net change in total soil water content from one year to the next (as monitored, say, either at the end of the wet season or at the end of the dry season) is close to zero, the *water balance* per unit area, disregarding surface runoff, is as follows:

$$V_i + V_r + V_g = V_{et} + V_h \quad (10.29)$$

Herein V_i, V_r, V_g, V_{et}, and V_h are total annual volumes of irrigation, atmospheric precipitation, capillary rise of groundwater, evapotranspiration, and drainage, respectively. Since evapotranspiration removes no salt and crops generally remove only a negligible amount, and if we disregard agricultural inputs and in situ precipitation and dissolution of salts, the salt balance corresponding to Eq. (10.29), assuming no accumulation, is

$$V_i c_i + V_r c_r = (V_h - V_g) c_h \quad (10.30)$$

where c_i, c_r, and c_h are the average concentrations of salt in the irrigation, atmospheric precipitation, and drainage waters, respectively. If the water table is kept deep enough so that no substantial capillary rise of groundwater into the root zone takes place, and since the salt content of atmospheric precipitation is usually negligible, Eq. (10.30) further simplifies to

$$V_i c_i = V_h c_d \quad (10.31)$$

I. Salt Balance of the Soil Profile

Such an overall "black box" approach disregards the mechanisms and rates of salt and water interactions in the root zone and is incapable of characterizing the internal distribution of salts within the soil profile and the manner in which it changes in time.

A more detailed treatment of the dynamic balance of salts in the root zone under irrigation was given by Bresler (1972), based on the following equation:

$$\frac{\partial}{\partial t} \int_0^z [c(z,t)\theta(z,t)] \, dz = q(o,t)c_0(t) - q(z,t)c(z,t) \tag{10.32}$$

wherein the flux q, soil wetness θ, and salt concentration c are considered as functions of depth z and time t, and c_0 is the salt concentration of the applied water. This equation was solved explicitly for $c(z,t)$, using numerical methods, for different initial and boundary conditions.

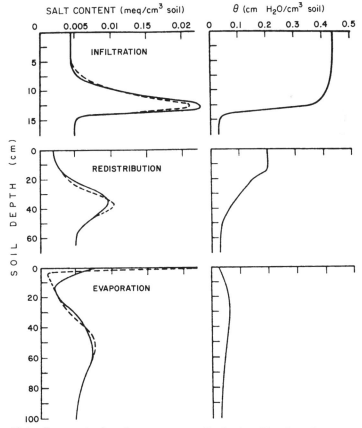

Fig. 10.4. Computed salt and water content distribution. The dispersion term was considered (solid lines) and omitted (dashed lines). (From Bresler, 1972.)

Figure 10.4 demonstrates the results obtained from these computations for a fallow (unvegetated) loam soil during cycles of infiltration, redistribution of soil moisture, and evaporation. The computation of salt and water dynamics in the presence of an active root system, while more complicated, is amenable to the same type of quantitative approach (e.g., Hillel *et al.*, 1975).

The problems encountered in any attempt to apply the theory in practice and to monitor the dynamic balance of salts in the field include difficulties of in situ measurement of concentration changes, of precipitation and dissolution of salts, and of water fluxes in the soil and subsoil. These difficulties are compounded by the heterogeneous nature of many fields.

J. Leaching of Excess Salts

The leaching of excess salts from the root zone is an essential aspect of salinity control. When irrigation is practiced in arid regions, particularly if the applied water contains an appreciable concentration of soluble salts, the twin processes of evaporation and transpiration tend to concentrate salts in the root zone. Without leaching, salt will accumulate in direct proportion to the amount of water applied and to its salt content.

It is a startling fact that 1 m depth of even reasonably good quality irrigation water (an amount normally applied in a single irrigation season) contains sufficient salt to salinize an initially salt-free soil (i.e., about 5000 kg/hectare).

To prevent salt residues from accumulating during repeated irrigation–evapotranspiration cycles, the obvious remedy is to apply water in an amount greater than evapotranspiration, so as to deliberately cause a significant fraction of the applied water to flow through and past the root zone and leach away the excess salts. However, unless the water table is very deep, or lateral groundwater drainage is sufficiently rapid, the excess irrigation can, and often does, cause a progressive rise of the water table. And, once the water table comes within a meter or two of the soil surface, groundwater tends to seep upwards into the root zone by capillary action between irrigations and thus to reinfuse the soil with salt.

From the foregoing it should be obvious that any attempt to leach without provision of adequate drainage is not merely doomed to fail but can indeed exacerbate the problem. It should be obvious, but alas, to too many people, it is not. In many areas where natural drainage is slow and artificial drainage is not provided, it becomes impossible to sustain irrigation and the land must sooner or later be abandoned owing to progressive salinization. This has happened repeatedly throughout the history of irrigated agriculture, and still continues to happen in great and small river valleys from the Jordan to the San Joaquin, and from the Indus to the Rio Grande.

J. Leaching of Excess Salts

Much attention has been devoted to the assessment of the optimal quantity of water which must be applied to effect leaching. Clearly, the application of too much water can be as harmful as the application of too little. Exaggerated leaching not only wastes water but also tends to remove essential nutrients and to impede aeration by waterlogging the soil. The *leaching requirement* concept was developed by the U. S. Salinity Laboratory (Richards, 1954). It has been defined as the fraction of the irrigation water that must be leached out of the bottom of the root zone in order to prevent average soil salinity from rising above some specifiable limit.[3] As such, the leaching requirement depends upon the salt concentration of the irrigation water, upon the amount of water extracted from the soil by the crop (the evapotranspiration), and upon the salt tolerance of the crop, which determines the maximum permissible concentration of the soil solution in the root zone.

Assuming steady-state conditions of through-flow (thus disregarding short-term changes in soil moisture content, flux, and salinity), and furthermore assuming no appreciable dissolution or precipitation of salts in the soil and no removal of salts by the crop or capillary rise of salt-bearing water from below, we obtain from Eq. (10.31)

$$V_h/V_i = c_i/c_h \tag{10.33}$$

in which V_h and V_i are the volumes of drainage and of irrigation, respectively, and c_h and c_i are the corresponding concentrations of salt. The water volumes are normally expressed per unit area of land as equivalent depths of water and the salt concentrations are generally measured and often reported in terms of electrical conductivity. Since the volume of water drained is the difference between the volumes of irrigation and evapotranspiration, i.e., $V_h = V_i - V_{et}$, we can transform the last equation as follows:

$$(V_i - V_{et})/V_i = c_i/c_h$$

Hence

$$V_i = [c_h/(c_h - c_i)]V_{et} \tag{10.34}$$

which is equivalent to the formulation by Richards (1954):

$$d_i = [E_h/(E_h - E_i)]d_{et} \tag{10.35}$$

where d_i is the depth of irrigation, d_{et} the equivalent depth of "consumptive use" by the crop (evapotranspiration), and E_h and E_i are the electrical conductivities of the drainage and irrigation waters, respectively.

---

[3] According to the standards of the 1954 Handbook of the U.S. Salinity Laboratory, the maximum concentration of the soil solution should be kept below 4 mmhos/cm (expressed in terms of electrical conductivity) for sensitive crops. Tolerant crops like beets, alfalfa, and cotton, may give good yields at values up to 8 mmhos/cm, while a very tolerant crop like barley may give good yields at 12 mmhos/cm.

The leaching requirement equation implies that by varying the fraction of applied water which is percolated through the root zone it is possible to control the concentration of salts in the drainage water and hence to maintain the concentration of the soil solution in the main part of the root zone at some intermediate level between c_i and c_h. In the limit, of course, as the volume of drainage approaches the volume of irrigation water applied, i.e., when irrigation greatly exceeds evapotranspiration, the concentration of the drainage water approaches that of the irrigation water.

The leaching requirement concept obviously disregards the distribution of salts within the root zone itself, as it is affected by the frequency of irrigation as well as by its quantity and water quality. In particular, the spatial and temporal variation of root zone salinity is affected by the degree to which soil moisture is depleted between irrigations. The less frequent the irrigation regime, the greater the build-up of salt concentration between successive irrigations. With the recent trend toward high-frequency irrigation (Rawlins and Raats, 1975), it has become possible to maintain a zone of soil near the surface with a soil solution concentration essentially equal to that of the irrigation water. This depth can be increased by increasing the volume of water applied. Beyond this depth, the salt concentration of the soil solution increases with depth to a salinity level which depends on the leaching fraction (defined as V_h/V_i). Because the effects of matric and osmotic potentials on crop growth are approximately additive (Wadleigh and Ayers, 1945) it is doubly important to maintain a higher soil moisture condition (and hence a higher level of both matric and osmotic potentials, i.e., lower suction levels) by frequent and sufficient irrigation whenever low quality or brackish water is used for irrigation.

In determining the optimal leaching fraction, a possible constraint must be taken into account, namely the limited hydraulic conductivity of the subsoil at the lower boundary of the root zone. The flow rate through this zone may limit the *attainable leaching fraction* (Rhoades, 1974). Unless this limitation is recognized and defined, a blind attempt to follow the leaching requirement concept may lead to waterlogging the soil and thus aggravating, rather than solving, the problem.

An interesting approach to the reclamation of saline–sodic soils having adequate drainage was developed by Reeve and Bower (1960). They recommended using saline rather than good-quality water for the initial leaching of such soils, then gradually decreasing the salinity of applied water. The higher concentration of salts in the applied water can help maintain the high conductivity of the soil and prevent the strong dispersion of clay which often results when salts are removed abruptly by high-quality water without immediately replacing the high percentage of exchangeable sodium with a divalent cation such as calcium.

A fundamental approach to leaching was attempted by van der Molen (1956), who used the theory of chromatography and found agreement between theoretical curves and measured leaching rates in Dutch soils which had been inundated by seawater. Gardner and Brooks (1956) developed a theory in which a distinction was made between immobile (detained) and mobile salt, the latter moving with the velocity of the leaching front. They observed that in several soils about 1.4 pore volumes of water are needed to reduce salinity by 80%, and held that the flow process itself, rather than diffusion, is responsible both for the diffuse boundary between the soil solution and the leaching water and for the subsequent removal of the temporarily bypassed salt.

Nielsen and Biggar (1961) conducted studies on miscible displacement and applied these to the leaching of excess salts from saline soils. They suggested that leaching soils at a water content below saturation (e.g., under low-intensity sprinkling irrigation or rainfall or under intermittent irrigation) could produce more efficient leaching and thereby reduce the amount of water required as well as reduce drainage problems in areas of high water tables. This corroborated observations that in a soil with large vertical cracks most of the water infiltrated under ponding moves through these cracks and is ineffective in leaching, whereas under rainfall, more of the water moves through the soil blocks and micropores, producing more efficient leaching per unit volume of water infiltrated.

Sample Problems

1. A soluble pollutant was inadvertently spilled on the ground. Suppose that it is nondegradable, nonvolatile, not taken up by plants, not adsorbed by the soil, nor immobilized by any other mechanism. If the annual rainfall is 1500 mm, the annual evapotranspiration 1250 mm, the water table is 20 m deep, and the so-called unsaturated zone underlying the soil has a constant volumetric wetness of 25%, estimate the residence time in the unsaturated zone and the time required for the pollutant to reach the groundwater.

As a rough approximation, we assume that the solute is transported only by the convective stream of water draining out of the soil and flowing vertically downward through the unsaturated zone toward the water table. The possible effects of diffusion and dispersion are thus disregarded. We further assume that this drainage occurs under steady state conditions, i.e., that temporal perturbations at the soil surface such as railfall and evapotranspiration are damped out in the soil.

To estimate the distance of travel of a solute per unit time, we consider the average apparent velocity $\bar{v}$ of the flowing solution to be [Eq. (10.2)]

$$\bar{v} = q/\theta$$

where $\bar{v}$ is the straight line length of the path traversed through the soil or subsoil by the solution in unit time, q is flux, and θ volumetric wetness. Now substituting the values given above, we obtain

$$\bar{v} = (1500 - 1250) \text{ mm/yr}/0.25 = 1000 \text{ mm/yr}$$

To estimate the residence time t_r of the solute within the unsaturated zone, we refer to Eq. (10.4):

$$t_r = L/\bar{v}$$

where L is the thickness of the zone considered. Thus,

$$t_r = 20 \text{ m}/1 \text{ m/yr} = 20 \text{ yr}$$

Thus, the bulk of the solute can be expected to reach the water table and enter the groundwater in about 20 years time. Since in actual fact diffusion and dispersion phenomena do operate, we can expect some of the solute to move faster than the bulk of the solution and some to lag behind. However, the calculation based on convective transport alone can serve as a useful rough estimation of the movement rate.

2. Estimate the rate of dissolution of a gypsum block placed in contact with the soil surface. Assume one-dimensional steady-state diffusion from the surface of the block through a soil column having a length of 20 cm at the end of which is a chemical *sink* which effectively removes all of the solute from the solution. Assume, furthermore, that soil water is immobile (i.e., no convective flow and hence no hydrodynamic dispersion), that the molecular diffusion coefficient in bulk water is 10^{-5} cm^2/sec, that the soil's volumetric wetness is 40%, and that the tortuosity factor is 0.65. Finally, assume that the solubility of gypsum (calcium sulfate) is 2.4×10^{-3} gm/cm^3 (i.e., a saturated solution contains 2400 parts per million).

We first estimate the diffusion coefficient in the soil (D_s). Recalling Eq. (10.8), which relates D_s to that in bulk water D_0,

$$D_s = D_0 \theta \xi$$

where θ is volumetric wetness and ξ the tortuosity factor, we have

$$D_s = 10^{-5} \text{ cm}^2/\text{sec} \times 0.4 \times 0.65 = 2.6 \times 10^{-6} \text{ cm}^2/\text{sec}$$

Sample Problems

We now use Fick's first law in the form of Eq. (10.10):

$$J_h = -D_s \, dc/dx$$

For a linear distribution of concentration, we can write:

$$J_d = D_s \Delta c/L$$

where Δc is the concentration differential across a column of length L.

Now assume that the concentration of the solute is 2.4×10^{-3} gm/cm³ at one end of the soil column (at the *source*) and zero at the other end (the *sink*). Thus we have

$$J_d = (2.6 \times 10^{-6}) \times \frac{2.4 \times 10^{-3} - 0}{20} = 3.12 \times 10^{-10} \text{ gm/cm}^2 \text{ sec}$$

The dissolution rate per year is therefore:

$$(3.12 \times 10^{-10} \text{ gm/cm}^2 \text{ sec})(86{,}400 \text{ sec/day})(365 \text{ day/yr})$$
$$\cong 9.84 \times 10^{-3} \text{ gm/cm}^2 \text{ yr}$$

Thus, the gypsum block is likely to lose about 10 mgm/cm² yr. If it is completely embedded in the soil under similar conditions, and if its surface area is, say, 100 cm² and initial mass is 50 gm, its dissolution rate will be nearly 1 gm/yr (i.e., 2% of its initial mass).

3. The following data were obtained in a field located in an arid zone:

(1) Rainfall occurred entirely in winter and amounted to 30 cm, with a total salt concentration of 40 ppm.

(2) Capillary rise from shallow, saline groundwater occurred during spring and autumn and amounted to 10 cm at a concentration of 1000 ppm.

(3) Irrigation was applied during the summer season and amounted to 90 cm (400 ppm salts).

(4) Drainage occurred only during the irrigation season and amounted to 20 cm, with a soluble salt concentration of 800 ppm.

(5) An additional increment 120 gm/m² of soluble salts was added in the form of fertilizers and soil amendments, whereas an amount of 100 gm/m² was removed by harvested crops.

Disregarding dissolution and/or precipitation of salts within the soil, compute the annual salt balance. Is there a net accumulation or release of salts by the soil?

We begin with a slightly modified version of Eq. (10.28) to calculate the salt balance of the soil's root zone per unit of land area:

$$\Delta M_s = \rho_w(V_r c_r + V_i c_i + V_g c_g - V_d c_d) + M_a - M_c$$

where M_s is the change in mass of salt in the root zone (in the liquid phase as well as in the solid phase; i.e., the sum of all dissolved, adsorbed, and precipitated salt); ρ_w is the density of water; V_r, V_i, V_g, and V_d are the volumes of rain, irrigation, groundwater rise, and drainage, respectively, with corresponding salt concentrations of c_r, c_i, c_g, and c_d; and M_a and M_c are the masses of salts added agriculturally and removed by the crops, respectively.

Using 1 cm² as the unit "field" area, calculating water volumes in terms of cm³ and masses in terms of gm (water density being 1 gm/cm³), we can substitute the given quantities into the previous equation to obtain the change in mass content of salt in the soil:

$$\Delta M_s = 1 \text{ cm}^2 \times [1 \text{ gm/cm}^3 \times (30 \text{ cm} \times 40 \times 10^{-6} + 90 \text{ cm} \times 400 \times 10^{-6}$$
$$+ 10 \text{ cm} \times 1000 \times 10^{-6} - 20 \text{ cm} \times 800 \times 10^{-6})$$
$$+ 120 \times 10^{-4} \text{ gm/cm}^2 - 100 \times 10^{-4} \text{ gm/cm}^2]$$
$$= 3.52 \times 10^{-2} \text{ gm}$$

Thus, the soil's root zone is accumulating salt at the rate of 3.52×10^{-2} gm/cm² yr, equivalent to 3520 kg/hectare yr.

4. Estimate the "leaching requirements" of a field subject to a seasonal evapotranspiration of 1000 mm, if the electrical conductivity of the irrigation water is 1 mmho/cm (approximately equivalent to a salt concentration of 650 ppm), and that of the drainage water is allowed to attain 4 mmho/cm (≈ 2600 ppm). What would be the leaching requirement if the irrigation water were half as concentrated? And what if the drainage water were allowed to become twice as concentrated? Finally, what would be the electrical conductivity of the drainage water if the amount of irrigation were 1500 mm?

Using Eq. (10.35):

$$d_i = [E_d/(E_d - E_i)]d_{et}$$

where d_i and d_{et} are the volumes of water per unit land area (in depth units) of irrigation and of evapotranspiration, respectively; and E_d and E_i are the electrical conductivities of the drainage and irrigation waters, respectively. Substituting the appropriate values, we get

$$d_i = [4/(4 - 1)] \, 1000 \text{ mm} = 1333 \text{ mm}$$

The "leaching depth,"

$$d_e = d_i - d_{et} = 1333 - 1000 = 333 \text{ mm}$$

If the irrigation water were half as concentrated, i.e., if E_i were 0.5 mmho, we would have

$$d_i = [4/(4 - 0.5)] \, 1000 \text{ mm} = 1143 \text{ mm}$$

Sample Problems

Thus, the leaching depth would be only 143 mm, which is less than half the previously required leaching volume.

If the drainage water were permitted to become twice as concentrated, i.e., if E_d were 8 mmho instead of 4, then

$$d_i = [8/(8 - 1)] \, 1000 \text{ mm} = 1143 \text{ mm}$$

In words: Doubling the allowable concentration of the drainage water is equivalent to halving the concentration of the applied irrigation water, in terms of its effect on reducing leaching requirements.

If the depth of irrigation water applied were 1500 mm and its concentration were of 1 mmho, the electrical conductivity of the drainage water would be

$$E_d = E_i/[1 - (d_{et}/d_i)]$$

which is obtained by simple transformation of Eq. (10.35). Thus,

$$E_d = [1 - (1000/1500)]^{-1} = 3 \text{ mmho}$$

Part IV:

THE GASEOUS PHASE

> Impressed with the transitoriness of objects, the ceaseless mutation and transformation of things, Buddha formulated a philosophy of change. He reduced substances, souls, monads, things to forces, movements, sequences and processes, and adopted a dynamic conception of reality.
>
> S. Radhakrishnan[1]

11 Soil Air and Aeration

A. Introduction

The process of *soil aeration* is one of the most important determinants of soil productivity. Plant roots adsorb oxygen and release carbon dioxide in the process of *respiration*. In most terrestrial plants (excepting such specialized plants as rice), the internal transfer of oxygen from the parts above the ground (leaves and stems) to those below the ground surface (roots) cannot take place at a rate sufficient to supply the oxygen requirements of the roots. Adequate root respiration requires that the soil itself be *aerated*, that is to say, that gaseous exchange take place between soil air and the atmosphere at such a rate as to prevent a deficiency of oxygen and an excess of carbon dioxide from developing in the root zone (Fig. 11.1). Soil microorganisms also respire, and, under conditions of restricted aeration, might compete with the roots of higher plants (Stotzky, 1965).

Gases can move either in the air phase (that is, in the pores which are drained of water, provided they are interconnected and open to the atmosphere) or in dissolved form through the water phase. The rate of diffusion of gases in the air phase is generally greater than in the water phase, hence soil aeration is dependent largely upon the volume fraction of air-filled pores.

Impeded aeration resulting from poor drainage and waterlogging, or from mechanical compaction of the soil, can strongly inhibit crop growth. In particular, the problem of soil compaction seems to have worsened in recent decades, along with the growing trend to use larger and heavier machinery and the tendency to tread over the field repeatedly for such purposes as

---

[1] In (Radhakrishnan, 1958).

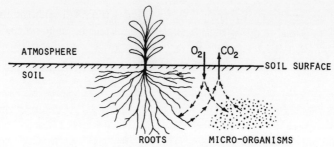

Fig. 11.1. Soil aeration as a process of O_2 and CO_2 exchange with the atmosphere (schematic). Among other gases involved in the soil atmosphere exchange are various volatile forms of nitrogen (e.g., N_2, NH_3, NO, NO_2), sulfur (e.g., H_2S, SO_2) and hydrocarbons (e.g., CH_4).

fertilization and pest control. Moreover, with greater use of fertilizers and irrigation, shortages of nutrients and water have been obviated in many places so that more and more it is aeration which has become a major limiting factor to the attainment of maximal productivity. It seems likely that root systems are commonly restricted in extent by the progressive decrease of aeration in the deeper regions of the soil profile. Poor aeration can decrease the uptake of water and induce early wilting (Stolzy et al., 1963). According to Kramer (1956), restricted aeration also causes a decrease in the permeability of roots to water.

Anaerobic conditions in the soil induce a series of reduction reactions, both chemical and biochemical. Included among these reactions are *denitrification* (the processes by which nitrate is reduced to nitrite, thence to nitrous oxide and eventually to elemental nitrogen: $NO_3^- \rightarrow NO_2^- \rightarrow N_2O \rightarrow N_2$; *manganese reduction* from the manganic to the manganous form); *iron reduction* from the ferric to the ferrous form; and *sulfate reduction* to form hydrogen sulfide. Some of the numerous products of these anaerobic processes are toxic to plants (e.g., ferrous sulfide, ethylene, as well as acetic, butyric, and phenolic acids).

The subject of soil aeration has been reviewed repeatedly over the years and the evolving state of our knowledge is reflected in the successive publications by Russell (1952), Stolzy and Letey (1964), Grable (1966), Letey et al. (1967), Wesseling (1974), Currie (1975), and Cannell (1977).

B. Volume Fraction of Soil Air

In most natural soils, the volume ratios of the three constituent phases, namely, the solids, water, and air, are continually changing as the soil as a whole undergoes wetting or drying, swelling or shrinkage, tillage or compaction, aggregation or dispersion, etc. Specifically, since the twin fluids—

B. Volume Fraction of Soil Air

water and air—compete for the same pore space, their volume fractions are so related that an increase of the one generally entails a decrease of the other. From Eq. (2.11), we get

$$f_a = f - \theta \tag{11.1}$$

wherein f_a is the volume fraction of air, f the total porosity (the fractional volume of soil not occupied by solids), and θ the volume fraction of water (volume wetness).

Since the volume fraction of air is a transient property, if we wish to use it as an index of soil aeration we must obviously specify some characteristic and reproducible wetness value. The wetness value generally chosen for this purpose is the so-called "field capacity," discussed in detail by Hillel (1980, Chapter 3). Although the term defies exact physical definition, suffice it to say at present that it is an approximation of the amount of water retained by a soil after the initially rapid stage of internal drainage. Objections to this concept notwithstanding for the moment, we might define an analogous soil aeration index, which we might call *"field air-capacity,"* definable as the fractional volume of air in a soil at the field-capacity water content. Though under different names (e.g., noncapillary porosity, air-filled porosity, air content at field capacity, etc.), this index, or something very similar to it, has been investigated for many years in relation to soil aeration and plant response. It has been found to depend on numerous factors, not all of which are amenable to human control.

In the first place, air capacity depends on soil texture. In sandy soils, it is of the order of 25% or more, in loamy soils it is generally between 15 and 20%, and in clayey soils, which tend to retain the most water, it is likely to fall below 10% of total soil volume. In fine-textured soils, however, soil structure, too, has much to do with determining the air capacity. Strongly aggregated soils, with macroaggregates of the order of 5 mm or more in diameter, generally have a considerable volume of *macroscopic* (interaggregate) pores which drain very quickly and remain air filled practically all of the time. Hence such soils exhibit an air capacity of 20–30%. As the aggregates are dispersed, or broken down by mechanical forces, the macroscopic pores tend to disappear so that a strongly compacted soil may contain less than 5% air by volume at its characteristic field capacity value of soil moisture. The effect of compaction on a soil's capacity to retain air and water is illustrated hypothetically in Fig. 11.2.

Investigators have long attempted to establish a value of the field air capacity, or air-filled porosity, at which soil aeration is likely to become limiting to root respiration and hence to plant growth (Vomocil and Flocker, 1961). The results, however, have been rather baffling. The reported limiting values have ranged between 5 and 20%, averaging around 10%.

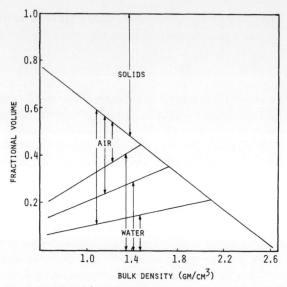

Fig. 11.2. Fractional air, water, and total pore volumes as related to soil bulk density, for different mass wetness w values (0.1, 0.2, and 0.3).

There are, of course, fundamental limitations to the air capacity index as a means of characterizing soil aeration. First, the value is difficult to determine with a satisfactory degree of accuracy, since it generally depends on prior determination of two highly variable parameters which are themselves rather cumbersome to measure—namely, the field capacity and the total porosity. Both determinations depend on the method of sampling and their results can be grossly inaccurate. An even more serious objection to the use of air capacity as an index of soil aeration is that in principle the *rate of exchange* of soil air rather than simply the *content* of soil air constitutes the decisive factor. At high wetness values, soils often contain isolated pockets of occluded air which, though forming a part of the air-filled volume, do not contribute to active gas exchange. At times, even a thin surface crust, if highly compact or saturated, can form a bottleneck limiting aeration to the entire soil profile, regardless of the air-filled porosity beneath. We are thus led to the necessity of characterizing aeration in more dynamic terms.

C. Composition of Soil Air

In a well-aerated soil, the composition of soil air is close to that of the external ("open") atmosphere, as the oxygen consumed in the soil is readily replaced from the atmosphere. Not so in a poorly aerated soil. Analyses of the actual composition of soil air in the field reveal it to be much more variable than the external atmosphere. Depending on such factors as time of year, temperature, soil moisture, depth below the soil surface, root growth, microbial activity, pH, and—above all—the rate of exchange of gases through the soil surface, soil air can differ to a greater or lesser degree from the composition of the external atmosphere. The greatest difference is in concentration of carbon dioxide (CO_2), which is the principal product of aerobic respiration by the roots of higher plants and by numerous macro- and microorganisms in the soil. The CO_2 concentration of the atmosphere is about 0.03%. In the soil, however, it frequently reaches levels which are ten or even one-hundred times greater.

As CO_2 is produced in the soil by oxidation of carbonaceous (organic) matter, an increase of CO_2 concentration is generally associated with a decrease in elemental oxygen (O_2) concentration (though not necessarily to an exactly commensurate degree, since additional sources of oxygen may exist in dissolved form and in easily reducible compounds). Since the O_2 concentration of air is normally about 20%, it would seem that even a hundredfold increase of CO_2 concentration, from 0.03 to 3%, can only diminish the O_2 concentration to about 17%. However, even before they begin to suffer from lack of oxygen per se, some plants may suffer from excessive concentrations of CO_2 and other gases in both the gaseous and aqueous phases. In more extreme cases of aeration restriction, the O_2 concentration can fall to near zero and prolonged anaerobic conditions can result in the development of a chemical environment characterized by reduction reactions such as denitrification, the evolution of such gases as hydrogen sulfide (H_2S), methane (CH_4), and ethylene, and the reduction of mineral oxides (such as those of iron and manganese).

Another interesting difference between the atmosphere and soil air is that the latter is characterized by a high relative humidity, which nearly always approaches 100% except at the soil surface during prolonged dry spells.

D. Convective Flow of Soil Air

The exchange of air between the soil and the atmosphere can occur by means of two different mechanisms: *convection* and *diffusion*. Each of these processes can be formulated in terms of a linear rate law, stating that the flux is proportional to the moving force. In the case of convection, also called mass flow, the moving force consists of a gradient of *total gas pressure*, and it results in the entire mass of air streaming from a zone of higher pressure to one of lower pressure. In the case of diffusion, on the other hand, the moving force is a gradient of *partial pressure* (or concentration) of any constituent member of the variable gas mixture which we call air, and it causes the molecules of the unevenly distributed constituent to migrate from a zone of higher to lower concentration even while the gas as a whole may remain isobaric and stationary. We propose to discuss convective flow of soil air in this section, and diffusion in the next.

A number of phenomena can cause pressure differences between soil air and the external atmosphere, thereby inducing convective flow into or out of the soil. Among these phenomena are barometric pressure changes in the atmosphere, temperature gradients, and wind gusts over the soil surface. Additional phenomena affecting the pressure of soil air are the penetration of water during infiltration, causing displacement of antecedent soil air, the fluctuation of a shallow water table pushing air upward or drawing air downward, and the extraction of soil water by plant roots. Short-term changes in soil air pressure can also occur during tillage or compaction by machinery.

The degree to which air pressure fluctuations and the resulting convective flow can contribute to the exchange of gases between the soil and the atmosphere has long been a subject of debate among soil physicists. The majority have tended to support the hypothesis that diffusion rather than convection is the more important mechanism of soil aeration (Keen, 1931; Penman, 1940; Russell, 1952). Recent evidence suggests, however, that previous analyses have been incomplete and that convection can in certain circumstances contribute significantly to soil aeration, particularly at shallow depths and in soils with large pores (Vomocil and Flocker, 1960; Grable, 1966; Farrel et al., 1966; Scotter and Raats, 1968, 1969; Kimball and Lemon, 1971, 1972). For example, Vomocil and Flocker (1961), assuming that the O_2 consumption rate is 0.2 ml/hr gm of root tissue (fresh weight), that rooting density is 0.1 gm/cm^3, that the water extraction rate is 7.5 mm/day, and that soil bulk density remains constant, calculated that the process of root water extraction by itself can draw into the soil as much as 70% of the oxygen required for respiration of crop roots.

The convective flow of air in the soil is similar in some ways to the flow of water, and different in other ways. The similarity is in the fact that the flow of both fluids is impelled by, and is proportional to, a pressure gradient. The dissimilarity results from the relative incompressibility of water in comparison with air, which is highly compressible so that its density and viscosity are strongly dependent on pressure (as well as temperature). Quite another difference is that water has the greater affinity to the surfaces of mineral particles (i.e., it is the *wetting fluid*) and is thus drawn into narrow necks and pores, forming capillary films and wedges (menisci). In a three-phase system,

D. Convective Flow of Soil Air

therefore, air tends to occupy the larger pores. The two fluids—water and air—coexist in the soil by occupying different portions of the pore space having different geometric configurations. For this reason, the soil exhibits toward the two fluids different conductivity or permeability functions, as these relate to the different effective diameters and tortuosities of the pore sets occupied by each fluid. Only when the soil is completely permeated by one or the other, be it water or air, should either flowing fluid encounter the same transmission coefficient of the medium. (See the discussion of *intrinsic permeability* in Chapter 8.)

Notwithstanding the differences between water flow and air flow, it is possible to formulate the convective flow of air in the soil as an equation analogous to Darcy's law for water flow, as follows:

$$\mathbf{q}_v = -(k/\eta)\nabla P \qquad (11.2)$$

where $\mathbf{q}_v$ is the *volume convective flux* of air (volume flowing through a unit cross-sectional area per unit time), k is permeability of the air-filled pore space, η is viscosity of soil air, and ∇P is the three-dimensional gradient of soil air pressure. In one dimension, this equation takes the form

$$q_v = -(k/\eta)(dP/dx) \qquad (11.3)$$

If the flux is expressed in terms of mass (rather than volume) per unit area and per unit time, then the equation is

$$q_m = -(\rho k/\eta)(dP/dx) \qquad (11.4)$$

wherein q_m is the mass convective flux and ρ the density of soil air.

Recalling that the density of a gas depends on its pressure and temperature, we shall now assume that soil air is an *ideal gas*, in which the relation of mass, volume, and temperature is given by the equation

$$PV = nRT \qquad (11.5)$$

where P is pressure, V volume, n number of moles of gas, R the universal gas constant per mole, and T absolute temperature. Since the density $\rho = M/V$, and the mass M is equal to the number of moles n times the molecular weight m, we have

$$\rho = (m/RT)P \qquad (11.6)$$

We now introduce the *continuity equation* (see Chapters 9 and 10) for a compressible fluid:

$$\partial \rho/\partial t = -\partial q_m/\partial x \qquad (11.7)$$

Substituting the expression for ρ from Eq. (11.6) and the expression for q_m from Eq. (11.4) into Eq. (11.7), we obtain

$$\frac{m}{RT}\frac{\partial P}{\partial t} = \frac{\partial}{\partial x}\left(\frac{\rho k}{\eta}\frac{\partial P}{\partial x}\right) \tag{11.8}$$

If $\rho k/\eta$ are more or less constant (i.e., the pressure differences are small), we can write

$$\partial P/\partial t = \alpha\, \partial^2 P/\partial x^2 \tag{11.9}$$

wherein $\alpha = RTk/m$, a composite constant. This is an approximate equation for the transient state convective flow of air in soil. An assumption underlying the use of this equation is that flow is laminar, which it was indeed shown to be for small pressure differences (Muskat, 1946).

Quite a different mechanism of convective movement of gases in the soil is the transfer of dissolved gases by rain or irrigation water infiltrating into and percolating through soils. Some investigators have speculated that dissolved O_2 in percolating water may allow growth, or at least survival, of crop plants in soils covered by flowing water (Russell, 1952) or even permit roots to thrive below the water table (Robinson, 1964). However, oxygen-saturated air at atmospheric pressure contains only 6 ml of O_2 per liter, which is only enough to provide for the respiration of 1 gm (dry weight) of active roots. Except in the case of sandy soils with unusually high infiltration rates, therefore, it seems unlikely that oxygen supply by infiltrating water can have anything but a very temporary or marginal effect. However, a sudden heavy rain or flood irrigation can sometimes trap large volumes of air between the penetrating surface water and the water table and this air can serve as an oxygen reservoir during a flooding period lasting several days (Kemper and Amemiya, 1957).

E. Diffusion of Soil Air

The diffusive transport of gases such as O_2 and CO_2 in the soil occurs partly in the gaseous phase and partly in the liquid phase. Diffusion through the air-filled pores maintains the exchange of gases between the atmosphere and the soil, whereas diffusion through water films of various thickness maintains the supply of oxygen to, and disposal of CO_2 from, live tissues, which are typically hydrated. For both portions of the pathway, the diffusion process can be described by Fick's law (see Chapter 10):

$$q_d = -D\, dc/dx \tag{11.10}$$

wherein q_d is the diffusive flux (mass diffusing across a unit area per unit

E. Diffusion of Soil Air

time), D is the diffusion coefficient (generally having the dimensions of area per time), c is concentration (mass of diffusing substance per volume), x is distance, and dc/dx is the concentration gradient. If partial pressure p is used instead of concentration of the diffusing component, we get

$$q_d = -(D/\beta)(dp/dx) \tag{11.11}$$

where β is the ratio of the partial pressure to the concentration.

Considering first the diffusive path in the air phase, we note that the diffusion coefficient in the soil D_s must be smaller than that in bulk air D_0 owing to the limited fraction of the total volume occupied by continuous air-filled pores and also to the tortuous nature of these pores. Hence we can expect D_s to be some function of the air-filled porosity, f_a. Different workers have over the years found different relations between D_s and f_a for various soils. For instance, Buckingham (1904) reported the following nonlinear relation:

$$D_s/D_0 = \kappa f_a^2 \tag{11.12}$$

On the other hand, Penman (1940) found a linear relation:

$$D_s/D_0 = 0.66 f_a \tag{11.13}$$

wherein 0.66 is a tortuosity coefficient, suggesting that the apparent path is about two-thirds the length of the real average path of diffusion in the soil. Tortuosity very probably depends on the fractional volume of air-filled pores (i.e., it stands to reason that the tortuous path length should increase as the air-filled pore volume decreases); hence we can expect Penman's constant coefficient to hold for only a limited range of variation of air-filled porosity or of volume wetness.

In fact, other investigators have reported different values for different soils and ranges of air and water contents. To mention just a few:

1. Blake and Page (1948) found that the ratio between (D_s/D_0) and air-filled porosity f_a varied between 0.62 and about 0.8, and that in an aggregated soil D_s approached zero when air content fell below about 10% of porosity, a phenomenon which they attributed to the discontinuity of blocked air-filled pores inside the aggregates.

2. Van Bavel (1952) found $D_s/D_0 = 0.61 f_a$.

3. Marshall (1959) took account of the size distribution of soil pores affecting diffusion and proposed the relationship $D_s/D_0 = f_a^{3/2}$, which accords with Penman's coefficient at high air-filled porosity values but falls below it at lower ones.

4. Millington (1959) proposed $D_s/D_0 = (f_a/f)^2 f_a^{4/3}$, where f is total porosity.

5. Wesseling (1962) proposed $D_s/D_0 = 0.9 f_a - 0.1$, which suggests

that D_s becomes zero at an air-filled porosity value of $0.1/0.9 = 11\%$. The confusion was explained by de Vries (1952), who showed on the basis of theory that the relation sought between the effective diffusion coefficient and air-filled porosity should be curvilinear and dependent on pore geometry and hence is not expected to be the same for different soils and water versus air contents. Consequently, Grable and Siemer (1968) found that as air-filled porosity fell to around 10% the ratio D_s/D_0 fell to about 0.02. At even lower f_a values of 4–5%, Lemon and Erickson found the D_s/D_0 ratio to be as low as 0.005.

Having established the variable nature of D_s, we now return to the mathematical formulation of diffusion processes. To account for transient conditions, we must once again introduce the continuity principle:

$$\partial c/\partial t = -\partial q_d/\partial x \tag{11.14}$$

which states that the rate of change of concentration with time must equal the rate of change of diffusive flux with distance. The foregoing assumes that the diffusing substance is conserved throughout. As O_2 and CO_2 diffuse through the soil, however, O_2 is taken up and CO_2 is generated by aerobic biological activity along the diffusive path. To take account of the amount of a diffusing substance added to or subtracted from the system per unit time we add a $\pm S$ term to the right-hand side of Eq. (11.14). Note that a positive sign represents an increment rate (source) and a negative sign represents a decrement rate (sink) for the substance considered. Accordingly,

$$\partial c/\partial t = -(\partial q_d/\partial x) \pm S(x,t) \tag{11.15}$$

The designation $S(x,t)$ implies that the source–sink term is a function of (i.e., varies with) both space and time.

We next substitute Eq. (11.10) into (11.15) and consider only the vertical direction z (depth) to obtain

$$\frac{\partial c}{\partial t} = \frac{\partial}{\partial z}\left(D_s \frac{\partial c}{\partial z}\right) \pm S(z,t) \tag{11.16}$$

Note that we use D_s, for which one may wish to substitute an expression such as the one by Penman ($D_s = 0.66 D_0 f_a$) or any other empirical or theoretically based function.

In the event that D_s is constant, Eq. (11.16) simplifies to

$$\frac{\partial c}{\partial t} = D_s \frac{\partial^2 c}{\partial z^2} \pm S(z,t) \tag{11.17}$$

In aggregated soils, gaseous diffusion can be expected to take place rapidly in the interaggregate macropores, which quickly drain out after a rain or

E. Diffusion of Soil Air

irrigation and form a network of continuous air-filled voids. On the other hand, the intraaggregate micropores can remain nearly saturated for extended periods and thus restrict the internal aeration of aggregates. It is common observation that plant roots are generally confined to the larger pores between aggregates and scarcely penetrate the aggregates themselves, whether because the small internal pores of the aggregates and their mechanical rigidity do not permit penetration or because of aeration restriction per se. However, microorganisms do penetrate aggregates and, by their demand for oxygen, affect soil aeration as a whole. The centers of large, dense crumbs can be anaerobic even while the larger pores all around indicate good aeration. Thus we can have pockets of anaerobiosis in the midst of a seemingly well-aerated soil.

Equations (11.16) and (11.17) consider diffusion in the soil profile as a whole. If we wish to take a closer look at diffusion into or out of individual soil aggregates (clods) we must recast our equations into an omni-directional (three-dimensional) form. Assuming an aggregate to be isotropic and approximately spherical, we can use polar coordinates to obtain:

$$\frac{\partial c}{\partial t} = \frac{1}{r^2}\frac{\partial}{\partial r}\left(D_s r^2 \frac{\partial c}{\partial r}\right) \pm S \qquad (11.18)$$

wherein r is radial distance from the center of the sphere. If we further assume steady state exchange (say, absorption of O_2 and release of CO_2) between the clod and its surroundings, we can simply set $\partial c/\partial t$ equal to zero and solve the equation for appropriate boundary conditions. For example, for a uniform aggregate of radius R, respiring steadily and uniformly throughout, the concentration difference Δc between the surface and center is given by

$$\Delta c = \pm SR^2/6D_s \qquad (11.19)$$

Note that the concentration difference (i.e., expressing the O_2 deficit or CO_2 excess inside the clod) is proportional directly to the respiration rate S and inversely to the diffusion coefficient D_s. As pointed out by Currie (1975) a knowledge of both S and D_s is required to understand aeration.

Thus far we have confined our attention almost entirely to diffusion in the air phase. The final segment of the diffusion path to a root must take place through the hydration film or shell surrounding the root. To describe this stage of the aeration process, we employ the cylindrical form of the diffusion equation:

$$\frac{\partial c}{\partial t} = \frac{1}{r}\frac{\partial}{\partial r}\left(rD_w \frac{\partial c}{\partial r}\right) \qquad (11.20)$$

in which r is radial distance from the root axis. Here we use the diffusion coefficient for the soluble diffusing substance in water D_w. Although this stage of the process takes place through a very short distance (the hydration film being, at most, only a few millimeters thick), it may be the rate limiting stage, as the diffusivity of O_2 in water is only about 1/10,000 its value in air (i.e., about 2.6×10^{-5} in water as against 0.226 cm^2/sec in air). The solubility of oxygen in water (only 4% of the solubility of carbon dioxide) may be limiting also, especially at relatively high temperatures.

At 20°C, the solubility of oxygen in water is about 4.3 gm/m^3. If rain water is saturated with oxygen, then a rainfall of, say, 25 mm could add about 0.1 gm of O_2 (equivalent to about 70 ml of gaseous oxygen at atmospheric pressure) to each square meter of soil. This amount of oxygen is but an insignificant fraction of the amount required daily by an actively growing crop. However, such a rain could conceivably have a temporary invigorating effect on a crop suffering from oxygen stress.

The diffusion of oxygen to roots was analyzed by Lemon (1962). His analysis divided the process between two regions, as illustrated in Fig. 11.3: an inner region representing the root itself, and an outer region representing the film of water surrounding the root. For steady-state diffusion, the two equations pertaining to the internal region (subscripted i) and the external one (subscripted e) are

$$D_i \left(\frac{\partial^2 c_i}{\partial r^2} + \frac{1}{r} \frac{\partial c_i}{\partial r} \right) - S = 0, \qquad D_e \left(\frac{\partial^2 c_e}{\partial r^2} + \frac{1}{r} \frac{\partial c_e}{\partial r} \right) = 0$$

If the diffusion coefficients for the root D_i and for the water film D_e are

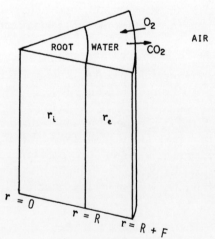

Fig. 11.3. Diffusion of O_2 from soil to root: Representation of root and hydration film as concentric cylinders. Note that r represents radial distance from the root axis, R the radius of the root, F thickness of the water film.

known, and if the oxygen consumption rate per length of root S is also known, it becomes possible to solve the pair of equations simultaneously for different conditions and to define the circumstances likely to limit root respiration.

There is, incidentally, a fundamental difference between the functional dependence of a soil's permeability (or conductivity) upon pore geometry and the corresponding function for the diffusion coefficient. As the permeability pertains to pressure-induced convective-viscous flow, it obeys Poiseuille's law which states that flow varies as the fourth power of the pore radius. Hence permeability is strongly dependent upon *pore-size distribution*. Diffusion, on the other hand, depends primarily on the total volume and tortuosity of continuous pores available for diffusion. The reason that diffusion does not depend on pore-size distribution is that the *mean free path* of molecules in thermal motion (i.e., the distance an "average" molecule in random motion travels before it collides with another) is of the order of 0.0001 to 0.0005 mm and thus much smaller than the radii of the pores which generally account for most of a soil's air-filled porosity.

F. Soil Respiration and Aeration Requirements

The overall rate of respiration due to all biological activity in the soil—i.e., the amount of oxygen consumed and the amount of carbon dioxide produced by the entire profile—determines the *aeration requirement* of the soil. It is important to acquire quantitative knowledge of the aeration requirements of different crops and soils in varying circumstances if we are to devise means to ensure that these requirements are indeed met. However, the information is difficult to obtain, as the rate and spatial distribution of soil respiration, as well as its temporal variation, depend on numerous factors, included among which are temperature, soil wetness, organic matter, and the time-variable respiratory activity of both macro- and microorganisms (Alexander, 1961).

Anaerobiosis, or oxygen stress, will occur in the soil whenever the rate of supply falls below the demand. This condition can develop quite quickly, since the storage of oxygen in the soil is generally rather low in relation to the quantity required for soil respiration. To illustrate, let us consider a soil with an effective root zone depth of 60 cm and 15% air-filled porosity, containing 90 liter of air under each square meter of soil surface. With an initial oxygen concentration of 20% in the gaseous phase, the storage of oxygen can be calculated to be 18 liters, equivalent to about 25 g.[2] If the oxygen require-

---

[2] An ideal gas at standard temperature and pressure occupies a volume of 22.4 liter/mole. The mass of a mole of oxygen (atomic weight = 16) is 32 gm. The mass contained in 18 liter is therefore $(18/22.4) \times 32 = 25.7$ gm.

ment for soil respiration is of the order of 10 gm/day, per square meter of ground, the initial oxygen reserve in the soil would last only $2\frac{1}{2}$ day. Oxygen stress symptoms would probably begin even earlier. However, these figures are given only to provide an order of magnitude, since in actual conditions the aeration rate probably varies between wide limits. Plant growth probably depends more upon the occurrence and duration of periods of oxygen deficiency than upon average conditions (Erickson and van Doren, 1960).

Soil respiration values reported by various investigators have varied widely. Papendick and Runkles (1965), working under laboratory conditions, found respiration rates of 1.7×10^{-11} mole/cm^3 sec, equivalent to 0.75 mole of oxygen per cubic meter per 12-hr day, or 24 gm/m^3 day. At the other extreme, Grable and Siemer (1968) reported values 10,000 times higher. Under field conditions, respiration values are likely to be smaller than those found in growth chambers. Greenwood (1971), working in England, measured average oxygen consumption rates of 1.3×10^{-7} ml oxygen per second per milliliter of a soil carrying a mature crop, and maximum rates three times as high. The average value is equivalent to 8 gm of oxygen per cubic meter of soil per 12-hr day. Dasberg and Bakker (1970) found very similar values in cropped soils in Holland.

Comprehensive data on soil respiration rates have been obtained in

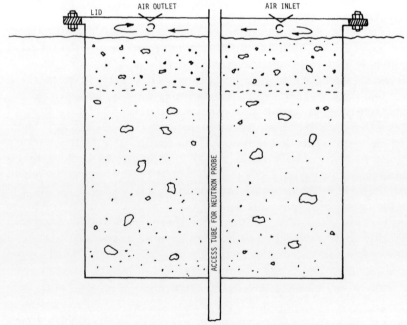

Fig. 11.4. Soil respirometer at Rothamsted. (After Currie, 1975.)

F. Soil Respiration and Aeration Requirements

England from *field respirometers*, first built at Wrest Park and later resited at the Rothamsted Experiment Station (Fig. 11.4). These installations have recently been described by Currie (1975). They consist of large containers (91 × 91 × 91 cm) filled with soil and sealed on top, with provisions to maintain normal atmospheric composition by continuously adding or removing measured amounts of oxygen and carbon dioxide as required. When the soil is cropped, the plants are grown through holes in the lid, using cold-setting silicone rubber to seal the gap around the stem. It is thus possible to measure daily carbon dioxide output and hourly oxygen uptake for soils with and without plants.

The results available to date demonstrate the primary importance of soil temperature. This is illustrated in terms of the seasonal variation of soil temperature and soil respiration in Fig. 11.5. Respiration rates in summer can be more than ten times as great as in winter. However, other factors can also have a strong influence. For example, respiration rates at a given temperature

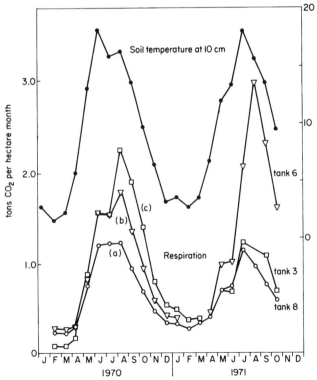

Fig. 11.5. Seasonal variation of soil respiration and soil temperature at Rothamsted. (After Currie, 1975.)

tend to be greater during spring than during autumn. The cause seems to be the more vigorous population of microorganisms and the greater availability of undecomposed organic residues in spring than in autumn. Soil respiration is obviously much greater in cropped than in fallow land, owing both to root respiration and to the enhanced microbial activity resulting from root exudates and root decay. The diurnal variation of soil respiration also follows the pattern of soil temperature, with oxygen uptake rates registering a more than twofold increase from early morning to mid afternoon. Thus, the respiration rates in soil vary from season to season, from day to day, and from hour to hour, and are related to crop growth stage and to microbial activity. The effect of waterlogging the soil is slight in winter, for although air diffusion is constricted, the oxygen requirements are slight. The same degree of waterlogging in summer, however, could be severely damaging to a crop at its most active growth stage.

G. Measurement of Soil Aeration

The complex group of processes which we lump together in the term soil aeration is altogether too elusive and incompletely understood to be definable unequivocally by any single measurement. We are reduced, therefore, to measuring soil attributes and component processes which we consider to be pertinent to the problem at hand and to be *indicative* of soil aeration as a whole.

An early approach to the problem of measuring aeration was to determine the fractional air space, or *air-filled porosity*, at some standardized value of soil wetness. This could be done by taking an "undisturbed" sample from a soil presumed to be at its field capacity (e.g., two days after a deep wetting) or by saturating a soil sample in the laboratory and then subjecting it to some arbitrarily specified water tension. For example, the term *aeration porosity* has been defined (Kohnke, 1958) as the pore space filled with air when the soil sample is placed on a porous plate and equilibrated with a 50 cm hanging column of water. At a tension head of 50 cm, all pores with an effective diameter wider than 0.06 mm are drained of water.[3] The air space as a fraction of total porosity can then be measured directly with the aid of an *air pycnometer* (Page, 1948; Vomocil, 1965) or by taking the difference between total porosity (obtained from the measurement of bulk density) and volumetric wetness. By either method, the determination of fractional air space is fraught with uncertainties and gives no indication of aeration dynamics.

[3] Students may wish to verify this statement independently. Hint: use the capillary tension equation $h = 2\gamma/\rho g R$ (Chapter 7).

G. Measurement of Soil Aeration

Another traditional approach is to measure the composition of soil air. Although still a static measurement, this appears to be a better diagnostic tool than measurement of air volume alone, for it can reveal more directly when a problem might exist—i.e., when the oxygen content of soil air falls significantly below that of the atmosphere owing to restricted gas exchange. The difficulty here is how to extract an air sample at once large enough to provide a reliable measurement, and yet small enough to represent the sampled point and to avoid disturbance and mixing of soil air, or even contamination from the atmosphere. The gas chromatograph technique, using a syringe to extract small samples (only 0.5 ml in volume) helps to make the measurement more reliable (Yamaguchi et al., 1962). An alternative method permitting repeated monitoring of O_2 concentrations in soil air without extraction of samples is based on the use of membrane-covered electrodes such as described by Willey and Tanner (1963) and by McIntyre and Philip (1964). Electrodes have also been developed for measuring CO_2 concentrations (Letey and Stolzy, 1964). Such electrodes must be kept dry to measure oxygen concentration. Hence they are usually sealed in a hollow tube which is inserted into the soil. The volume of air surrounding an electrode must be equilibrated with soil air prior to the measurement. If the soil is very wet, this equilibration may require many hours or even several days.

A quite different approach to characterizing soil aeration is to measure the *air permeability*, i.e., the coefficient governing convective transmission of air through the soil in response to a total pressure gradient. This measurement can provide useful information on the effective sizes and the continuity of air-filled pores. The techniques which have been proposed include constant-pressure and falling-pressure devices (Fig. 11.6). The method has been applied in the field (Grover, 1956) and found useful for assessing the "openness" of the surface layer to the entry of air, as affected by such cultural practices as tractor traffic and tillage (Fig. 11.7).

Numerous techniques have been proposed for measuring diffusion processes in the soil, both in the gas phase and in the liquid phase. One of the earliest (Raney, 1950) was to fill a cavity in the soil with nitrogen gas, then sample the cavity periodically to determine the rate of diffusion of gases, particularly oxygen, from the surrounding soil. An alternative technique is to introduce some volatile substance into the cavity and measure the rate of its dissipation by diffusion into the surrounding soil's air phase. However, the measurement of diffusion rates through the air-filled pores alone fails to provide any indication of the possible impedance of the liquid envelope surrounding a root. As pointed out by Kristensen and Lemon (1964), it is possible to have high O_2 concentration in the network of large pores open to the atmosphere and yet have an inadequate supply to the root if the root is thickly hydrated.

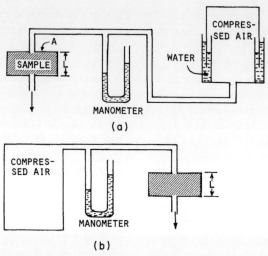

Fig. 11.6. Measurement of air permeability with (a) constant-pressure (variable volume) permeameter ($k = (L\eta/Af)(\Delta V/t)$), and (b) a falling-pressure permeameter (constant volume; $k = (2.3L\eta V/AP_a)[\log(p_1/p_2)/\Delta t]$). *Note:* k is air permeability (cm^2), L sample length (cm), A sample cross-sectional area (cm^2), V volume of the air cell (cm^3), P_a barometric pressure, p cell air pressure; t time, and p_1 and p_2 cell pressure at start and end of time step Δt.

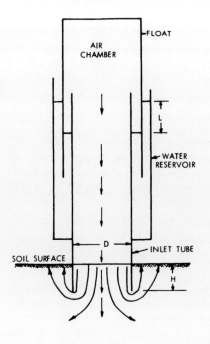

Fig. 11.7. Air permeameter for field use. (After Grover, 1956.)

An interesting method for measuring oxygen diffusion to a rootlike probe was introduced by Lemon and Erickson (1952). The process measured is the chemical reduction of elemental oxygen by a thin platinum electrode maintained at a constant potential.[4] The resulting current measures the flux of oxygen to the moisture-covered electrode which acts as a sink, thus simulating the action of a respiring root. In practice, one inserts the probe into a moist soil and waits until the reading becomes steady, at which time the flux of oxygen to the probe is taken to represent the oxygen-supplying power of the soil, and is commonly called ODR (for O_2 diffusion rate). The technique fails in relatively dry soils which, however, are unlikely to present aeration problems. Descriptions of the technique and the results obtainable by it have been published by Birkle *et al.* (1964), Letey and Stolzy (1964), and McIntyre (1970). Numerous investigations have shown correlations between ODR and plant response in soils of restricted aeration. According to Stolzy and Letey (1964) the roots of many plant species will not grow in an environment with an ODR of less than 0.2 mgm/cm^2 min. Note that the method is not based on solution of the diffusion equation for definable boundary conditions, and the results do not provide a value for the effective diffusion coefficient for oxygen in either soil air or soil water. Rather, what is measured is a flux, which, of course, depends on the size and shape of the electrode and the location of its insertion, as well as on the diffusion coefficients of the surrounding porous medium, the temperature, etc.

Additional methods for measuring soil aeration include determination of redox (reduction–oxidation) potential (Grable and Siemer, 1968; Dasberg and Bakker, 1970) and the already mentioned use of soil respirometers.

Sample Problems

1. Consider a cropped field with an effective root zone depth of 80 cm, a daily transpiration rate of 6 mm, and a daily soil respiration rate of 10 gm O_2/m^2. Calculate what fraction of the oxygen requirement is supplied by convection if air is drawn from the atmosphere in immediate response to the pressure deficit created in the soil by the extraction of soil moisture.

Calculation: Volume of water extracted per square meter: 1 m^2 × 0.006 m = 0.006 m^3 = 6 liter per day.

---

[4] Birkle *et al.* (1964) recommended use of a potential of 0.65 V. The reaction which takes place at the negatively charged electrode is that dissolved oxygen molecules reaching the electrode take up four electrons and react with hydrogen ions to form water in an acid solution ($O_2 + 4H^+ + 4e^- \rightarrow H_2O$, or react with water to form hydroxyl in an alkaline solution ($O_2 + 2H_2O + 4e^- \rightarrow 4OH^-$).

Volume of air drawn from the atmosphere = volume of water withdrawn from soil = 6 liter.
Volume of oxygen drawn from the atmosphere = 6 × 21% = 1.26 liter.
Mass of oxygen drawn from the atmosphere = (1.26/22.4) × 32 = 1.8 gm.
Percentage of daily oxygen requirement supplied by convection = (1.8/10) × 100 = 18%.

Note: 32 is the molecular weight of oxygen (O_2), and 22.4 liter is the volume of 1 mole of gas at standard temperature and pressure.

2. Consider a soil profile in which the air-phase oxygen concentration diminishes linearly from 21% at the soil surface to half of that at 100 cm depth. If the total porosity is a uniform 45% and the volume wetness 35%, calculate the diffusion rate using Penman's coefficient for the effective diffusion coefficient of oxygen in the soil (D_s). Assume steady-state diffusion. Use a value of 1.89×10^{-1} cm²/sec for the bulk-air diffusion coefficient.

Our first step is to estimate the effective diffusion coefficient D_s using Penman's linear relation between D_s and the air-filled porosity f:

$$D_s = 0.66 f_a D_0$$

where f_a is air-filled porosity, D_0 is the diffusion coefficient in bulk air, and 0.66 is the tortuosity factor (assumed by Penman to be constant). Substituting the given values, we have

$$D_s = 0.66 \times (0.45 - 0.35) \times 0.189 \text{ cm}^2/\text{sec} = 0.0126 \text{ cm}^2/\text{sec}$$

We now use Fick's first law to calculate the steady state one-dimensional diffusive flux q_d of oxygen through the soil profile from the external atmosphere to a plane at which the oxygen concentration is 50% of the atmosphere's at a depth of 100 cm:

$$q_d = D_s \Delta c / \Delta x$$

Recall that the concentration of oxygen in the external atmosphere is about 0.32×32 gm per 22.4 liter, i.e., 3×10^{-4} gm/cm³. Therefore

$$q_d = 1.26 \times 10^{-2} \times (3 \times 10^{-4} - 1.5 \times 10^{-4})/100$$
$$= 1.89 \times 10^{-8} \text{ gm/cm}^2 \text{ sec}$$

This quantity can be multiplied by 10^4 cm²/m² and by 8.64×10^4 sec/day to obtain the flux in grams per square meter of soil surface per day. Thus,

$$(q_d)_{\text{daily}} = 1.89 \times 10^{-8} \times 10^4 \times 8.64 \times 10^4 = 15.83 \text{ gm/m}^2 \text{ day}$$

Part V:

COMPOSITE PROPERTIES AND BEHAVIOR

> I hear, and I forget;
> I see, and I remember;
> I do, and I understand.
> Kung Fu-tse (Confucius)
> 551–479 B.C.E.

12 Soil Temperature and Heat Flow

A. Introduction

Soil temperature, its value at any moment and the manner with which it varies in time and space, is a factor of primary importance in determining the rates and directions of soil physical processes and of energy and mass exchange with the atmosphere—including evaporation and aeration. Temperature also governs the types and rates of chemical reactions which take place in the soil. Finally, soil temperature strongly influences biological processes, such as seed germination, seedling emergence and growth, root development, and microbial activity.

Soil temperature varies in response to changes in the radiant, thermal, and latent energy exchange processes which take place primarily through the soil surface. The effects of these phenomena are propagated into the soil profile by a complex series of transport processes, the rates of which are affected by time-variable and space-variable soil properties. Hence the quantitative formulation and prediction of the soil thermal regime can be a formidable task. Even beyond passive prediction, the possibility of actively controlling or modifying the thermal regime requires a thorough knowledge of the processes at play and of the environmental and soil parameters which govern their rates. The pertinent soil parameters include the specific heat capacity, thermal conductivity, and thermal diffusivity (all of which are strongly affected by bulk density and wetness), as well as the internal sources and sinks of heat operating at any time.

Present day theory can provide at least a semiquantitative interpretation of observed influences of soil surface conditions, including the presence of mulching materials and various tillage treatments, on the soil's thermal regime. Moreover, available theory can help to explain why the annual temperature variation penetrates into the soil much more deeply than the diurnal variation; it can account for the obvious difference in temperature distribution among soils of differing constitution, such as sand, clay, or peat. It can also explain why the surface of a dry soil exhibits high maxima and low minima temperatures, and suggest how these extremes may be moderated when the soil moisture content is changed. Theories are now being developed to deal with freezing and thawing phenomena in cold-region soils. Finally, the important interaction of heat flow and water flow is being examined in an effort to understand how the transports of matter and energy occur simultaneously and interdependently in the soil.

Reviews of soil temperature and heat flow have been published over the years by Kersten (1949), Hagan (1952), van Rooyen and Winterkorn (1959), Smith et al. (1964), Taylor and Jackson (1965), van Wijk (1963), Chudnovskii (1966), van Bavel (1972), and de Vries (1975).

B. Modes of Energy Transfer

We begin with some basic physics. In general, there are three principal modes of energy transfer: radiation, convection, and conduction.

By *radiation*, we refer to the emission of energy in the form of electromagnetic waves from all bodies above 0°K. According to the *Stephan–Boltzmann* law, the total energy emitted by a body J_t integrated over all wavelengths, is proportional to the fourth power of the absolute temperature T of the body's surface. This law is usually formulated

$$J_t = \varepsilon \sigma T^4 \tag{12.1}$$

where σ is a constant and ε is the *emissivity coefficient* which equals unity for a perfect emitter (generally called a *black body*). The absolute temperature also determines the wavelength distribution of the emitted energy. *Wien's law* states that the wavelength of maximal radiation intensity λ_m is inversely proportional to the absolute temperature:

$$\lambda_m = 2900/T \tag{12.2}$$

where λ_m is in microns. The actual intensity distribution as a function of wavelength and temperature is given by *Planck's law*:

$$E_\lambda = C_1/\lambda^5 [\exp(C_2/\lambda T) - 1] \tag{12.3}$$

where E_λ is the energy flux emitted in a particular wavelength range, and

C_1, C_2 are constants. Since the temperature of the soil surface is generally of the order of 300°K (though it can range, of course, from below 273°K, the freezing point, to 330°K or even higher), the radiation emitted by the soil surface has its peak intensity at a wavelength of about 10 μm and its wavelength distribution over the range of 3–50 μm. This is in the realm of *infrared*, or *heat, radiation*. A very different spectrum is emitted by the sun, which acts as a black body at an effective surface temperature of about 6000°K. The sun's radiation includes the visible light range of 0.3–0.7 μm, as well as some infrared radiation of greater wavelength (up to about 3 μm) and some ultraviolet radiation (λ <0.3 μm). Since there is very little overlap between the two spectra, it is customary to distinguish between them by calling the incoming solar spectrum *short-wave* radiation, and the spectrum emitted by the earth *long-wave* radiation.

The second mode of energy transfer, called *convection*, involves the movement of a heat-carrying mass, as in the case of ocean currents or atmospheric winds. An example more pertinent to soil physics would be the infiltration of hot waste water (from, say, a power plant) into an initially cold soil.

Conduction, the third mode of energy transfer, is the propagation of heat within a body by internal molecular motion. Since temperature is an expression of the kinetic energy of a body's molecules, the existence of a temperature difference within a body will normally cause the transfer of kinetic energy by the numerous collisions of rapidly moving molecules from the warmer region of the body with their neighbors in the colder region. The process of heat conduction is thus analogous to the process of diffusion, and in the same way that diffusion tends in time to equilibrate a mixture's composition throughout, heat conduction tends to equilibrate a body's internal distribution of molecular kinetic energy—that is to say, its temperature.

In addition to the three modes of energy transfer described, there is a composite phenomenon which one may recognize as a fourth mode, namely the *latent heat transfer*. A prime example is the process of *distillation*, which includes the heat-absorbing stage of evaporation, followed by the convective or diffusive movement of the vapor, and ending with the heat-releasing stage of condensation. A similar catenary process can also occur in transition back and forth from ice to liquid water.

The transfer of heat through the soil surface may occur by any or all of the above mechanisms. Within the soil, however, heat transfers by radiation, convection, and distillation are generally of secondary importance, and the primary process of heat transport is by molecular conduction.

We end this section with a word to the wise. The terms transport, movement, or even flow of heat are metaphorical, as they suggest an analogy with material flow. In fact, heat is not a material substance at all, but a form of energy. However, the poverty of our language is not, in this case at least, a serious hindrance. Though not to be taken literally, the analogy between heat and fluid transfers is useful, as it enables us to deal with heat in familiar terms and it facilitates our speculations on the forces, rates, and directions of heat transfer. The visual image of something flowing leads us to think of gradients and of a natural tendency to flow from a higher to a lower level. This has turned out to be a helpful approach to heat transfer in soils. Beyond the conceptual analogy, there is indeed a strong mathematical resemblance between the equations describing heat flow (Carslaw and Jaeger, 1959) and those describing water flow as well as diffusion processes in porous media (e.g., Bear, 1969).

C. Energy Balance for a Bare Soil

Hillel (1980) devotes a chapter to the energy and water balance of fields, with a detailed discussion of the concepts involved. Here we wish only to outline the overall energy regime of a bare (unvegetated) soil as a preliminary stage in the elucidation of soil heat conduction.

We begin with the radiation balance of a bare soil surface, which can be written (van Bavel and Hillel, 1976)

$$J_n = (J_s + J_a)(1 - \alpha) + J_{li} - J_{le} \qquad (12.4)$$

Here J_n is the *net radiation*, that is, the sum of all incoming minus outgoing radiant energy; J_s the incoming flux of short-wave radiation from the sun and J_a the short-wave radiation from the atmosphere (sky); J_{li} the incoming long-wave radiation flux from the sky and J_{le} the long-wave radiation emitted by the soil; and finally, α is the *albedo*, or *reflectivity coefficient*, which is the fraction of incoming short-wave radiation reflected by the soil surface rather than absorbed by it.

We shall disregard in the present context all terms which do not pertain to the soil, namely J_s, J_a, and J_{li}. The albedo α is an important characteristic of soil surfaces, and it can vary widely in the range of 0.1–0.4, depending upon the soil's basic color (whether dark or light colored), the surface's roughness and inclination (Sellers, 1965), and, in the short run, upon the changing wetness of the exposed soil (Jackson *et al.*, 1974). The drier the soil, the smoother its surface, and the brighter its color—the higher its albedo. To a certain extent, the albedo can be modified by various surface treatments such as tillage and mulching. Apart from the reflected short-wave radiation, governed by the albedo, we have another soil-dependent process, namely the emitted long-wave radiation J_{le}. In accordance with Eq. (12.1), J_{le} depends primarily on soil surface temperature but is also affected by the soil's emissivity ε. This parameter also depends on soil wetness, but its range of variation is generally small, i.e., between 0.9 and 1.0.

The net radiation received by the soil surface is transformed into heat which warms the soil and air and vaporizes water. We can thus write the surface energy balance as follows:

$$J_n = S + A + LE \tag{12.5}$$

where S is the soil heat flux (the rate at which heat is transferred from the surface downward into the soil profile), A is the "sensible" heat flux transmitted from the surface to the air above, and LE is the evaporative heat flux, a product of the evaporative rate E and the *latent heat* per unit quantity of water evaporated L.

The total surface energy balance [Eq. (12.4) and (12.5)] is therefore

$$(J_s + J_a)(1 - a) + J_{li} - J_{le} - S - A - LE = 0 \tag{12.6}$$

Conventionally, all components of the energy balance are taken as positive if directed toward the surface and negative otherwise.

D. Conduction of Heat in Soil

The conduction of heat in solids was analyzed as long ago as 1822 by Fourier, whose name is associated with the linear transport equations which have been used ever since to describe heat conduction. These equations are mathematically analogous to the diffusion equations (Fick's laws) as well as to Darcy's law for the conduction of fluids in porous media. An analogy can also be drawn between these laws and Ohm's law for the conduction of electricity (see Chapter 8). A definitive text on the mathematics of heat conduction was published by Carslaw and Jaeger (1959).

The first law of heat conduction, known as *Fourier's law*, states that the flux of heat in a homogeneous body is in the direction of and proportional to the temperature gradient:

$$\mathbf{q}_h = -\kappa \nabla T \tag{12.7}$$

where $\mathbf{q}_h$ is the *thermal flux* (i.e., the amount of heat conducted across a unit cross-sectional area in unit time), κ is *thermal conductivity*, and ∇T the spatial gradient of temperature T. In one-dimensional form, this law is written

$$q_h = -\kappa_x \, dT/dx \quad \text{or} \quad q_h = -\kappa_z \, dT/dz \tag{12.8}$$

Here dT/dx is the temperature gradient in any arbitrary direction designated

x, and dT/dz is, specifically, the vertical direction representing soil depth ($z = 0$ being the soil surface). The subscripts attached to the thermal conductivity term are meant to account for the possibility that this parameter may have different values in different directions. The negative sign in these equations is due to the fact that heat flows from a higher to a lower temperature (i.e., in the direction of a negative temperature gradient).

If q_h is expressed in calories per square centimeter per second and the temperature gradient in degrees Kelvin per centimeter, κ has the units of calories per centimeter-degree-second. If, on the other hand, the thermal flux is given in watts per meter and the gradient in degrees per meter, the thermal conductivity assumes the units of watts per meter-degree.

Equation (12.7) is sufficient to describe heat conduction under steady-state conditions, that is to say where the temperature at each point in the conducting medium and the flux remain constant in time. To account for nonsteady or transient conditions, we need a second law analogous to Fick's second law of diffusion as embodied in Eq. (10.13). To obtain the second law of heat conduction, we invoke the principle of *energy conservation* in the form of the *continuity equation*, which states that, in the absence of any sources or sinks of heat, the time rate of change in heat content of a volume element of the conducting medium (in our case, soil) must equal the change of flux with distance:

$$\rho c_m \, \partial T/\partial t = -\nabla \cdot \mathbf{q}_h \tag{12.9}$$

where ρ is mass density and c_m *specific heat capacity* per unit mass, (called simply specific heat and defined as the change in heat content of a unit mass of the body per unit change in temperature). The product ρc_m (often designated C) is the specific heat capacity per unit volume, and $\partial T/\partial t$ is the time rate of temperature change. Note that the symbol ρ represents the total mass per unit volume, including the mass of water in the case of a moist soil. The symbol ∇ (del) is the shorthand representation of the three-dimensional gradient. An equivalent form of Eq. (12.9) is

$$\rho c_m \, \partial T/\partial t = -(\partial q_x/\partial x + \partial q_y/\partial y + \partial q_z/\partial z)$$

where x, y, z are the orthogonal direction coordinates.

Combining Eqs. (12.9) and (12.7), we obtain the desired *second law of heat conduction*:

$$\rho c_m \, \partial T/\partial t = \nabla \cdot (\kappa \nabla T) \tag{12.10}$$

which, in one-dimensional form, is

$$\rho c_m \frac{\partial T}{\partial t} = \frac{\partial}{\partial x}\left(\kappa \frac{\partial T}{\partial x}\right) \tag{12.11}$$

E. Volumetric Heat Capacity of Soils

Sometimes we may need to account for the possible occurrence of heat sources or sinks in the realm where heat flow takes place. Heat sources include such phenomena as organic matter decomposition, wetting of initially dry soil material, and condensation of water vapor. Heat sinks are generally associated with evaporation. Lumping all these sources and sinks into a single term S, we can rewrite the last equation in the form

$$\rho c_m \frac{\partial T}{\partial t} = \frac{\partial}{\partial x}\left(\kappa \frac{\partial T}{\partial x}\right) \pm S(x,t) \tag{12.12}$$

in which the source–sink term is shown as a function of both space and time.

The ratio of the thermal conductivity κ to the volumetric heat capacity $C(=\rho c_m)$ is called the *thermal diffusivity*, designated D_T. Thus,

$$D_T = \kappa/C \tag{12.13}$$

Substituting D_T for κ, we can rewrite Eq. (12.8) and (12.11):

$$q_h = -D_T C\, dT/dx \tag{12.14}$$

and

$$\frac{\partial T}{\partial t} = \frac{\partial}{\partial x}\left(D_T \frac{\partial T}{\partial x}\right) \tag{12.15}$$

In the special case where D_T can be considered constant, i.e., not a function of distance x, we can write

$$\partial T/\partial t = D_T\, \partial^2 T/\partial x^2 \tag{12.16}$$

To solve the foregoing equations so as to obtain a description of how temperature varies in both space and time, we need to know, by means of measurement or calculation, the pertinent values of the three parameters just defined, namely, the volumetric heat capacity C, thermal conductivity κ, and thermal diffusivity D_T. Together, they are called the *thermal properties* of soils.

E. Volumetric Heat Capacity of Soils

The *volumetric heat capacity* C of a soil is defined as the change in heat content of a unit bulk volume of soil per unit change in temperature. Its units are calories per cubic centimeter per degree (Kelvin), or joules per cubic meter per degree. As such, C depends on the composition of the soil's solid phase (the mineral and organic constituents present), bulk density, and the soil's wetness (see Table 12.1).

The value of C can be calculated by addition of the heat capacities of the various constituents, weighted according to their volume fractions. As given by de Vries (1975),

$$C = \sum f_{si}C_{si} + f_w C_w + f_a C_a \qquad (12.17)$$

Here, f denotes the volume fraction of each phase: solid (subscripted s), water (w), and air (a). The solid phase includes a number of components, subscripted i, such as various minerals and organic matter, and the symbol $\sum$ indicates the summation of the products of their respective volume fractions and heat capacities. The C value for water, air, and each component of the solid phase is the product of the particular density and specific heat per unit mass (i.e., $C_w = \rho_w c_{mw}$, $C_a = \rho_a c_{ma}$, $C_{si} = \rho_{si} c_{mi}$).

Most of the minerals composing soils have nearly the same values of density (about 2.65 gm/cm³ or 2.65×10^3 kg/m³) and of heat capacity (0.48 cal/cm³ °K or 2.0×10^6 J/m³ °K). Since it is difficult to separate the different kinds of organic matter present in soils, it is tempting to lump them all into a single constituent (with an average density of about 0.5 gm/cm³ or 1.3×10^3 kg/m³, and an average heat capacity of about 0.6 cal/cm³ °K or 2.5×10^6 J/m³ °K). The density of water is less than half that of mineral matter (1 gm/cm³ or 1.0×10^3 kg/m³) but the specific heat of water is more than twice as large (1 cal/cm³ °K or 4.2×10^6 J/m³ °K). Finally, since the density of air is only about 1/1000 that of water, its contribution to the specific heat of the composite soil can generally be neglected.

Thus, Eq. (12.17) can be simplified as follows:

$$C = f_m C_m + f_o C_o + f_w C_w \qquad (12.18)$$

Table 12.1

DENSITIES AND VOLUMETRIC HEAT CAPACITIES OF SOIL CONSTITUENTS (AT 10°C) AND OF ICE (AT 0°C)

Constituent	Density ρ		Heat capacity C	
	(gm/cm³)	(kg/m³)	(cal/cm³ °K)	(W/m³ °K)
Quartz	2.66	2.66×10^3	0.48	2.0×10^6
Other minerals (average)	2.65	2.65×10^3	0.48	2.0×10^6
Organic matter	1.3	1.3×10^3	0.6	2.5×10^6
Water (liquid)	1.0	1.0×10^3	1.0	4.2×10^6
Ice	0.92	0.92×10^3	0.45	1.9×10^6
Air	0.00125	1.25	0.003	1.25×10^3

F. Thermal Conductivity of Soils

where subscripts m, o, w refer to mineral matter, organic matter, and water, respectively. Note that $f_m + f_o + f_w = 1 - f_a$, and the total porosity $f = f_a + f_w$. The reader will recall that in our preceding chapters we designated the volume fraction of water f_w as θ. Knowing the approximate average values of C_m, C_o, and C_w (0.46, 0.60, and 1.0 cal/gm, respectively), we can further simplify Eq. (12.18) to give

$$C = 0.48 f_m + 0.60 f_o + f_w \tag{12.19}$$

The use of Eq. (12.18) must be qualified in the case of frozen or partially frozen soils, since the properties of ice differ somewhat from those of liquid water ($\rho = 0.92$ gm/cm^3 or 0.92×10^3 kg/m^3, and $C = 0.45$ cal/cm^3 °K or 1.9×10^6 J/m^3 °K).

In typical mineral soils, the volume fraction of solids is in the range of 0.45–0.65, and C values range from less than 0.25 cal/cm^3 °K (about 1 MJ/m^3 °K) in the dry state to about 0.75 cal/cm^3 °K ($\approx$ 3MJ/m^3 °K) in the water-saturated state.

Apart from the method described for calculating a soil's volumetric heat capacity, it is, of course, also possible to measure it, using calorimetric techniques (Taylor and Jackson, 1965).

F. Thermal Conductivity of Soils

Thermal conductivity, designated κ, is defined as the amount of heat transferred through a unit area in unit time under a unit temperature gradient. As shown in Table 12.2, the thermal conductivities of specific soil constituents differ very markedly (see also Table 12.3). Hence the space-average (macroscopic) thermal conductivity of a soil depends upon its mineral composition and organic matter content, as well as on the volume fractions of water and air. Since the thermal conductivity of air is very much

Table 12.2

THERMAL CONDUCTIVITIES OF SOIL CONSTITUENTS (AT 10°C) AND OF ICE (AT 0°C)

Constituent	mcal/cm sec °K	W/m °K
Quartz	21	8.8
Other minerals (average)	7	2.9
Organic matter	0.6	0.25
Water (liquid)	1.37	0.57
Ice	5.2	2.2
Air	0.06	0.025

Table 12.3
AVERAGE THERMAL PROPERTIES OF SOILS AND SNOW[a]

Soil type	Porosity f	Volumetric wetness θ	Thermal conductivity (10^{-3} cal/cm sec °C)	Volumetric heat capacity C_v (cal/cm^3 °C)	Damping depth (diurnal) d (cm)
Sand	0.4	0.0	0.7	0.3	8.0
	0.4	0.2	4.2	0.5	15.2
	0.4	0.4	5.2	0.7	14.3
Clay	0.4	0.0	0.6	0.3	7.4
	0.4	0.2	2.8	0.5	12.4
	0.4	0.4	3.8	0.7	12.2
Peat	0.8	0.0	0.14	0.35	3.3
	0.8	0.4	0.7	0.75	5.1
	0.8	0.8	1.2	1.15	5.4
Snow	0.95	0.05	0.15	0.05	9.1
	0.8	0.2	0.32	0.2	6.6
	0.5	0.5	1.7	0.5	9.7

[a] After van Wijk and de Vries (1963).

smaller than that of water or solid matter, a high air content (or low water content) corresponds to a low thermal conductivity. Moreover, since the proportions of water and air vary continuously, κ is also time variable. Soil composition is seldom uniform in depth, hence κ is generally a function of depth as well as of time. It also varies with temperature, but under normal conditions this variation is ignored. The factors which affect thermal conductivity κ are the same as those which affect the volumetric heat capacity C, but the measure of their effect is different so that the variation in κ is much greater than of C. In the normal range of soil wetness experienced in the field, C may undergo a threefold or fourfold change, whereas the corresponding change in κ may be hundredfold or more. One complicating factor is that, unlike heat capacity, thermal conductivity is sensitive not merely to the volume composition of a soil but also to the sizes, shapes, and spatial arrangements of the soil particles.

The problem of expressing the overall thermal conductivity of a soil as a function of the specific conductivities and volume fractions of the soil's constituents is very intricate, as it involves the internal geometry of soil structure and the transmission of heat from particle to particle and from phase to phase.

Two relatively simple alternative cases can be envisaged: a dry soil or a

F. Thermal Conductivity of Soils

water-saturated soil with the same particle configuration. In either case, we have a two-phase system in which the particles are dispersed in a continuous medium of fluid (air or water) with a volume fraction f_o and thermal conductivity κ_0. The particles then occupy a volume fraction $f_1 = 1 - f_0$ and have a thermal conductivity κ_1. A composite thermal conductivity for the medium as a whole can be defined as follows: Consider a representative cube of soil with side l, large in comparison with the diameters of the particles and pores. Assume that the upper face is at a temperature T_1 and the bottom face at a lower temperature T_2. A constant heat flux q_h will then pass through the cube, proportional to the overall temperature gradient, with κ_c as the factor of proportionality for the composite medium:

$$q_h = -\kappa_c \, dT/dx = \kappa_c(T_1 - T_2)/l$$

Since the cube is a mixture of two phases, the composite thermal conductivity κ_c will be intermediate between κ_0 and κ_1. According to Burger (1915),

$$\kappa_c = (f_0 \kappa_0 + k f_i \kappa_i)/(f_0 + k f_1) \tag{12.20}$$

wherein the factor k is the ratio of the average temperature gradient in the particles to the corresponding gradient in the continuous fluid:

$$k = \frac{(dT/dz)_2}{(dT/dz)_1}$$

According to de Vries (1975) the value of k depends not only on the ratio κ_1/κ_0, but also on the particle sizes, shapes, and mode of packing. These variables are difficult to characterize quantitatively in the case of particles of irregular shapes, distribution of sizes, and packing arrangements.

If there are several types of particles with different shapes or conductivities, Eq. (12.20) can be generalized:

$$\kappa_c = \sum_{i=1}^{n} k_i f_i \kappa_i \Big/ \sum_{i=1}^{n} k_i f_i \tag{12.21}$$

Here n is the number of particle classes within which all particles have about the same shape and conductivity.

As shown by de Vries (1975), the thermal conductivity of soils of widely differing compositions can be estimated by Eq. (12.21) with a fair degree of accuracy. The deviations between measured and estimated values of thermal conductivity were reported to be less than 10% for κ_1/κ_0 ratios smaller than 10 (i.e., for saturated soils) and about 25% for κ_1/κ_0 ratios of the order of 100 (i.e., dry soils). In moist but unsaturated soils, water can be considered as a continuous medium in which soil particles and air are dispersed. Since the ratio κ_1/κ_0 is greater than unity for mineral particles and less than unity for organic matter and air, the errors caused by these different constituents may compensate each other, at least partially. Thus, Eq. (12.21) may still provide a fair estimate of the composite thermal conductivity even of a three-phase soil. The de Vries model can also be applied in two steps to calculate the thermal conductivity of a soil with aggregated particles. In the first step the conductivity of the aggregates is calculated, in the second step that of the soil as a whole. There are insufficient data to verify this model, however.

The following form of Eq. (12.21) for an unsaturated soil was used by van Bavel and Hillel (1975, 1976):

$$\kappa_c = \frac{f_w \kappa_w + k_s f_s \kappa_s + k_a f_a \kappa_a}{f_w + k_s f_s + k_a f_a}$$

wherein κ_w, κ_a, and κ_s are the specific thermal conductivities of each of the soil constituents (water, air, and an average value for the solids, respectively). The factor k_s represents the ratio between the space average of the temperature gradient in the solid relative to the water phase. The factor k_s depends on the array of grain shapes as well as on mineral composition and organic matter content. The k_a factor represents the corresponding ratio for the thermal gradient in the air and water phases.

The dependence of thermal conductivity and diffusivity on soil wetness is illustrated in Fig. 12.1. The influence of latent heat transfer by water vapor in the air-filled pores is proportional to the temperature gradient in these pores. It can be taken into account (van Bavel and Hillel, 1976; Hillel, 1977) by adding to the thermal conductivity of air an apparent conductivity due to evaporation, transport, and condensation of water vapor (the so-called vapor enhancement factor). This value is strongly temperature dependent, and rises rapidly with increasing temperature.

The flux of sensible heat associated with liquid water movement in the soil is generally considered negligible.

Given the complexities involved in any attempt to predict a soil's thermal conductivity by calculation based on theory, one might be justified in asking, "Why bother? Why not simply take measurements?" Indeed, one should never depend on theory alone. Measurements are necessary, if only to validate (or invalidate) theory. However, the task of measuring thermal conductivity presents its own difficulties and complexities. Because of the fact that soil water potential depends on temperature, the development of a temperature gradient generally induces the movement of water as well as of heat. Hence techniques for measuring heat transfer through a soil sample based on steady-state heat flow between two planes maintained at a constant temperature differential involve the risk of changing the sample's internal moisture distribution, and therefore its thermal properties, during the process

F. Thermal Conductivity of Soils

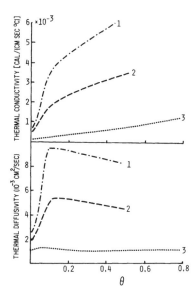

Fig. 12.1. Thermal conductivity and thermal diffusivity as functions of volume wetness (volume fraction of water) for (1) *sand* (bulk density 1.46 gm/cm^3; volume fraction of solids 0.55), (2) *loam* (bulk density 1.33 gm/cm^3; volume fraction of solids 0.5), and (3) *peat* (volume fraction of solids 0.2). (After de Vries, 1975.)

of measurement: the soil near the warmer plane becomes drier, while that near the cooler plane becomes wetter. Early attempts to measure thermal conductivity (e.g., Smith, 1932) failed to recognize this pitfall as they purported to maintain constant soil moisture conditions during prolonged steady-state heat flow. Hence their results can only be considered approximations at best. While steady-state methods may be sufficiently accurate for measuring thermal conductivity of dry soils, short-term transient heat-flow techniques are preferable, in principle, for moist soils.

The principal advantages of using transient-state methods in the measurement of thermal conductivity are that water movement in the soil volume of interest is obviated, or at least minimized, and that the long wait required for attainment of steady-state heat flow is avoided. In addition, whereas steady-state methods are confined practically entirely to the laboratory, transient-state methods are applicable both in the laboratory and in the field. At present one of the most practical of these methods for measuring thermal conductivity in situ is the cylindrical-probe heat source, which can be inserted into the soil to any depth in the field, and can be used in the laboratory as well (de Vries and Peck, 1958; Woodside, 1958). Its use is based on solution of the equation for heat conduction in the radial direction from a line source

(Carslaw and Jaeger, 1959):

$$\frac{\partial T}{\partial t} = \kappa \frac{\partial^2 T}{\partial r^2} + \frac{1}{r}\frac{\partial T}{\partial r} \tag{12.22}$$

where T is temperature, t time, r radial distance from the line source of heat, and κ, as before, thermal conductivity.

In practice, a cylindrical probe containing a heating wire is embedded in the soil, an electrical current is then supplied, and the rate of temperature rise is measured with a thermocouple or thermistor placed next to the wire. For a short distance from the line source, the rise in temperature $T - T_0$ is given by

$$T - T_0 = (q_h/4\pi\kappa)(c + \ln t) \tag{12.23}$$

wherein T is the measured temperature, T_0 the initial temperature, q_h the heat generated per unit time and unit length of heating wire, κ the conductivity, c a constant, and t the time. A plot of temperature versus the logarithm of time permits a calculation of κ. A correction factor may be necessary to account for the dimensions of the probe (Jackson and Taylor, 1965).

The thermal diffusivity D_h, instead of the conductivity κ, is sometimes desired. It can be defined as the change in temperature produced in a unit volume by the quantity of heat flowing through the volume in unit time under a unit temperature gradient. An alternative definition, easier to perceive, is that the thermal diffusivity is the ratio of the conductivity to the product of the specific heat and density:

$$D_h = \kappa/c_s\rho = \kappa/C_v \tag{12.24}$$

where C_v is the volumetric heat capacity. As shown in the preceding section, the specific heat and density of both solids and water must be considered when calculating the volumetric heat capacity:

$$C_v = \rho_s(c_s + c_w w) \tag{12.25}$$

where ρ_s is the density of dry soil, c_s the specific heat of dry soil, c_w the specific heat of water, and w the ratio of the mass of water to the mass of dry soil.

The thermal diffusivity can thus be calculated from prior measurements of thermal conductivity and volumetric heat capacity, or it can be measured directly as described by Jackson and Taylor (1965).

G. Simultaneous Transport of Heat and Moisture

The flows of water and of thermal energy in the soil are interactive phenomena: the one entails the other. Temperature gradients affect the moisture

G. Simultaneous Transport of Heat and Moisture

potential field and induce both liquid and vapor movement. Reciprocally, moisture gradients move water which carries heat. The simultaneous occurrence of temperature gradients and of moisture gradients in the soil therefore brings about the combined transport of heat and moisture. This combined transport can generally be ignored in the extreme cases of a relatively wet soil and a nearly dry soil: in the former, the influence of temperature gradients on liquid water flow is generally small in comparison with the influence of moisture gradients; in the latter, the movement of heat can entail no significant movement of either liquid water or vapor. Thus, we are left with the problem of how to deal with situations in which transport of liquid water and of vapor are quantitatively similar, and in which thermal gradients are more important than other moisture potential gradients.

Two separate approaches to the combined transfer of heat and moisture have been attempted: (1) a mechanistic approach, based on a physical model of the soil system, and (2) a thermodynamic approach, based on an attempt to formulate the phenomenology of irreversible processes in terms of coupled forces and fluxes. Though starting from different points of view, the two approaches have been shown to be related and, properly formulated, can be cast into an equivalent mold (Groenevelt and Bolt, 1969; Jury, 1973). Yet neither approach has yet been developed sufficiently to encompass the full complexity of the interactive set of transport processes involved in simultaneous heat and moisture transport.

The mechanistic approach is exemplified by the work of Philip and de Vries (1957). Their model is based on the concept of viscous flow of liquid water under the influence of gravity and of capillary and adsorptive forces and on the concept of vapor movement by diffusion. Local "microscopic-scale" thermodynamic equilibrium between liquid and vapor is assumed to exist at each point within the soil. The general differential equation describing moisture movement in a porous system under combined temperature and moisture gradients for one-dimensional vertical flow is, accordingly,

$$\partial \theta / \partial t = \mathbf{V} \cdot (D_T \mathbf{V} T) + \mathbf{V} \cdot (D_w \mathbf{V} \theta) - \partial K / \partial z \qquad (12.26)$$

where θ is volumetric wetness, t time, T absolute temperature, D_T the water diffusivity under a temperature gradient (the sum of the liquid and vapor diffusivities), D_w the water diffusivity under a moisture gradient, K hydraulic conductivity, and z the vertical space coordinate. The last term on the right-hand side is due to the gravity gradient [see Eq. (9.13)] and becomes positive if z is taken to be increasing downwards.

The heat transfer equation is similarly

$$C_v \, \partial T / \partial t = \mathbf{V} \cdot (\kappa \mathbf{V} T) - L \mathbf{V} \cdot (D_{w,\text{vap}} \mathbf{V} \theta) \qquad (12.27)$$

where C_v is volumetric heat capacity, κ apparent thermal conductivity of the soil, L latent heat of vaporization of water, and $D_{w,\text{vap}}$ the diffusivity for heat conveyed by water movement (mostly vapor). The above simultaneous equations are both of the diffusion type, involving θ- and T-dependent diffusivities as well as gradients of both θ and T. Taken together, Eq. (12.26) and (12.27) describe the coupled transport of moisture and heat in soils. The mechanistic nature of the theory and of the coefficients involved was explained by de Vries (1975).[1] However, the difficulty encountered in making the theory operational is in the actual measurement of the diffusivities (Dirksen and Miller, 1966). A more fundamental problem is that, since the two mechanisms of flow represented in each equation do interact, they are not, strictly speaking, simply additive.

To consider the approach based on irreversible thermodynamics, we must first understand in principle the difference between this relatively new branch of science and the older, "classical" thermodynamics, which deals with reversible processes and equilibrium states. Classical thermodynamics can predict whether, and in what direction (but not at what rate) a spontaneous process will occur in a system not at equilibrium. However, in a natural system any number of different forces might be operating simultaneously to produce mutually interacting fluxes in a combination of irreversible processes. For instance, a concentration gradient causes diffusion, a pressure gradient causes convection, and a temperature gradient results in the transfer of heat, with each of these fluxes involving the others in the same system. If the system is not too far from equilibrium, the fluxes are taken to be related linearly to the forces causing them.

In application to simultaneous water and heat flow, as an example, the approach based on the thermodynamics of irreversible processes formulates a pair of *phenomenological equations* in which the fluxes of moisture q_w and heat q_h are expressed as linear functions of the moisture potential gradient dP/dz and the temperature gradient dT/dz:

$$q_w = -L_{ww}\frac{1}{T}\frac{dP}{dz} - L_{wh}\frac{1}{T^2}\frac{dT}{dz}, \qquad q_h = -L_{hw}\frac{1}{T}\frac{dP}{dz} - L_{hh}\frac{1}{T^2}\frac{dT}{dz} \quad (12.28)$$

The four phenomenological coefficients occurring in these equations (L_{ww}, L_{wh}, L_{hw}, L_{hh}, relating water flow to the water potential gradient, water flow to the thermal potential gradient, heat flow to the water potential gradient, and heat flow to the thermal potential gradient, respectively) are unknown functions of P (or θ) and T. According to *Onsager's theorem* (Katchalsky and Curran, 1965), the *cross-coupling coefficients* L_{wh} and L_{hw}

---

[1] The assumption of local thermodynamic equilibrium links the vapor pressure p_v to the matric potential ψ by the relation $p_v = p_{vs}h = p_{vs}\exp(Mg\psi/RT)$, where p_{vs} is the saturated vapor pressure at the particular temperature T, h relative humidity, M molar mass, g acceleration of gravity, and R the universal gas constant. The diffusivities for water and heat by vapor transport are obtained by use of this relationship.

are equal when the fluxes and forces are properly formulated. Thus, the number of coefficients which must be measured is reduced.

The irreversible thermodynamics approach is exemplified in the works of Cary and Taylor (1962), Cary (1963), and Taylor and Cary (1960, 1964). An extensive discussion is given in the book by Nielsen *et al.* (1972). An apparent advantage of the irreversible thermodynamics approach is that it makes no a priori assumptions regarding the mechanisms of the transport phenomena formulated. Hence it would seem to be less restrictive than a physical theory whose validity is constrained at the outset by its mechanistic assumptions. The disadvantage of the approach, however, is precisely in its failure to address itself to, and provide insight into, the nature and internal workings of the phenomena considered. For instance, it treats the phase change of water involved in the transfer of heat by vapor in an implicit way only. At the present state of our knowledge, the actual, apart from the merely formalistic, advantages inherent in the application of irreversible thermodynamics to the problem at hand remain to be proven.

H. Thermal Regime of Soil Profiles

In nature, soil temperature varies continuously in response to the ever-changing meteorological regime acting upon the soil–atmosphere interface. That regime is characterized by a regular periodic succession of days and nights, and of summers and winters. Yet the regular diurnal and annual cycles are perturbed by such irregular episodic phenomena as cloudiness, cold waves, warm waves, rainstorms or snowstorms, and periods of drought. Add to these external influences the soil's own changing properties (i.e., temporal changes in reflectivity, heat capacity, and thermal conductivity as the soil alternately wets and dries, and the variation of all these properties with depth), as well as the influences of geographic location and vegetation, and you can expect the thermal regime of soil profiles to be complex indeed. Yet not altogether unpredictable.

The simplest mathematical representation of nature's fluctuating thermal regime is to assume that at all depths in the soil the temperature oscillates as a pure harmonic (sinusoidal) function of time around an average value. Since nature's actual variations are not so orderly, this may be a rather crude approximation. Nevertheless, it is an instructive exercise in itself, and when used in conjunction with field data it can lead to a better understanding, and perhaps even provide a basis for the prediction, of a soil's thermal regime.

Now let us assume that, although soil temperature varies differently at different depths in the soil, the average temperature is the same for all depths. We next choose a starting time ($t = 0$) such that the surface is at the average temperature. The temperature at the surface can then be expressed as a function of time (Fig. 12.2):

$$T(0,t) = \bar{T} + A_0 \sin \omega t \tag{12.29}$$

where $T(0,t)$ is the temperature at $z = 0$ (the soil surface) as a function of time t, T is the average temperature of the surface (as well as of the profile), and A_0 is the amplitude of the surface temperature fluctuation [the range from maximum (or minimum) to average temperature]. Finally, ω is the radial frequency, which is 2π times the actual frequency. In the case of diurnal variation the period is 86,400 sec (24 hr) so $\omega = 2\pi/86,400 = 7.27 \times 10^{-5}$/sec. Note that the argument of the sine function is expressed in radians rather than in degrees.

The last equation is the boundary condition for $z = 0$. For the sake of convenience, let us assume that at infinite depth ($z = \infty$) the temperature is constant and equal to $\bar{T}$. Under these circumstances, temperature, at any depth z and time t, is also a sine function of time, as shown in Fig. 12.3 (Lettau, 1962; van Wijk, 1963):

$$T(z,t) = \bar{T} + A_z \sin[\omega t + \varphi(z)] \tag{12.30}$$

in which A_z is the amplitude at depth z. Both A_z and $\varphi(z)$ are functions of z but not of t. They can be determined by substituting the solution of Eq. (12.29) in the differential equation $\partial T/\partial t = \kappa(\partial^2 T/\partial z^2)$. This leads to the solution

$$T(z,t) = \bar{T} + A_0[\sin(\omega t - z/d)]/e^{z/d} \tag{12.31}$$

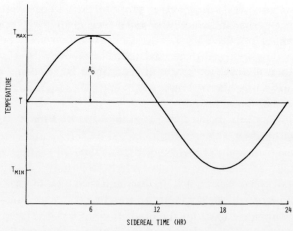

Fig. 12.2. Idealized daily fluctuation of surface soil temperature, according to the equation: $T = \bar{T} + A_0 \sin(\omega t/p)$, where T is temperature, $\bar{T}$ average temperature, A_0 amplitude, t time, and p period of the oscillation (in this case, p refers to the diurnal period of 24 hr).

H. Thermal Regime of Soil Profiles

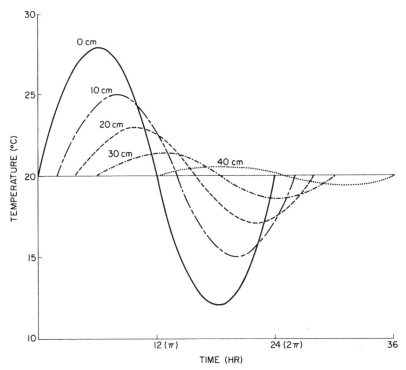

Fig. 12.3. Idealized variation of soil temperature with time for various depths. Note that at each succeeding depth the peak temperature is damped and shifted progressively in time. Thus, the peak at a depth of 40 cm lags about 12 hr behind the temperature peak at the surface and is only about $\frac{1}{16}$ of the latter. In this hypothetical case, a uniform soil was assumed, with a thermal conducitivity of 4×10^{-3} cal/cm sec deg and a volumetric heat capacity of 0.5 cal/cm^3 deg.

The constant d is a characteristic depth, called the *damping depth*, at which the temperature amplitude decreases to the fraction $1/e$ ($1/2.718 = 0.37$) of the amplitude at the soil surface A_0. The damping depth is related to the thermal properties of the soil and the frequency of the temperature fluctuation as follows:

$$d = (2\kappa/c\omega)^{1/2} = (2D_h/\omega)^{1/2} \tag{12.32}$$

It is seen that at any depth z the amplitude of the temperature fluctuation A_z is smaller than A_0 by a factor $e^{z/d}$ and that there is a phase shift (i.e., a time delay of the temperature peak) equal to $-z/d$. The decrease of amplitude and increase of phase lag with depth are typical phenomena in the propagation of a periodic temperature wave in the soil.

The physical explanation for the damping and retarding of the temperature waves with depth lies in the fact that a certain amount of heat is absorbed or released along the path of propagation when the temperature of the conducting soil increases or decreases, respectively. The damping depth is related inversely to the frequency, as can be seen from Eq. (12.32). Hence it depends directly on the period of the temperature fluctuation considered. The damping depth is $\sqrt{365} \approx 19$ times larger for the annual variation than for the diurnal variation in the same soil. For example, van Wijk and de Vries (1963) calculated the damping depth for a soil with $\kappa = 2.3 \times 10^{-3}$ cal/cm sec deg and obtained $d = 12$ cm for the diurnal temperature fluctuation and $d = 229$ cm for the annual fluctuation. Whereas at depth $z = d$ the amplitude is 0.37 as great as the amplitude at the surface, it is only about 0.05 of the surface amplitude at $z = 3d$ ($=36$ cm for the diurnal variation in the case of the soil used by these authors). When an arbitrary zero point t_0 is introduced into the time scale, Eq. (12.31) becomes

$$T(z,t) = \bar{T} + A_0[\sin(\omega t + \varphi_0 - z/d)]/e^{z/d} \qquad (12.33)$$

The constant $\varphi_0 = -\omega t_0$ is called the *phase constant*.

The annual variation of soil temperature down to considerable depth causes deviations from the simplistic assumption that the daily average temperature is the same for all depths in the profile. The combined effect of the annual and diurnal variation of soil temperature can be expressed by

$$\begin{aligned} T(z,t) = \bar{T}_y &+ A_y[\sin(\omega_y t + \varphi_y - z/d_y)]/e^{z/d_y} \\ &+ A_d[\sin(\omega_d t + \varphi_d - z/d_y)]/e^{z/d_d} \end{aligned} \qquad (12.34)$$

where the subscripted indices y and d refer to the yearly and daily temperature waves, respectively. Thus $\bar{T}_y$ is the annual mean temperature. The daily cycles are now seen to be short term perturbations superimposed upon the annual cycle. Vagaries of weather (e.g., spells of cloudiness or rain) can cause considerable deviations from simple harmonic fluctuations, particularly for the daily cycles. Longer term climatic irregularities can also affect the annual cycle, of course. Also, since the annual temperature wave penetrates much more deeply, the assumptions of soil homogeneity in depth and of the time constancy of soil thermal properties are clearly unrealistic.

An alternative theroretical approach is now possible, with fewer constraining assumptions. It is based on numerical, rather than analytical, methods for solving the differential equations of heat conduction. With the brute force of increasingly powerful digital computers, it is possible to construct and solve mathematical simulation models which allow soil thermal properties to vary in time and space (e.g., in response to periodic changes in soil wetness), so as to account for alternative surface saturation and desiccation and for profile layering, and which also allow boundary conditions (e.g., climatic inputs) to follow more realistic and irregular

H. Thermal Regime of Soil Profiles

patterns. The surface amplitude of temperature need no longer be assumed to be an independent variable, but one which itself depends on the energy balance of the surface and thus is affected by both soil properties and above-soil conditions. Examples of the numerical approach can be found in the published works of Wierenga and deWit (1970), van Bavel and Hillel (1975, 1976), and Hillel (1977).

Other recent innovations of practical importance involve the development of techniques for monitoring the soil thermal regime more accurately and precisely than previously available techniques. One such innovation is the infrared *radiation thermometer* for the scanning or remote sensing of surface temperature for both fallow and vegetated soils without disturbance of the measured surface. Knowledge of the surface temperature and its variation in time is important in assessing energy exchange between soil and atmosphere as well as in determining boundary conditions for in-soil heat transfer.

Another important technique is the use of *heat flux plates*. These are flat and narrow plates or disks of constant thermal conductivity which allow precise measurement of the temperature difference between their two sides so as to yield the heat flux through them. When embedded horizontally in the soil at regular depth intervals, a series of such heat flux plates can provide a continuous record of heat transfer throughout the profile. There are problems, however. The presence of heat flux plates can distort the flow of heat in the surrounding medium if their thermal conductivity is very different from that of the soil. The experimental error can be minimized by con-

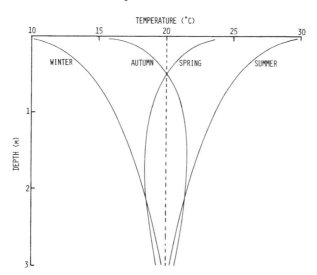

Fig. 12.4. The soil temperature profile as it might vary from season to season in a frost-free region.

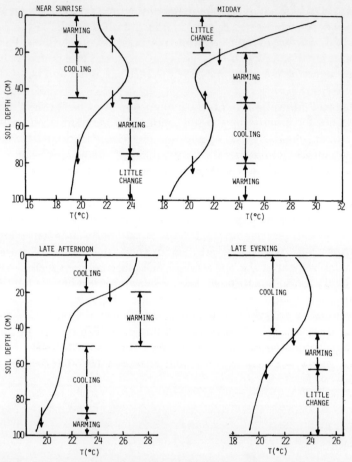

Fig. 12.5. Typical variation of temperature with depth at different times of day in summer. (From Sellers, 1965, based on data given by Carson, 1961.)

structing plates of maximal thermal conductivity and minimal thickness, and by calibrating them in a medium with a thermal conductivity close to that of the soil in which they are to be placed. Another problem is that such plates do not allow vapor flow, which can be an important component of heat transfer. Studies demonstrating the use of heat flux plates were reported by Fuchs and Tanner (1968) and by Hadas and Fuchs (1973).

The soil temperature profile as it might vary from season to season in a frost-free region is illustrated in Fig. 12.4. The diurnal variation of temperature and the directions of heat flow within a soil profile are illustrated in Fig. 12.5.

I. Modification of the Soil Thermal Regime

Although a comprehensive, fundamental understanding of the soil system's thermal regime has been late in coming, and in fact is still in the process of evolving, agriculturists and other practitioners have long attempted to manage the system by devising empirical methods to modify the thermal regime so as to obtain more advantageous conditions (e.g., to ensure or hasten the germination and growth of plants). Empirical experience is often a good guide, but any uninformed attempt to apply it beyond the specific circumstances in which it was obtained may be misleading.

Since it is at the soil surface that energy is partitioned and transformed into alternative fluxes, and since the soil surface is the most accessible part of the system and the one most amenable to manipulation, the majority of the methods aimed at affecting soil heat are surface treatments. These include covering, or mulching, the surface with a bewildering array of materials (such as peat, straw, sand, gravel, paper, asphalt, wax, chalk, charcoal, and various types of plastic sheeting—black, white, or transparent) so as to warm or cool the soil and/or to reduce evaporation. Other methods are based on the mechanical manipulation of the soil's top layer by land-shaping or tillage machinery. Still other methods, not specifically designed to modify soil temperature but having such an effect incidentally, include irrigation, drainage, weed control, and repeated traffic over the field for such purposes as harvesting or controlling pests. Apart from any direct intervention by man, the growth of plants can affect the soil thermal regime by shading the soil surface from solar radiation and by modifying the soil moisture regime. The effect is to moderate the temperature amplitude at the soil surface, i.e., to reduce the maximum and increase the minimum recorded at any depth and often to cause a small overall decrease in average soil temperature (Monteith, 1973).

Drainage can have a very strong effect on soil temperature. At springtime, in particular, wet soils tend to be cold soils, at least at the surface. The high heat capacity resulting from the high volume fraction of water reduces the temperature rise resulting from the absorption of a unit quantity of heat. Another effect of a high water content is to increase the soil's thermal conductivity, thus enhancing the downward conduction of heat rather than its retention in the surface zone. Moreover, the evaporation of water from wet soils consumes energy which might otherwise go to warm the soil surface zone. For all these reasons combined, the removal of excess moisture from the soil's top layer by both surface and subsurface drainage, and the consequent increase of the soil's air-filled porosity, can help to warm the soil and thus to enhance seed germination, root growth, tillering and shoot extension, and microbial activity in early spring.

Since the inclination and direction of a sloping surface affect the amount of solar radiation intercepted per unit area (i.e., slopes which face the sun receive more energy and hence are warmer than those which face away from it), methods have been devised to shape agricultural fields by ridging or bedding the portion of the land to be planted. The optimal direction and height of the ridges depends on geographic latitude and time of year. Studies on modification of soil microrelief and its effects on soil temperature and plant response have been reported by Shaw and Buchele (1957) and by Burrows (1963).

Mulching has already been mentioned as a method to modify a soil's thermal regime. The number of investigators who have experimented with the application of mulches is indeed legion. Yet the results are baffling. So varied have been the circumstances and the selection of mulching materials, that no generalization can be made regarding their effect. Consider the following example: The application of a highly reflective material such as chopped straw can have the effect of lowering soil temperature by reducing the radiant flux reaching the soil surface; on the other hand, a dense and less reflective mulch may tend to increase soil temperature by inhibiting evaporation. Another example is the use of thin plastic sheets to cover the soil surface. As shown by Waggoner et al. (1960), plastic materials of different opacity or transparency have very different effects. Black plastic absorbs most of the incoming radiant energy but transmits little of it to the soil, because of the insulating effect of the intermediate layer of still air. The plastic cover becomes warm during the day but the soil underneath remains cool. Transparent plastic is somewhat more reflective and much less absorptive, as it transmits short-wave (visible-light) radiation to the soil while reducing the upward escape of emitted long-wave (infrared) radiation because of vapor condensation on the underside. Thus, a "greenhouse effect" is produced, warming the soil.

Still another example is the use of powdered chalk (Stanhill, 1965) or charcoal to either lighten or darken the soil surface. The black powder absorbs more radiation, hence it causes more rapid evaporation at first. It is only after the surface desiccates that further evaporation is retarded, and the soil surface becomes very warm. White powder, on the other hand, reflects much of the incoming radiation and thus reduces the amount of energy available for evaporation, but allows the evaporation process to continue much longer (Hillel, 1968). Hence the relative effects of the two alternative treatments, or of any intermediate one, depends on the time of measurement and on antecedent conditions (e.g., whether the soil was initially dark or light, wet or dry, fallow or vegetated). The simulated effect of soil surface albedo on the soil thermal regime is illustrated in Figs. 12.6 and 12.7.

In recent years, interest has arisen in the possibility of using the soil for the

I. Modification of the Soil Thermal Regime

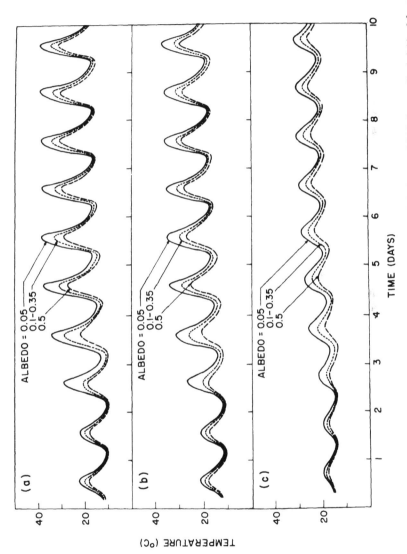

Fig. 12.6. Time course of soil temperature at three depths [(a at surface; (b) in seedbed, 0.03 m; and (c) below seedbed, 0.12 m] for three values of surface reflectivity (low, high, and intermediate variable). (From Hillel, 1977.)

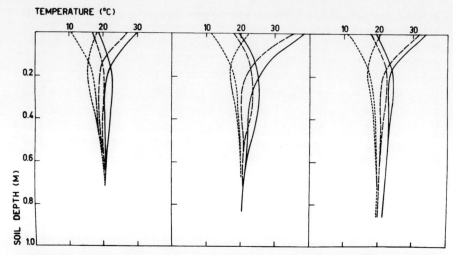

Fig. 12.7. Soil temperature profiles at midday and at midnight of second, sixth, and tenth days of simulated evaporation under low (0.5), high (0.05), and intermediate variable (0.1–0.35) albedo conditions. (From Hillel, 1977.)

disposal of waste heat from power plants and other industrial facilities. Warm water, ciculated in underground pipes, can warm the soil sufficiently to hasten the growth of crops during the transition seasons of spring and autumn. The problem here, as in the other instances given, is to be able to predict the effect quantitatively and thus to optimize the installation so as to obtain the greatest possible economic and environmental benefit. The optimal spacing and depth of the heat sources in the soil obviously depends on soil thermal properties and meteorological conditions, and hence would vary from place to place.

A fundamental investigation into the influence of tillage on soil temperature was published by van Duin (1956). Despite the passage of nearly 25 years, his work can still be considered definitive and deserves reviewing. Tillage establishes a loose upper layer of soil with physical properties different from those of the underlying soil. Macroporosity is greatly increased, at least temporarily. The effect of tillage on the soil's temperature regime is primarily a consequence of the change in thermal properties of this layer. Since surface temperature depends on the rate at which energy enters and leaves the soil surface, van Duin studied the influence of tillage on the energy balance, as affected by changes in the surface reflectivity and in the upper layer's volumetric heat capacity and thermal conductivity. He was thus able to show that the amplitude of temperature variation at the surface is greater in tilled than in untilled soil, but that this amplitude decreases with depth more steeply in the tilled than in the untilled soil (because of the former's lowered thermal conductivity resulting from its higher air content). This means that the daily maximum temperature at the surface of a tilled soil will be several degrees higher and the minimum similarly lower. At certain critical times, such as during germination or when night frost might occur, even small changes in the soil's thermal regime may be crucial. The dangers to seeds and seedlings of excessive temperature and desiccation in daytime, and of frost at nighttime, are increased by tillage. The increase in diurnal variation due to tillage may have a noticeable effect on the germination of species or varieties that are sensitive to extremes of temperature, even though the *average* temperature may remain nearly the same. Moreover, since tilled soil can be warmer at the surface but cooler below a certain depth, the rate of seedling root growth may be affected even after germination.

It follows from the foregoing that a comprehensive analysis and interpretation of experimental data is only possible when temperature measurements are taken at various depths both in the tilled and in the undisturbed soil so as to provide complete data on the dynamics of the respective temperature profiles. The fact that tillage can have an opposite effect at the surface and at some depth may help explain the inconsistent conclusions drawn by different investigators in the past. A comparison of temperature measurements taken at one arbitrary depth or time is of dubious value and may be misleading. Unfortunately, examples of comprehensive experiments providing sufficient data on soil thermal properties and temperature regimes are still too few. All too often, results of field experiments cannot be generalized owing to lack of vital data. Once again we are reminded of the importance of comprehensive theoretical knowledge in the planning, implementation, and interpretation of field experiments.

Sample Problems

1. Consider the energy balance of a bare-surface soil. The daytime (12-hr) average global (sun and sky) radiation is 0.8 cal/cm² min; the albedo is 0.15. The average soil surface temperature during the diurnal period is 27°C. Advection in daytime balances the outflow of sensible heat during the night so that diurnal net sensible heat exchange with the atmosphere is negligible. Evaporation is 2 mm/day: Assume that the emissivity is 0.9 and that the atmosphere returns 60% of the long-wave radiation emitted by the ground. Estimate the soil heat transfer term. Is it positive or negative (i.e., is the soil gaining or losing heat)?

Following Eq. (12.6), the diurnal energy balance can be written

$$J_s(1 - \alpha) - J_1 - S - A - \text{LE} = 0$$

where J_s is incoming global short-wave radiation, α albedo, J_1 net long-wave emitted radiation, S heat flow into the soil, A sensible heat transfer to the air, and LE latent heat loss. To obtain S, we rearrange this equation to read

$$S = J_s(1 - \alpha) - J_1 - A - LE$$

For the net short-wave radiation, we have

$$\begin{aligned}J_s(1 - \alpha) &= 0.8 \text{ cal/cm}^2 \text{ min} \times 60 \text{ min/hr} \times 12 \text{ hr/day} \times (1 - 0.15) \\ &= 489.6 \text{ cal/cm}^2 \text{ day}\end{aligned}$$

The outgoing long-wave radiation (using the Stefan–Boltzmann law) is

$$\begin{aligned}J_1 &= 0.4 \times (\varepsilon \sigma T^4) \\ &= 0.4 \times 0.9 \times 1.17 \times 10^{-7} \text{ cal/cm}^2 \text{ day }°K^4 \times (273 + 27)^{4} °K \\ &= 341.50 \text{ cal/cm}^2 \text{ day}\end{aligned}$$

Here we used a value of 1.17×10^{-7} cal/cm^2 day $°K^4$ for Stefan–Boltzmann constant (equivalent to 8.14×10^{-11} cm^2 min $°K^4$). The net sensible heat transfer A is negligible.

For the latent heat loss term, we have

$$LE = 580 \text{ cal/gm} \times 0.2 \text{ gm/cm}^2 \text{ day} = 116 \text{ cal/cm}^2 \text{ day}$$

Note: An evaporation of 2 mm/day is equivalent to 0.2 cm^3/cm^2 day. Given the density of water (1 gm/cm^3), this equals 0.2 gm/cm^2 day.

Finally, we can sum up all of these quantities to obtain the soil heat flow:

$$S = 489.6 - 341.50 - 0 - 116 = 32.1 \text{ cal/cm}^2 \text{ day}$$

Ergo, the soil is gaining heat. If, for example, this amount of heat is distributed uniformly in the top 20 cm of the soil having a specific heat capacity of about 0.5 cal/gm deg and a bulk density of 1.6 gm/cm^3, it will raise the temperature by about 2 deg. This example, however, is completely hypothetical, as is the entire exercise.

2. Assuming steady-state conditions, calculate the one-dimensional thermal flux and total heat transfer through a 20-cm thick layer if the thermal conductivity is 3.6×10^{-3} cal/cm sec deg and a temperature differential of 10°C is maintained across the sample for 1 hr.

Using Eq. (12.8) in discrete form, we can write

$$\begin{aligned}q_h &= \kappa \, \Delta T/\Delta x = 3.6 \times 10^{-3} \text{ cal/cm sec deg} \times 10 \text{ deg}/20 \text{ cm} \\ &= 1.8 \times 10^{-3} \text{ cal/cm}^2 \text{ sec}\end{aligned}$$

Total heat transfer is

$$q_h t = 1.8 \times 10^{-3} \text{ cal/cm}^2 \text{ sec} \times 3600 \text{ sec} = 6.48 \text{ cal/cm}^2$$

3. A thermal flux of 10^{-3} cal/cm^2 sec is maintained into the upper surface of a 10-cm thick sample, the bottom of which is thermally insulated. Calculate

Sample Problems

the time rate of temperature change and the total temperature rise per hour if the bulk density is 1.2 gm/cm³ and the specific heat capacity is 0.6 cal/gm deg.

For this case of heat flow, we use a discrete form of Eq. (12.9):

$$dT/dt = (\Delta q_h/\Delta x)(1/\rho_b c_m)$$

Using the data provided, the time rate of temperature change

$$dT/dt = (10^{-3} \text{ cal/cm}^2 \text{ sec}/10 \text{ cm})(1.2 \text{ gm/cm}^3 \times 0.6 \text{ cal/gm deg})^{-1}$$
$$= 1.39 \times 10^{-4}$$

Total temperature rise is 1.39×10^{-4} deg/sec $\times$ 3600 sec/hr $= 0.5°C/\text{hr}$.

4. Calculate the volumetric heat capacity C of a soil with a bulk density of 1.46 gm/cm³ when completely dry, when completely saturated. Assume that the density of solids is 2.60 gm/cm³ and that organic matter occupies 10% of the solid matter (by volume).

First calculate the volume fraction of pores (the porosity):

$$f = (\rho_s - \rho_b)/\rho_s = (2.60 - 1.46) \text{ gm/cm}^3/2.60 \text{ gm/cm}^3 = 0.44$$

Hence the volume fraction of solids is $1 - 0.44 = 0.56$. Since organic matter constitutes 10% of the soil's solid phase, the volume fraction of mineral matter is

$$f_m = 0.56 \times 0.9 = 0.504$$

The volume fraction of organic matter is

$$f_o = 0.56 \times 0.1 = 0.056$$

The volumetric heat capacity can be calculated using Eq. (12.18):

$$C = f_m C_m + f_o C_o + f_w C_w$$

where f_m, f_o, and f_w, are the volume fractions of mineral matter, organic matter, and water, respectively; and C_m, C_o, and C_w refer to heat capacities of the same constituents (namely, 0.48 cal/cm³ deg for mineral matter, 0.6 cal/cm³ deg for organic matter, and 1 cal/cm³ deg for water). Accordingly, when the soil is completely dry,

$$C = (0.48 \times 0.504) + (0.60 \times 0.05) = 0.24 + 0.03 = 0.27 \text{ cal/cm}^3 \text{ deg}$$

When the soil is saturated, its volume fraction of water equals the porosity. Thus

$$C = 0.27 \text{ cal/cm}^3 \text{ deg} + 0.44 \times 1 \text{ cal/cm}^3 \text{ deg} = 0.71$$

Note: We have completely neglected the heat capacity of air, since it is too small to make any significant difference (see Table 12.1).

5. The daily maximum soil-surface temperature is 40°C and the minimum is 10°C. Assuming that the diurnal temperature wave is symmetrical, that the mean temperature is equal throughout the profile (with the surface temperature equal to the mean value at 6 A.M. and 6 P.M.), and that the "damping depth" is 10 cm, calculate the temperatures at noon and midnight for depths 0, 5, 10, and 20 cm.

Since the temperature range is 30°C and the mean ($\bar{T}$) 25°C, the amplitude at the surface A_0, the maximum value above the mean, is 15.

We use Eq. (12.31) to calculate the temperature T at any depth z and time t:

$$T(z,t) = \bar{T} + A_0[\sin(\omega t - z/d)]/e^{z/d}$$

where ω is the radial frequency ($2\pi/24$ hr) and d is the "damping depth" at which the temperature amplitude is $1/e (= 0.37)$ of A_0. *Note:* The radial angle is expressed in radians rather than in degrees.

At depth zero (the soil surface): Noontime temperature (6 hr after $T = \bar{T}$):

$$T(0,6) = 25 + 15 \times [\sin(\pi/2 - 0)]/e^0 = 25 + 15 = 40°C$$

Midnight temperature (18 hr after $T = \bar{T}$)

$$T(0,18) = 25 + 15 \times [\sin(3\pi/2 - 0)]/e^0 = 25 - 15 = 10°C$$

At depth 5 cm: Noontime temperature:

$$\begin{aligned}T(5,6) &= 25 + 15 \times [\sin(\pi/2 - 5/10)]/e^{5/10} \\ &= 25 + 15 \times [\sin(1.57 - 0.5)]/e^{1/2} \\ &= 25 + 15 \times [\sin(1.07)]/1.65 \\ &= 25 + 15 \times (0.87720/1.65) = 32.97°C\end{aligned}$$

Midnight temperature:

$$\begin{aligned}T(5,18) &= 25 + 15 \times [\sin(3\pi/s - 5/10)]/1.65 \\ &= 25 + 15 \times [\sin(4.71 - 0.5)]/1.65 \\ &= 25 + 15 \times (-0.87720/1.65) = 17.03°C\end{aligned}$$

At depth 10 cm (the damping depth): Noontime temperature:

$$\begin{aligned}T(10,6) &= 25 + 15 \times [\sin(\pi/2 - 1)]/e^1 = 25 + 15 \times \sin(0.57)/e \\ &= 25 + 15 \times (0.53963/2.718) = 27.98°C\end{aligned}$$

Midnight temperature:

$$\begin{aligned}T(10,18) &= 25 + 15 \times [\sin(3\pi/2 - 1)]/e \\ &= 25 + 15 \times [\sin(4.71 - 1)]/2.718 \\ &= 25 + 15 \times (-0.53963/2.718) = 22.02°C\end{aligned}$$

Sample Problems

At depth 20 cm: Noontime temperature:

$$T(20,6) = 25 + 15 \times [\sin(1.57 - 20/10)]/e^{20/10}$$
$$= 25 + 15 \times [\sin(-0.43)]/e^2$$
$$= 25 + 15 \times (-0.41687/7.39) = 25 - 0.85 = 24.15°C$$

Midnight temperature:

$$T(20,18) = 25 + 15 \times [\sin(4.71 - 2)]/7.39$$
$$= 25 + 15 \times (0.41687/7.39 = 25.85°C$$

Note: At a depth of 20 cm the phase shift is so pronounced that the temperature at midnight is actually higher than at noontime.

A useful exercise for students at this point is to calculate and to plot the sinsusoidal variation of temperature at each depth so as to observe how the phase shift (time lag of maximum and minimum values) increases and the amplitude decreases with depth.

τα παντα ρεi
(everything flows.)
Heraclitus
540–475 B.C.

13 Stress–Strain Relations and Soil Strength

A. Introduction

The fundamental concepts of *soil rheology* and *soil dynamics* and the applied science of *soil mechanics* based upon them all pertain to the behavior of soil bodies under applied forces. Civil engineers have long been using these concepts in solving problems related to the stability of slopes and foundations. Recently and increasingly the same fundamental concepts have been found to be applicable to soil behavior and management in tillage and traction. The important property known as *soil strength*, the ability of a soil body to bear or withstand stresses without collapsing or deforming excessively, derives from the mutual bonding of interlocking particles and from their frictional resistance to deformation.

To deal with soil deformation and strength, we must first review a few of the fundamental principles of a scientific discipline known as rheology. The term *rheology* originates from a Greek word meaning flow. As a branch of science, however, it is concerned with all types of deformations, of which flow is merely one. Broadly speaking, rheology can be described as the mechanics of deformable bodies. Classical Newtonian mechanics deals with the relation between forces and the conditions of equilibrium or motion of bodies but disregards the internal reactions of the bodies to the forces acting on them. For example, classical mechanics tells us that a drop of water, a ball of wet clay, and a spherical stone will all be subject to uniform gravitational acceleration as they fall side by side, but it fails to describe their different behaviors as they strike the ground. At this point, we turn to rheology to

B. The Concept of Strain and Stress

define the basic mechanical properties of a liquid, a plastic solid, and an elastic solid and to describe the different relationships they exhibit among the variables of force, deformation, and time.

If the three spheres are subjected to a uniform pressure everywhere perpendicular to the surface (i.e., an *isotropic pressure*, also called hydrostatic), they will react similarly by compressing without changing shape. When the pressure is removed, all three bodies will reassume their original volumes. In other words, under isotropic pressure all materials behave uniformly (i.e., elastically). Where materials differ is in their reaction to *shearing forces*, or to combined shearing and *tensile* or *compressive forces*. In general, such forces can produce changes in both shape and volume of the body to which they are applied. How materials in general, and soils in particular, react to applied forces under different conditions is the subject of this chapter.

B. The Concept of Strain and Stress

The reaction of a body to a force or a combination of forces can be characterized in terms of its *relative deformation*, i.e., the ratio of its deformation to its initial dimensions. Consider for example, a rectangular parallelepiped of metal or rubber subjected to a force F_a directed along its main axis as in Fig. 13.1. If the initial length is l_0 and the change in length Δl, then the relative

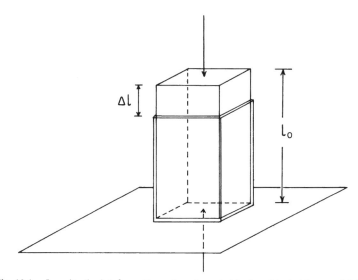

Fig. 13.1. Longitudinal deformation of an elongated body subjected to an axial load.

axial deformation is

$$\varepsilon = \Delta l/l_0 \tag{13.1}$$

where ε is called the *longitudinal strain*. The same expression applies whether the body is compressed or stretched, but of course the two directions of strain have opposite signs.

Another form of deformation occurs when the angles rather than the dimensions of a body change. The simplest such case, called simple shear, is illustrated in Fig. 13.2. The measure of *shear strain* γ (also called tangential strain) is the relative change of the initially right angle:

$$\gamma = u/h \tag{13.2}$$

where u is lateral (or tangential) displacement and h the height of the body. The ratio u/h is thus the tangent of the deformation angle. The dimensions of l, Δl, u, and h are those of length. Hence the strain, being a ratio of lengths, is dimensionless and is often expressed as a fraction or percent of the original dimension.

Now consider what possible types of deformation can occur within a cubic element of a material when subjected to various forces. A cube can elongate or contract in three principal directions, corresponding to the three orthogonal coordinates, and can also exhibit three possible directions of shear, affecting the angles on three pairs of parallel planes. Deformation therefore has six possible components. Though some of these components may be negligible in certain special cases, in general deformation can involve a composite of compression, tension, and shear strains. For example, if a beam is held at both ends and subjected to torque in opposite directions (Fig. 13.3), then the beam bends so that the material on the concave side is compressed, the material on the convex side is stretched, and the angles of any unit volume are deformed.

Bodies subjected to externally applied forces tend to resist deformation by mobilizing internal reactive forces. Although these internal forces generally cannot be measured directly, they can be evaluated by defining the internal distribution of the externally applied forces under *equilibrium conditions*. Equilibrium can be assumed when the body is at rest or in uniform linear motion, so that no uncompensated forces exist tending to cause acceleration,

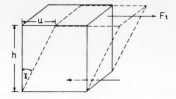

Fig. 13.2. Angular deformation in simple shear of a cube.

B. The Concept of Strain and Stress

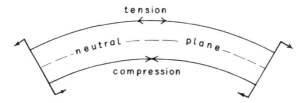

Fig. 13.3. Deformation of a beam subjected to bending.

rotation, or further deformation. Equilibrium thus implies that the vector sums of all forces F and all moments M (with respect to the center of mass) are zero:

$$\sum F = 0, \qquad \sum M = 0 \tag{13.3}$$

Experience shows that the effect of a force in causing deformation is related directly to the magnitude of the force and inversely to the area over which it acts. The ratio of force to area, i.e., a force per unit area, is called *stress*. A *normal stress* is caused by a force whose direction is perpendicular to the area over which it acts. Such a stress, equivalent to a pressure, is usually designated σ. Thus,

$$\sigma = F_n/A \tag{13.4}$$

where F_n is the force acting normal to the surface area A. When the direction of a force is not perpendicular but parallel to the surface area, it causes a *tangential stress*, also called *shearing stress*, and is usually designated:

$$\tau = F_t/A \tag{13.5}$$

where F_t is the tangential force acting on area A.

Another important factor in rheology is the time rate of stress application and the time dependence of strain. Thus,

$$\dot{\varepsilon} = \frac{d\varepsilon}{dt} = \frac{1}{l}\frac{d\,\Delta l}{dt} \tag{13.6}$$

where $\dot{\varepsilon}$ is the time rate of elongation (or, when negative, the time rate of contraction), and

$$\dot{\gamma} = \frac{d\gamma}{dt} = \frac{1}{h}\frac{du}{dt} = \frac{v}{h} \tag{13.7}$$

where v is the velocity du/dt and $\dot{\gamma}$ is the velocity gradient in the direction perpendicular to that of the shearing displacement. The stress–strain–time relationships of a material determine its rheological character, i.e., whether it be an elastic solid, a plastic solid, a Newtonian or non-Newtonian liquid, or any other of the various types of materials recognized and characterized by rheologists.

C. Elasticity and Plasticity

Let us now examine the rheological behavior of solids. According to the conventional physical definition, a solid is a crystalline material with an ordered internal structure, or lattice, of mutually bonded ions or atoms. The rheological criterion for a solid depends not on internal structure, important as it is, but on mechanical behavior. Accordingly, an elastic solid is one which, when subject to a stress, deforms instantaneously and retains its new form as long as the stress is maintained constant, then returns to its original form when the stress is released. Strictly speaking, there are no such materials, since most real bodies exhibit some residual deformation after release of stress. Over geological periods of time, even hard rocks are known to flow and deform irreversibly under pressure. Real materials often exhibit progressive deformation—albeit at a slowing rate—even under a constant pressure; that is to say, they creep. Whether the creep is appreciable or not depends on a property called *relaxation*, i.e., the tendency of a material to relieve stresses gradually through internal structural adjustments. For the moment, however, we choose to disregard creep and relaxation and to describe the behavior of perfectly elastic solids as if they do indeed exist.

Three hundred years ago, Robert Hooke found that if different weights were hung from springs, the springs stretched in proportion to the load applied. Based on this discovery, a law was formulated many years later to describe elastic bodies and termed *Hooke's Law*. This law states that *strain is proportional to stress*. It furthermore implies that strain occurs instanta-

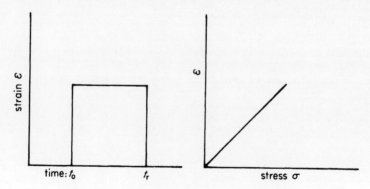

Fig. 13.4. Deformation of an ideally elastic body.

C. Elasticity and Plasticity

neously when stress is applied, and that it disappears immediately and completely when the stress is relieved (Fig. 13.4). The mathematical expression embodying this law is

$$\varepsilon = \sigma/E \qquad (13.8)$$

where E is a constant of proportionality known as *Young's modulus*. For most elastic materials (metals, wood, concrete, glass), this modulus is of the order of 10^5 or 10^6 bar.

Uniaxial tension and compression result not only in changes of a body's length l but also entail corresponding changes in its width or lateral dimension d. The ratio of elongation along one axis to the corresponding contraction of another axis is called *Poisson's ratio*, and is generally indicated by v:

$$v = -\frac{\Delta d/d_0}{\Delta l/l_0} = -\frac{\Delta d/\Delta l}{d_0/l_0} \qquad (13.9)$$

Poisson's ratio (dimensionless) is close to zero for small deformations of a porous material like cork and about 0.49 for a material like rubber, which can deform with practically no change in volume. In soils it depends on porosity and degree of saturation.

The equation of elasticity for shearing stress, analogous to Eq. (13.8), is

$$\gamma = \tau/G \qquad (13.10)$$

where G is the *modulus of shearing*, or rigidity.

We have stated at the outset that when any body is subjected to an isotropic pressure, its volume decreases in proportion to the pressure applied. The constant of proportionality in this case is termed the *bulk modulus* κ.

$$\varepsilon_v = P/\kappa \qquad (13.11)$$

where ε_v is the volume compression or expansion—relative to the original volume. The value of the bulk modulus varies from 10^4 bar for alcohol, through about $0.2-2 \times 10^5$ bar for pure water[1] (depending on temperature), to about 1.8×10^6 bar for steel.

In the case of *nonisotropic materials* (e.g., wood which can be expected to exhibit different properties in directions parallel to, and perpendicular to, its fibers), the various coefficients or moduli just defined can have different values for each of the three principal axes. There may also be different relationships between stress and deformation of different planes. Since theoretically there are six independent components of strain, and six corre-

---

[1] Bulk modulus is the reciprocal of the *coefficient of compressibility*, $C = \rho^{-1} d\rho/dP$, which is the relative change in density ρ with change in pressure P. The compressibility of pure water at 20°C and atmospheric pressure is about 4.6×10^{-11} cm²/dyn.

spondingly independent components of stress (three for normal stresses and three for shear stresses), we might expect as many as 36 elastic constants. However, in practice, we find that even for highly anisotropic materials, it is enough to define nine independent constants, namely, three values of G, three of E, and three of v.

The four elastic constants E, v, G, and κ are not independent of each other, but are in fact interrelated. In an isotropic body, the interrelations among the four parameters are as follows:

$$E = 9\kappa G/(3\kappa + G) \tag{13.12}$$

$$v = (3\kappa - 2G)/(6\kappa + 2G) \tag{13.13}$$

$$G = E/2(1 + v) \tag{13.14}$$

$$\kappa = E/3(1 - 2v) \tag{13.15}$$

For a completely incompressible body, if such were to exist, we would have $\varepsilon_v = 0$ for all values of ρ. Since $\varepsilon_v = \rho/\kappa$, $\varepsilon_v = 0$ requires $\kappa = \infty$. In such a case, $E = 3G$, and $v = \frac{1}{2}$.

Plasticity is the property of a body to deform progressively when under stress and to retain its deformed shape when the stress disappears (Fig. 13.5). Many materials under stress exhibit an elastic phase until some point at which elasticity gives way to plastic behavior. This is illustrated in Fig. 13.6.

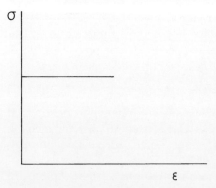

Fig. 13.5. Stress–strain relation for ideally plastic behavior.

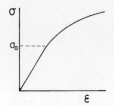

Fig. 13.6. Idealized stress–strain relation for an initially elastic material which becomes plastic when stressed beyond its yield strength σ.

D. Rheology of Liquids

The stress value σ_0 at which this happens is called the *yield point*. The transition from purely elastic to purely plastic behavior may not be as abrupt as shown in the idealized representation of Fig. 13.6. Moreover, it is often affected by the rate of strain and by a phenomenon known as *strain hardening*. Such behavior is typically encountered in the case of ductile metals. The increase in strength of a metal during or as a consequence of deformation is usually due to internal structure changes and recrystallization. An analogous effect may be observed in soils due to compaction.

So-called *plastic flow*, in analogy to viscous flow of fluids, exhibits a positive relation between stress and the rate of strain. In plastic flow, however, the "coefficient of proportionality" is not a true constant, but a function of the yield strength σ_0 and of the rate of straining.

D. Rheology of Liquids

Liquids, from the rheological point of view, are materials which, when under the influence of a shearing stress, undergo continuous deformation (flow). A liquid placed in a container is subject to a hydrostatic pressure which at each point is equal in all directions and hence isotropic. A liquid at rest experiences no shearing stresses. The pressure, however, is not equal at all points. To illustrate this, let us visualize two horizontal planes inside a body of liquid, separated from each other by a vertical distance dz. If the volume element of liquid having the dimensions $dx\ dy\ dz$ is at rest (i.e., equilibrium), then the forces acting on it must be balanced. The forces acting in a vertical direction are

1. the gravity force acting downward,

$$F_g = \rho g\ dx\ dy\ dz$$

2. the downward force due to the pressure P acting on the upper plane,

$$F_d = P\ dx\ dy$$

3. the upward force due to the pressure $P + dP$ acting on the lower plane

$$F_u = (P + dP)\ dx\ dy$$

Equilibrium requires that the vector sum of all forces be zero:

$$\sum F = 0 = (P + dP)\ dx\ dy - P\ dx\ dy - \rho g\ dx\ dy\ dz \quad (13.16)$$

Hence,

$$dP/dz = \rho g \quad (13.17)$$

Integration gives

$$P = \rho g z + c \tag{13.18}$$

If z is measured from the surface of the liquid downward, then it can be taken as equal to the depth, which we shall designate h. For $z = 0$, P is the atmospheric pressure P_0, so that

$$P = P_0 + \rho g h \tag{13.19}$$

It is seen that the pressure increases linearly with depth, so that h can be termed the *pressure head*. As pressure increases, the liquid undergoes compression. However, in the integration we assumed a constant ρ. Hence Eq. (13.19) is correct only for relatively incompressible liquids and for small depths. This equation does not apply in the case of gases and other highly compressible fluids.

As previously stated, liquids cannot bear shearing stresses and yet remain at rest. However, when shearing stresses are applied to a liquid (that is, when forces are applied tending to cause a portion of the liquid to slide over an adjacent portion), a resistance appears, caused by the collision of molecules across the surface of shearing, thus inhibiting slippage. The *viscous friction* transferred through a plane of shear *is proportional to the velocity gradient*. If the velocity of the liquid $u(y)$ is everywhere parallel to the x axis and varies in some way with y, then [see Eq. (8.1)]

$$\tau = \eta \, du/dy \tag{13.20}$$

where τ is the shearing stress. The factor of proportionality characterizes the internal friction, and is called the *coefficient of viscosity*. The value of η depends on the nature of the fluid and changes very strongly with temperature. In liquids, increasing temperature makes η decrease, whereas in gases at low density the opposite effect is observed.

The dimensions of the velocity gradient are T^{-1}, and since τ is a stress, the dimensions of η are those of the product of stress and time, namely, $ML^{-1}T^{-1}$. The unit dyn/cm^2 sec is named *poise* (after Poiseuille).

In some cases, it is convenient to have a symbol to represent the viscosity divided by the mass density of the fluid. Thus,

$$\eta_k = \eta/\rho \tag{13.21}$$

where η_k is called the *kinematic viscosity*, with dimensions of L^2T^{-1} (cm^2/sec).

Liquids which behave in accordance with Eq. (13.20) are termed *Newtonian liquids*. Those which do not obey this simple law (e.g., pastes, slurries, and high polymers) are called *non-Newtonian*. In a Newtonian liquid, the coefficient of viscosity is independent of the rate of shear. If viscosity decreases with increasing rate of shear, the liquid is termed *pseudoplastic*. If it increases

E. Rheological Models

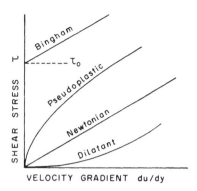

Fig. 13.7. Rheological behavior of various types of liquids.

with the rate of shear the liquid is termed *dilatant*. If the material remains rigid when the shear stress is smaller than a threshold value τ_0 (called the *yield stress*) but flows somewhat like a Newtonian liquid when shear stress exceeds τ_0, then it is called a *Bingham liquid*. This last mode characterizes the behavior of many liquid suspensions and pastes. These various types of liquids are illustrated in Fig. 13.7.

E. Rheological Models

An interesting technique in rheology is the formulation of idealized analog models presumed to simulate the behavior of real materials. A study of these idealized models may yield insights into the way different types of materials can be expected to react to variously applied stresses. The three primary elements of rheological models are (1) a *spring*, which reacts to stress instantaneously and in a perfectly elastic manner, in accordance with Hooke's law; (2) a *dashpot*, which is a piston moving in a Newtonian fluid of constant viscosity at a time rate proportional to the stress; and (3) a *friction block*, allowing no motion until a characteristic yield stress is attained, after which irreversible deformation takes place instantly. These three elements, depicted in Fig. 13.8, can be combined in numerous ways to describe alternative modes of rheological behavior. For example, a spring and a dashpot in series portray a *Maxwell-type* material, whereas the same elements in parallel represent a *Kelvin-type* material. Another example is a series combination of a dashpot, a friction block, and a spring, simulating the pattern for a so-called *Bingham fluid*. As illustrated in Fig. 13.9, the latter model is one in which the force exerted on the friction block is analogous to a shear stress, and the block's displacement to shear strain. If the force is applied gradually, at first only the spring stretches so that the body's reaction is elastic. As the force increases, it eventually exceeds the static resistance of the friction block, after

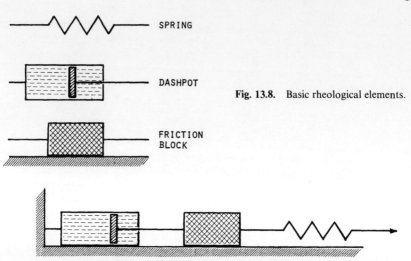

Fig. 13.8. Basic rheological elements.

Fig. 13.9. Rheological model of a Bingham body.

which the dashpot comes into play and exerts a viscous drag proportional to the rate of motion caused by the force. More complex models formed by combining rheological elements are beyond the scope of our treatise. Models applicable to soils were described by Kravtchenko and Sirieys (1966), Suklje (1969), and Kezdi (1974).

F. Stress Distribution in Soil

Forces are usually applied to the soil over a limited or finite section of its boundary. However, the soil itself in its natural location is typically semi-infinite; i.e., it extends in at least one direction beyond any possible influence of the forces applied at the body's boundary. The concept of force per unit area can become obscure inside a three-dimensional semi-infinite medium in which it is difficult to define finite areas and directions accurately. Some way is therefore needed to describe the forces acting at each point within the medium. A concept known as the state of stress at a point answers that need. Strictly speaking, this concept applies rigorously only to internally homogeneous and continuous materials. Although the soil is not continuous, consisting as it does of particles and pores, the assumption of a continuum may nevertheless be tenable if it is applied in a macroscopic sense, that is to say, as long as the smallest area considered is physically much greater than the microscopic discontinuities present. The concept of stress distribution has been successfully applied to soil by Nichols and Randolph (1925), Terzaghi and Peck (1948), and many others.

F. Stress Distribution in Soil

To analyze the stress concept in the hypothetical case of a continuous body, imagine a plane surface cutting through the body and consider the interaction between the two parts of the body across this surface. The material on one side of the plane can be assumed to exert a force on the material on the other side. Now consider a small area ΔA containing point 0, and the vector sum of all forces **F** acting on that area. The limit of the ratio of **F** to ΔA as ΔA approaches zero is defined as the stress vector **S** at the point 0:

$$\mathbf{S} = \lim_{\Delta A \to 0} \left(\frac{\mathbf{F}}{\Delta A}\right) \qquad (13.22)$$

In Fig. 13.10 such a surface of separation has been drawn normal to the x axis, and the two parts of the body have been disjoined for the sake of clarity in the drawing. We assume the x surface in this figure to have the shape of a rectangle with the sides Δy, Δz.

When the forces considered are parallel to the x direction, we speak of a normal stress σ_x, equal to the magnitude of the sum of the forces divided by the area $\Delta y \, \Delta z$. In general, however, the total force may not be perpendicular to the x surface but oblique to it, in which case we can resolve the force and the associated stress into components in the x, y, and z directions. On the plane shown in Fig. 13.10, the y and z components are recognized as tangential or shear stresses and denoted by τ_{xy}, τ_{xz}. This notation indicates the orientation of the surface (orthogonal to the x axis) by the first subscript and the direction of the stress by the second.

To completely define all stresses acting at point 0, we must be able to specify the stresses on all possible planes which can be passed through that point.

The plane shown in Fig. 13.10 is only one of an infinite number of possible planes which can be passed through point 0. To be able to resolve the stresses acting on any plane of any orientation, we need first of all to specify the stress vectors on three mutually perpendicular planes. This is shown in Fig. 13.11, which illustrates the possible distribution of stresses on the surfaces of a small volume element. Imagining this volume element to shrink to zero at

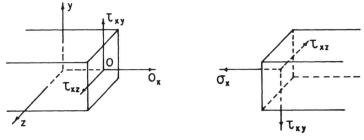

Fig. 13.10. Normal and shear stresses acting on a plane surface inside a continuous body.

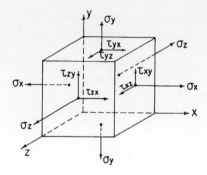

Fig. 13.11. Stress components on surfaces of a cubic element of volume.

point 0, we have at the limit a point of intersection of three orthogonal planes having the same stresses as shown.

Since the surfaces of our volume element are taken to be orthogonal to the three coordinate axes x, y, z, we can define the three normal stresses, namely, σ_x, σ_y, σ_z. Likewise, we can define six shear stresses in three pairs: τ_{xy}, τ_{yx}; τ_{zy}, τ_{yz}; τ_{xz}, τ_{zx}. If the volume element is in a state of equilibrium, as we assume it to be, the following conditions must hold:

$$\sum F_x = 0, \quad \sum F_y = 0, \quad \sum F_z = 0,$$
$$\sum M_x = 0, \quad \sum M_y = 0, \quad \sum M_z = 0 \quad (13.23)$$

wherein F_x is the component of force acting in the x direction and M_x is the moment of the force about the x axis. We now compute the components of all forces about the z axis. The forces are obtained from the products of the forces and the areas upon which they act, and the moments are obtained from the products of the forces and their arms. Thus,

$$\sum M_z = (\tau_{xy}\, dy\, dz)\, dx - (\tau_{yx}\, dx\, dz)\, dy = 0 \quad (13.24)$$

(wherein clockwise action is taken as negative). From this equation we obtain

$$\tau_{yx} = \tau_{xy}$$

and, similarly,

$$\tau_{zy} = \tau_{yz}, \quad \tau_{xz} = \tau_{zx} \quad (13.25)$$

We thus remain with three normal and three tangential stresses which need to be specified to define the state of stress at a point.

We next consider the resolution of stresses on an arbitrarily inclined plane inside our volume element of Fig. 13.11. For the sake of simplicity, we assume that the plane is inclined only to the x and y axes, but parallel to the z axis, as shown in Fig. 13.12. We further assume that σ_x is the only stress (i.e., τ_{xy} and τ_{xz} are zero).

The inclined plane is orthogonal to the direction **n**. As the figure shows,

G. The Mohr Circle of Stresses

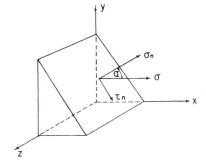

Fig. 13.12. The resolution of τ_x on an inclined plane.

σ_x can be resolved into two components on this plane: one normal to the plane (σ_n), the other tangential (τ_n). The relative magnitude of these two components depends on the angle α between the direction **n** and the x axis. There are two possible directions in which τ_n is zero: planes perpendicular to the x axis, and planes parallel to it. Planes upon which the tangential stresses disappear are called *principal planes*, and, conversely, the normal stresses acting on such planes are called *principal stresses*.

G. The Mohr Circle of Stresses

Those of us who tire occasionally of the seemingly endless sequence of equations encumbering scientific texts may yearn from time to time for an enlightening graphic representation to help us visualize the relationships which all too often remain obscure in the abstract mathematical formulation. Fortunately, such a representation of stress distribution within an homogeneous body is available to us in the form of a simple circle, called the *Mohr circle*. In order to grasp it, however, we must endure just a few more geometric and trigonometric manipulations (Fig. 13.13):

$$\sum F_{\sigma_\alpha} = 0$$

Therefore,

$$\sigma_\alpha \, dA = (\sigma_1 \, dA \cos \alpha) \cos \alpha + (\sigma_3 \, dA \sin \alpha) \sin \alpha$$

where $dA \cos \alpha$ is the area upon which σ acts, $\sigma_1(dA \cos \alpha)$ is the force exerted by σ_1 on this principal plane, and $(\sigma_1 \, dA \cos \alpha) \cos \alpha$ is the component of this force in a direction normal to plane dA. Similarly, $\sigma_3 \, dA \sin \alpha \sin \alpha$ is the component of force resulting from σ_3 acting in a direction normal to dA. Thus,

$$\sigma_\alpha = \sigma_1 \cos^2 \alpha + \sigma_3 \sin^2 \alpha \tag{13.26}$$

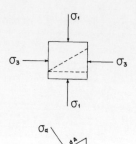

Fig. 13.13. Resolution of stresses on an inclined plane dA in a two-dimensional element of soil.

From basic trigonometry we know that

$$\cos^2 \alpha = \tfrac{1}{2}(1 + \cos 2\alpha), \qquad \sin^2 \alpha = \tfrac{1}{2}(1 - \cos 2\alpha)$$

Hence,

$$\sigma_\alpha = \tfrac{1}{2}(\sigma_1 + \sigma_3) + \tfrac{1}{2}(\sigma_1 - \sigma_3)\cos 2\alpha$$

or

$$\sigma_\alpha - \tfrac{1}{2}(\sigma_1 + \sigma_3) = \tfrac{1}{2}(\sigma_1 - \sigma_3)\cos 2\alpha \tag{13.27}$$

Similarly, equilibrium requires that all forces associated with τ_α balance out:

$$\sum F_{\tau_\alpha} = 0$$

Therefore,

$$\tau_\alpha\, dA = (\sigma_1\, dA \cos \alpha)\sin \alpha - (\sigma_3\, dA \sin \alpha)\cos \alpha$$

or

$$\tau_\alpha = \sigma_1 \sin \alpha \cos \alpha - \sigma_3 \sin \alpha \cos \alpha$$

$$\begin{aligned}\tau_\alpha &= \sigma_1 \sin \alpha \cos \alpha - \sigma_3 \sin \alpha \cos \alpha \\ &= (\sigma_1 - \sigma_3)\sin \alpha \cos \alpha\end{aligned} \tag{13.28}$$

Again, we recall from trigonometry that

$$\sin \alpha \cos \alpha = \tfrac{1}{2}\sin 2\alpha$$

Therefore,

$$\tau_\alpha = \tfrac{1}{2}(\sigma_1 - \sigma_3)\sin 2\alpha \tag{13.29}$$

G. The Mohr Circle of Stresses

Now we square both Eq. (13.27) and (13.29):

$$[\sigma_\alpha - \tfrac{1}{2}(\sigma_1 + \sigma_3)]^2 = [\tfrac{1}{2}(\sigma_1 - \sigma_3) \cos 2\alpha]^2$$

$$\tau_\alpha^2 = [\tfrac{1}{2}(\sigma_1 - \sigma_3) \sin 2\alpha]^2$$

Finally, we add the last two equations (remembering from trigonometry that $\sin^2 2\alpha + \cos^2 2\alpha = 1$) to obtain

$$[\sigma_\alpha - \tfrac{1}{2}(\sigma_1 + \sigma_3)]^2 + \tau_\alpha^2 = [\tfrac{1}{2}(\sigma_1 - \sigma_3)]^2 \qquad (13.30)$$

This corresponds to the equation of a circle

$$(x - a)^2 + y^2 = b^2$$

whose radius is b and whose center is at $(a, 0)$. As we mentioned before, the circle which can be drawn on the hypothetical σ–τ plane is called the *Mohr circle*. In the general case, as shown in Fig. 13.14, the center of the circle is at $(\tfrac{1}{2}(\sigma_1 + \sigma_3), 0)$ and the radius of the circle is $\tfrac{1}{2}(\sigma_1 - \sigma_3)$. This graphic representation enables us to determine the values of σ and τ for any angle of inclination α.

As Eq. (13.29) and Fig. 13.14 show, τ_α is maximal where $\sin 2\alpha = 1$, i.e., where $2\alpha = 90°$ ($\alpha = 45°$). At this point $\tau_\alpha = \tfrac{1}{2}(\sigma_1 - \sigma_3)$. The minimal value of τ_α occurs where $\alpha = 0°$. Maximal σ occurs where $\alpha = 0°$ (i.e., $\cos 2\alpha = 1$), whereas minimal σ occurs at $\alpha = 90°$ ($\cos 2\alpha = -1$). The values of τ are equal on orthogonal planes. In a uniform body, failure can be expected to occur on the plane over which the ratio of the shearing stress τ_α to the normal stress σ_α is maximal.

The Mohr circle also shows very clearly that the tangential (shearing) stress on any plane in a body depends upon the magnitude of the difference between the two principal stresses, σ_1 and σ_3. For a purely isotropic stress ($\sigma_1 = \sigma_3$) the Mohr circle shrinks to a point and no shearing stresses occur

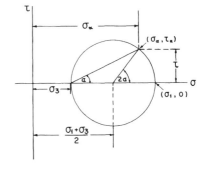

Fig. 13.14. The Mohr circle of stress.

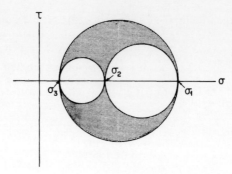

Fig. 13.15. The Mohr representation of the three-dimensional stress state at a point. The dotted area includes the stress components on all possible planes passing through the point.

on any plane. The difference between the major and minor principal stresses is called the *deviator stress*, and its magnitude determines whether a body under stress will maintain itself or whether it will fail. As the deviator stress increases, the shearing stress increases, and as the latter exceeds the strength of the body on any particular plane, the body experiences failure (i.e., breakdown or collapse).

The two-dimensional Mohr representation of stresses shown in Fig. 13.14 can be generalized to three dimensions (e.g., Jaeger, 1956) so as to describe the normal and shearing components of the stress vector which act on all possible planes passing through a point. Arranging the principal stresses so that σ_1 represents the largest principal stress and σ_3 the smallest gives three circles, as shown in Fig. 13.15. The shaded area in Fig. 13.15 describes the normal and shearing stress components on every possible plane. In the special case where two of the principal stresses are equal, the three circles coalesce and the shaded area becomes the circumference of a single circle, as in Fig. 13.14.

H. Stress–Strain Relations and Failure of Soil Bodies

In general, whenever a body of soil is subjected to stress, some strain will take place. If the magnitude and duration of the stress application are small, strain may be practically unnoticeable and disappear after removal of the stress. If, however, the magnitude of the stress is large, considerable strain may result with only partial recovery or none at all, so that permanent deformation results.

The mathematical description of strain is rather difficult. To describe strain at a point, the lengths and positional angles of lines or planes containing that point must be determined before and after the deformation.

The relative change in length of a line element is a measure of longitudinal strain, whereas the relative change in angle between two initially perpendicular line elements on a plane is a measure of shearing strain. Unfortunately, complex deformations may take place which cannot be defined in such simple terms (e.g., a line element may become curved during strain).

The conventional definitions of strain given by Eq. (13.1) and (13.2) apply to small strains only (normally confined to values of less than 0.1%). The usefulness of the strain concept can be extended to larger deformations by redefining it in terms of *natural strain*, namely, the incremental change of length or angle divided by the instantaneous value rather than by the original prestrained value.

In soil, strain is often very large. For example, an increase in bulk density from 1.1 to 1.4 gm/cm^3 represents a volume change of 27%. In small and confined test samples of soil, strain may be fairly uniform. Under natural soil conditions, though, deformation can take place in a very complicated geometric pattern.

There is as yet no way to predict a priori the amount and type of strain expectable from a given form and magnitude of stress (Gill and Vanden Berg, 1967). Hence the stress–strain relations of a soil must in each case be determined experimentally. One possible idealization is to assume that the soil obeys the *theory of elasticity*, based upon the linearity of stress versus strain, and the complete recovery of shape and volume after stress ceases. The *theory of plasticity* treats with permanent deformations and relates stress to strain through a parameter that is itself a function of the total strain. *Elastoplastic* materials are intermediate in that they exhibit partial recovery.

Soil deformation is often time dependent in a manner that does not conform to either elastic or plastic theories. More complex models are therefore needed to describe soil behavior. For instance, McMurdie (1963) applied the theory of *viscoelasticity* to soil in a study of creep phenomena. On the whole, progress in developing stress–strain–time relations which adequately characterize soil behavior has been slow, apparently owing to the inherent difficulty of the subject as well as to soil variability in space and time as regards composition, moisture, compaction, and aggregation.

The concept of *failure of a body under stress* underlies the very important concept of *strength*, which we shall discuss in our next section. "Failure" is sometimes used interchangeably with "yield," but this can be misleading. The concept of yield originally served to characterize the behavior of ductile metals. As stated in a previous section, when a rod of a ductile metal is stressed by tension, the strain is at first elastic. As the stress is increased, a point is eventually reached, called the yield point, beyond which plastic behavior is observed. In common engineering usage, the yield stress is taken to be the highest stress which can be borne by the metal member of a structure. In actual fact, there may be a considerable difference between the yield stress and what is known as the *ultimate stress,* or *failure stress,* at which fracture or separation actually occurs. Failure in soil is more complex than in most ductile or brittle materials. Depending on such variables as moisture and clay content, soils can vary their properties from those resembling a viscous liquid to those resembling a brittle solid.

Tensile failure of soil specimens has been measured in an attempt to establish the cohesive component of shear strength. Such measurements have been made on briquettes of soil formed in molds and are rather simple to carry out (Gill and Vanden Berg, 1967). Failure by compression is still more difficult to define than failure by tension, as the nonisotropic compression of a soil specimen may cause it to fail by shear. Nevertheless, values have been reported for compression failure (e.g., Winterkorn, 1936), and these generally indicate that soil bodies are apparently stronger in compression than in tension.

Failure of soil by shear is clearly definable when the soil is in the rigid and brittle state and exhibits a distinct fracture plane. An example is the shear fracture which often takes place in the so-called unconfined compression test, in which a solid cylinder of soil is loaded axially as shown in Fig. 13.16.

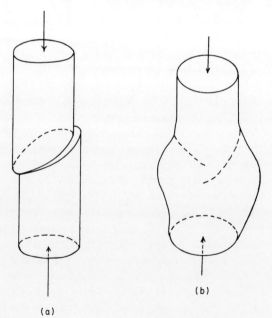

Fig. 13.16. Shear failure which occurs when a cylinder of soil is subjected to axial stress: (a) rigid soil—failure by fracture; (b) soft soil—plastic flow.

I. The Concept of Soil Strength

This type of failure is observed in cohesive, relatively dry soils. If the soil is moist and therefore less rigid (i.e., "soft" rather than "hard"), no distinct fracture is discernible. Rather, the soil specimen gradually thickens and bulges out, as shear failure occurs by plastic flow.

I. The Concept of Soil Strength

Stated qualitatively, *soil strength* is the capacity of soil to withstand forces without experiencing failure, whether by rupture, fragmentation, or flow. In quantitative terms, soil strength can be defined as the maximal stress which can be induced in a given soil body without causing the body to fail.

Though seemingly easy to define, soil strength is not at all easy to measure, being a highly variable property that often changes during the very process of measurement as the deformed body of soil might either decrease or increase in resistance to further deformation. In unsaturated soil, for instance, strength may increase as the soil becomes more compact; whereas in saturated or nearly saturated soils vibratory stresses may sometimes cause loss of cohesion and even liquefaction (a phenomenon known as *thixotropy*). The manner and rate of stressing can thus influence both the pattern of deformation and the mode of failure.

Since in most actual cases soil bodies fail by shear, it is useful at this point to concentrate specifically upon the ability of a soil to withstand shearing stress, known as *shearing strength*. A very useful approach to shearing strength is the theory of Mohr, based on the functional relationship between σ and τ. If a series of stress states just sufficient to cause failure is imposed on the same material and these states are plotted as a set or family of Mohr circles, as in Fig. 13.17, the line tangent to these circles, called the *envelope of the family of circles*, can be used as a criterion of shearing strength.

Where this envelope is found to be a straight line, or nearly so, it can be described mathematically by the equation

$$\tau = a + b\sigma \tag{13.31}$$

where the constant a, indicated in the figure as τ_0, is the intercept of the envelope line on the τ axis, and constant b is the tangent of angle ϕ which the envelope line makes with the horizontal line.

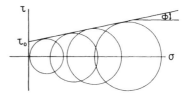

Fig. 13.17. Mohr's envelope of stresses.

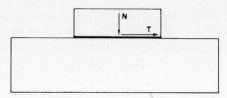

Fig. 13.18. Illustration of Coulomb's law: The tangential force T needed to overcome frictional resistance to the sliding of planar bodies over each other is proportional to the normal force N acting on the plane. Thus: $T = \mu N$, where μ is the coefficient of friction.

The linearity of τ versus σ pertaining to the internal shearing strength of bodies is analogous to that of *Coulomb's law* for sliding friction between bodies. Coulomb's law states that the frictional resistance toward a tangential stress tending to slide one planar body over another is proportional to the normal force pressing the bodies together (Fig. 13.18). Because of this analogy, the angle ϕ is often termed the angle of internal friction, Moreover, the intercept τ_0, visualized as the shear stress needed to cause failure at zero normal load, is generally called soil cohesion,[2] and designated by the symbol c. The relationship of Eq. (13.31) is generally expressed in these terms as

$$\tau = c + \sigma \tan \phi \qquad (13.32)$$

The Mohr theory of the straight line envelope has been widely accepted in engineering soil mechanics, although it does not necessarily represent shear failure conditions under all circumstances. As we shall see in the following section, the Mohr envelope can be determined experimentally by means of the *direct shear test*, or, preferably, by means of the so-called *triaxial shear test*. In principle, all points *under* the envelope represent states of stress which the soil can bear without failure; all points *on* the envelope are stress states just sufficient to cause failure, while points *above* the envelope are *impossible* stress states; that is to say, they represent stress states which the soil cannot stably bear (Barber, 1965).

J. Measurement of Soil Strength

The measurement of soil strength is generally based on some procedure for determining the stresses acting at incipient failure. As stated previously, soil resistance to applied stresses can be characterized in terms of two

---

[2] The term cohesion in this context, though prevalent, is not proper usage. Fundamentally, cohesion refers to the mutual attraction of like molecules, whereas the mutual adherence of heterogeneous particles and substances in the soil is more in the nature of *adhesion*. Hence the soil property designated as c in Eq. (13.32) would be more properly termed cohesiveness rather than cohesion.

J. Measurement of Soil Strength

parameters: *cohesiveness* c presumably representing the adherence or bonding of soil particles which must be broken if the soil is to be sheared and the *angle of internal friction* ϕ considered to represent the frictional resistance encountered when soil is forced to slide over soil. It should be understood that this interpretation of the two parameters as representing separate phenomena is merely a conceptual model, while in real cases they may not be truly independent of each other.

Soils which exhibit appreciable cohesiveness, called *cohesive soils*, generally contain a fair amount of clay which has the effect of bonding or cementing the soil internally. Dry sand, on the other hand, is generally *noncohesive*, hence the only resistance to shear is due to interparticle friction which results from the sliding and rolling of particles over each other. Dense sand might actually expand during shear, a phenomenon known as *dilatancy* and illustrated in Fig. 13.19. In the case of a moist (unsaturated) sand, the surface tension effect of the water menisci between grains imparts an *apparent cohesiveness* to the soil, which disappears upon either drying or saturation. This apparent cohesiveness is the reason why moist sand along the beach is fairly rigid and able to permit traffic even while the saturated sand under the water and the dry sand farther away from it will fail to support vehicles.

Two principal methods have been used to measure failure by shear in soil and thus to obtain quantitative values of c and ϕ (Sallberg, 1965). In the simpler of the two methods, the failure surface is controlled over a small area of soil, and the shearing stress necessary to cause failure is measured on several samples with different applied values of normal stress. These values are plotted directly on a τ–σ coordinate system. The line connecting the points is an envelope, and c and ϕ can be read directly. The devices used for these determinations fall into the general category of direct shear.

The direct shear apparatus is illustrated in Fig. 13.20, showing a sectioned container packed with soil to a known bulk density. Normal stress is applied to the shear plane by loading a pistonlike porous stone placed over the soil surface. Shearing is induced by gradually increasing the lateral, or tangential,

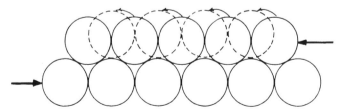

Fig. 13.19. Illustration of dilatancy (volume increase) during shearing of a dense sand, due to the interparticles rolling effect. It is the reason why a person stepping on saturated sand will observe that the sand around his foot seems to turn dry suddenly, only to become saturated again when one's foot is lifted off.

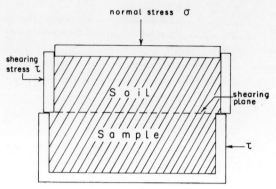

Fig. 13.20. Direct shear apparatus.

stress until failure occurs. The test can be carried out either at a constant rate of stress increase, or at a constant rate of deformation (the so-called stress-controlled versus strain-controlled methods). The direct shear test has several drawbacks. The shearing plane does not remain constant during the test, so the stresses can vary even if the normal and tangential forces remain constant. Moreover, the size and shape of the container influence the test results.

The alternative and more fundamentally based method for determination of shear strength is known as the *triaxial shearing test*. In this test, the failure surface is not predetermined but allowed to form within a specimen as successive combinations of lateral and axial stresses are applied to a series of samples of the same soil. Thus a series of Mohr circles representing shear failure states are determined, and the envelope of these circles yields the desired *c* and *b* values (Bishop and Henkel, 1964).

The triaxial shear apparatus[3] is illustrated in Fig. 13.21. A cylindrical sample of soil is wrapped in a thin sleevelike rubber membrane, tightly sealed to parallel metallic plates fitted with porous plugs at the top and bottom of the sample. All this is placed in a water-filled pressurized cell which provides a constant and uniform stress σ_3. An additional axial stress $\sigma_1 - \sigma_3$ is applied through a loading ram attached to a proving ring, and the stress is gradually increased until failure occurs, at which moment the axial stress as well as the lateral stress is noted. In a saturated sample the volume change can be determined by the quantity of water expelled from the sample during the

---

[3] The term triaxial for the apparatus described in Fig. 13.21 and the associated method of shear testing is really a misnomer, since stress is not controlled independently on three principal axes. With the lateral pressure imposed uniformly on the periphery of the sample, $\sigma_2 = \sigma_3$. Hence the apparatus and method should properly be called the cylindrical rather than the triaxial shearing test.

J. Measurement of Soil Strength

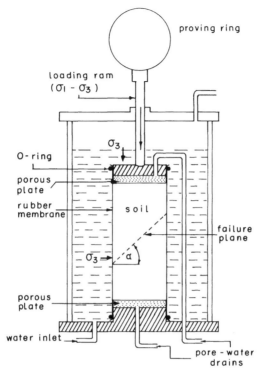

Fig. 13.21. The "triaxial" shear apparatus.

test. For pore water pressure measurements, the drains are connected to a manometer. In the case of unsaturated samples, tensiometric measurements can be made of matric suction and its changes during the test.

The internal stress σ acting on any plane inside a soil body consists of two components: the interparticle pressure, called the *effective stress*; and the *pore-water pressure*, which is neutral in the sense that it is isotropic and has no effect upon the shearing stress. We shall say more about effective versus total stress in our next chapter. For the present, however, it should be clear that increasing the lateral stress σ_3 acting on a *saturated* sample that is not allowed to drain can have no effect upon the deviatoric stress needed to cause failure, since it merely increases the hydrostatic pressure of pore water without having any effect on the soil matrix. Hence, the Mohr envelope in the case of an *undrained test* (also called the quick test) is horizontal. On the other hand, if the shearing test in the triaxial apparatus is carried out slowly and drainage is allowed so that at all times the pore-water pressure is zero (i.e., equal to atmospheric pressure), then the envelopes generally rise linearly, as shown in Fig. 13.22.

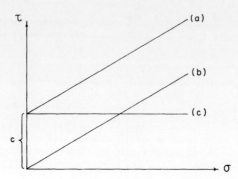

Fig. 13.22. Mohr envelopes for "triaxial" shearing of saturated soils: (a) drained test, cohesive soil; (b) drained test, noncohesive soil; (c) undarained test, cohesive soil.

A special case of the cylindrical shearing test is the so-called *unconfined compression test*, in which no lateral pressure is applied. Obviously, such a test can only be carried out on a soil specimen if the soil is cohesive. With σ_3 being zero, the axial stress σ_1 needed to cause failure is equal to the diameter of the Mohr circle, and hence to twice the value of soil cohesiveness c_u:

$$c_u = \sigma_1/2 \qquad (13.33)$$

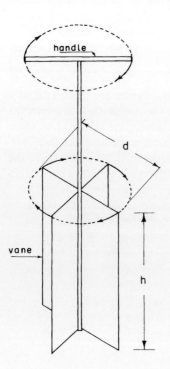

Fig. 13.23. The vane shear apparatus.

J. Measurement of Soil Strength

Another method of measuring soil cohesiveness, suitable for undisturbed in situ determinations, is the *vane shear test*, illustrated in Fig. 13.23 (ASTM, 1956). In this test, the vane is driven into the soil to the desired depth, and then rotated. The torque required to shear the soil, along the surface of a cylinder generated by the blade edges, is measured. The theoretical relation between torque T, soil cohesiveness c, and the dimensions of the blades is

$$T = c\pi(\tfrac{1}{2}d^2 h + \tfrac{1}{3}d^3) \tag{13.34}$$

For the standard height-to-radius ratio of 4:1, cohesiveness can be computed from the equation

$$c = 3T/28\pi r^3 \tag{13.35}$$

The advantage of this method is that measurements can be made at succeeding depths without extracting the apparatus, to obtain a complete strength profile of the natural soil (Evans and Sherratt, 1948). An alternative method for measuring shear strength of top soils in the field was developed by Fountaine and Payne (1951).

Measurement of soil cohesiveness is possible not only by shear but by tension as well. The *tensile strength* of soil is the normal force per unit area required to detach or pull apart one section of soil from another. A method for in situ measurement of tensile strength was proposed by Sourisseau (1935). The principle is shown in Fig. 13.24. Laboratory methods for determining tensile strength of soil specimens were reported by Gill (1959) and by Vomocil *et al.* (1961). A commonly used laboratory method of determining cohesive strength of dry soil briquettes is by measuring the *modulus of rupture* (Richards, 1953; Reeve, 1965b). This method consists of centrally loading a small beam of soil supported at both ends, until failure occurs (as shown in Fig. 13.25). The equation for determining the modulus of rupture σ_b is

$$\sigma_b = 3fl/2bd^2 \tag{13.36}$$

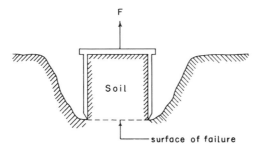

Fig. 13.24. Method for determining tensile strength of soil in the field.

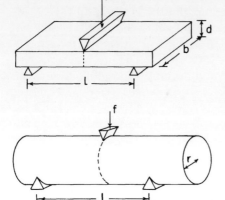

Fig. 13.25. Measurement of the modulus of rupture on a rectangular and on a cylindrical specimen of soil.

where f is the force applied to cause failure, l is the length of the beam, b the breadth, and d the depth or thickness, of the beam.

If the beam is cylindrical, the equation is

$$\sigma_b = fl/r^3 \tag{13.37}$$

where r is the radius of the cylinder of soil.

Modulus of rupture has been used to characterize the tendency of certain unstable soils to slake and form a dense surface crust that inhibits germination. The method was adapted to measurement of natural crust samples by Hillel (1959).

Still another method for measuring the strength of a cylindrical specimen of soil was proposed by Kirkham et al. (1959). In this method, the sample is loaded laterally until failure occurs and the cylinder is split in two, as shown in Fig. 13.26. The stress at failure is given by

$$\sigma = f/\pi lr \tag{13.38}$$

where l, r, f are as before.

A different technique for characterizing soil strength in the field is the use of a *penetrometer*. This instrument, illustrated in Fig. 13.27, is designed to evaluate soil resistance to the penetration of a narrow probe. What is measured is not soil strength per se, but a composite parameter considered to be related to soil strength, though the quantitative nature of its relation to soil strength is not generally known in any exact sense. Nevertheless, this technique has certain advantages, particularly in the ease and simplicity of in situ measurement (Davidson, 1965).

J. Measurement of Soil Strength

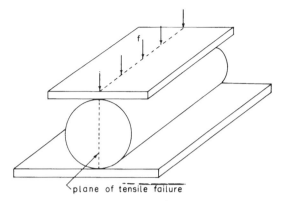

Fig. 13.26. Indirect tension method for measuring tensile strength.

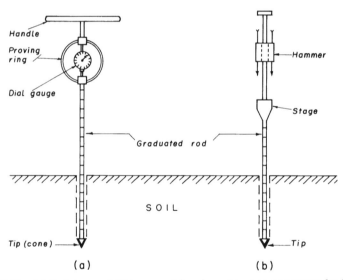

Fig. 13.27. (a) Push-type and (b) hammer-driven impact-type penetrometers for field characterization of soil strength.

When a metallic probe is inserted into the soil, several processes or effects may occur in combination, including cutting or separation of the soil, shear failure, plastic flow, compression, and metal-to-soil as well as soil-to-soil friction. The particular combination of processes occurring in any specific case depends upon the shape and type of instrument used, as well as on soil conditions. Many types of penetrometers have been designed, and some are available commercially, but there is as yet no standardization of shape or size of tip or of the mode of driving it into the soil. The results obtained from various penetrometers and the possibility of reducing them to a common dimensionless basis were analyzed by Kondner (1958).

The penetration resistance of the soil can be represented in at least two ways: (1) as the force required to cause penetration, per unit area of the probe tip or (2) as the energy required to cause penetration, per unit depth of soil. The penetrometer tips most often employed are conical, but disks have also been used.

Zelenin (1950) correlated data obtained by two types of penetrometers: a hammer-driven impact type versus a "static" type which is pushed slowly into the soil. His data indicate that the two methods give mutually proportional results. The static type of penetrometer has been fitted by several investigators (e.g., Terry and Wilson, 1953; McClelland, 1956; Hanks and Harkness, 1956) with a recording device to provide a continuous record of resistance with depth.

Penetrometers have been used very extensively to estimate the *draft* (force) required to plow soils under various conditions as well as to indicate soil *trafficability* (the ability of a soil to provide traction and allow unobstructed overland passage of vehicles such as tractors, automobiles, tanks, etc.). Other uses of penetrometers have also been proposed; e.g., Morton and Buchele (1960) used a penetrometer to simulate seedling emergence through a surface crust.

The modes of deformation and failure of various types of soil during tillage are illustrated schematically in Fig. 13.28.

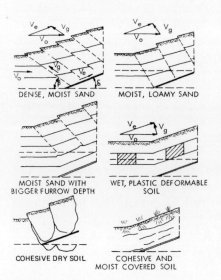

Fig. 13.28. Modes of deformation and shearing planes in different soils cut by a flat inclined blade. (From Söhne, 1966.)

K. Soil Consistency

Early attention to the various modes by which a soil may react to externally imposed forces led the forerunners of modern soil physics to develop a concept known as *soil consistency*. Like its sister term soil structure, however, the term soil consistency has resisted all attempts at exact quantitative definition. Typical is the definition proposed by Russell and Russell (1950), that soil consistency "designates the manifestations of the physical forces of cohesion and adhesion acting within the soil at various moisture contents... including the behavior toward gravity, pressure, thrust and pull... the tendency to adhere to foreign bodies (and) the sensations which are evidenced as feel by the fingers of the observer." In other words, the idea is to define how "consistent" a soil body can remain under stress, or to what extent it can maintain its shape when subjected to forces tending to cause deformation. No matter how difficult to define, the concept of soil consistency has spurred constructive efforts to find at least semiquantitative criteria to characterize what we now call the soil's rheological behavior. Although from a fundamental point of view its principal importance seems to lie in having served as a precursor of soil rheology, the study of soil consistency has in itself yielded significant practical benefits. From a practical point of view, it seems to have stood the test of time and its criteria are still used almost universally, especially by soil engineers.

The factor long recognized as the major determinant of soil consistency is the soil's degree of wetness, generally expressed as the mass of water present per unit mass of solids. Consider, for the sake of argument, a hypothetical body of soil of medium texture, e.g., a loam. When dry, such a body is likely to be relatively hard and brittle and exhibit a high degree of cohesiveness (or internal cementation) and a high resistance to tillage. If tilled nevertheless, a dry soil may break into hard and massive chunks, or clods, and if tilled excessively these clods may be pulverized into dust. When moist (but not too moist) the soil becomes typically *friable*, i.e., when tilled it tends to crumble easily and form a loose assemblage of relatively small, soft clods. In this state, the soil is at or near its "optimal wetness' for tillage, as it can be tilled to best advantage with the least investment of energy. As the degree of wetness is increased, however, the soil typically loses its friability and becomes puttylike, or *plastic*. When worked, instead of crumbling into clods, it tends to be molded ("puddled") into lumps which, upon drying, become extremely hard. In the plastic state, the soil is obviously to wet to be tilled effectively,[4] and if tilled nevertheless the soil will probably experience

---

[4] In fact, old farmers often test whether the soil has dried sufficiently to be tilled by kneading and rolling a pinch of the moist topsoil in the palm of the hand, then casting it on the ground. If the roll crumbles on striking ground, the soil is ready. If it remains a lump, the soil is still too wet. Here is a shining example of soil rheology reduced to its essentials.

structural degradation through mechanical destruction of its natural aggregates. If soil moisture is increased beyond the plastic range, the saturated soil becomes sticky and if worked it will become a muddy paste and tend to behave as a *viscous liquid*. In the extreme, as even more is added and the mixture churned, the soil will enter into a state of *suspension*.

Thus, in progressive transition from a dry state to a moist, then to a wet, then saturated, and finally to a supersaturated state, the soil by turns undergoes a rather dramatic series of changes in consistency, from a hard and brittle solid, to a soft and friable solid, to a moldable plastic semisolid, and then to a sticky and viscous liquid. These transitions were codified about 70 years ago by a Swedish soil scientist named Atterberg (1911, 1912), who devised simple empirical criteria, or testing procedures, known universally as the *Atterberg limits*. These procedures were designed to determine the mass wetness values at which a soil apparently changes from one consistency state to another. Students are often startled to discover how simplistic these testing procedures are, yet how universal in application.

Realistically, we know that the transition from one rheological state to another is not abrupt but gradual, and that much depends on the mechanical manipulation of the sample. The tested samples are not soil in the natural state, but soil material subjected to arbitrary treatment. We therefore cannot expect the consistency limits to be anything more than *indexes* of the workability or firmness of artificial mixtures of soil and water. Yet the concept underlying the Atterberg limits has proven to be extremely useful and has stood the test of time. In fact it characterizes the whole range of rheological transformations which a soil may undergo, and its actual results can be correlated with many of the fundamental properties and mechanisms of soil behavior that we have since learned to measure with great precision (Seed *et al.*, 1964), including swelling and shrinkage, compressibility, permeability, and strength. It is a remarkable fact that the Atterberg limits are still considered standard determinations in most laboratories of soil mechanics.

We shall give only a sketchy description of the Atterberg limits (illustrated schematically in Figs. 13.29 and 13.30), as the detailed procedures are given in the appropriate ASTM manual (1958) and in other standard handbooks (e.g., Sowers, 1965).

1. *Flocculation limit*: the mass wetness at which a soil suspension is transformed from a liquid state to a semiliquid state with appreciable increase in viscosity.

2. *Liquid limit*: the mass wetness at which the soil–water system changes from a viscous liquid to a plastic body. This limit is also called the *upper plastic limit*. It is measured in a special apparatus sketched in Fig. 13.31. The cup is filled with soil at different water contents. A groove is made with a

K. Soil Consistency

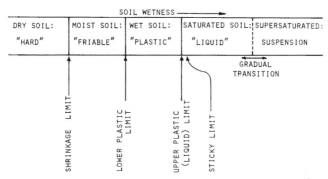

Fig. 13.29. The Atterberg consistency limits (schematic).

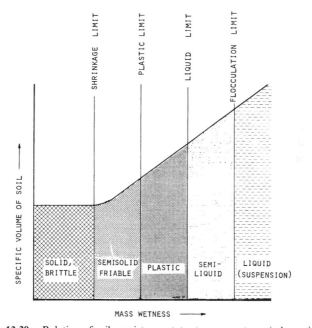

Fig. 13.30. Relation of soil consistency states to mass wetness (schematic).

grooving tool. The cup is then lifted and dropped from a standard height of 1 cm until the soil has flowed sufficiently to close the groove along a 12 mm length. The number of impacts and the corresponding water content are plotted. The liquid limit is defined and interpolated as the water content at which the groove is closed by 25 impacts.

3. *Plastic limit*: the mass wetness at which the soil stiffens from a plastic

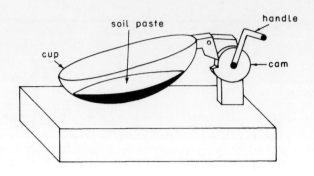

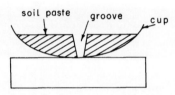

Fig. 13.31. Liquid limit apparatus for determination of the Atterberg "upper plastic limit."

to a semirigid and friable state. This is also called the *lower plastic limit*. In practice, the plastic limit is defined as the specific water content by mass at which a sample of soil can just be rolled into a thread of 3 mm diameter without breaking. It thus characterizes the lower end of the range over which a clayey soil is in a plastic state.

4. *Shrinkage limit*: the mass wetness at which the soil changes from a semirigid to a rigid solid with no additional change in specific volume as drying proceeds still further. Usually a cylindrical specimen of soil in the plastic state is obtained, and its weight and change of volume with continuing loss of water is determined. When there is no further volume change the specimen is dried out and its final volume and dry weight are found. Shrinkage of a soil specimen is illustrated in Fig. 13.32.

5. *Sticky limit* (less frequently used): the minimum mass wetness value at which a soil paste will adhere to a steel spatula drawn over its surface.

These are not entirely objective procedures. In fact, the ability to perform these tests in a reproducible and dependable way is an acquired skill which comes with experience, and is still something of an art rather than an exact science. We should note also that the tests are not very suitable for coarse-textured soils, which do not exhibit much plasticity, and are really meant primarily for soils with an appreciable content of clay. Sooner or later, the

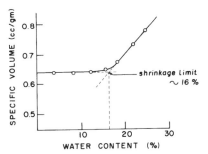

Fig. 13.32. The shrinkage behavior of a clayey soil (hypothetical).

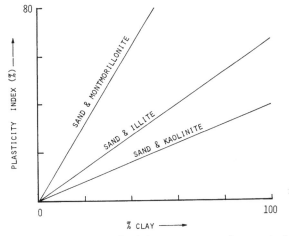

Fig. 13.33. Dependence of the plasticity index upon type and concentration of clay in various clay-sand mixtures (schematic).

Atterberg limits seem destined to be replaced by more rigorous and objective procedures for characterizing soil materials on the basis of their stress–strain–time behavior. The fact that the Atterberg system has survived so long in engineering practice attests to the inherent difficulty of defining exact soil properties and to the need for simple (if not simplistic) criteria.

A useful index derived from the consistency limits is the *plasticity index*, defined as the difference between the liquid and the plastic limits. It is generally taken to be an indication of a soil's "clayeyness" or potential plasticity, and is used as such in engineering classification systems for soils (Casagrande, 1948). The plasticity index depends, however, not only on the clay content of a soil but also on the nature of the clay present, whether of the swelling or nonswelling type (see Fig. 13.33), as well as on the adsorbed cations, the organic matter content, and, as already mentioned, on pretreatment of the sample.

Sample Problems

1. A triaxial shearing test was performed on a cohesive soil. Three samples of the soil were mounted in an apparatus of the type shown in Fig. 13.21. Each sample was subjected to a different cell pressure and loaded axially at a slow rate, so as to allow free drainage and dissipation of pore-water pressure. The axial stress at which failure occurred was recorded in each case. The following data were obtained:

Cell pressure, σ_3: 1.0 2.0 3.0 bar
Axial load, σ_1:   5.9 8.5 11.2 bar

Plot the set of Mohr circles and their envelope, and determine the soil's "cohesiveness" and "angle of internal friction." Predict the deviator and axial stresses, as well as the normal and shear stresses on the failure plane, at a lateral stress of 2.5 bar.

The three stress circles are drawn in a graph, Fig. 13.34. Note that the diameter of each circle is equal to the deviator stress $\sigma_1 - \sigma_3$. A tangent is drawn to the three circles and is seen to be a straight line. When this line is extrapolated leftward, it intercepts the τ axis. According to the Mohr theory of stress, this line can be expressed in terms of Eq. (13.32):

$$s = c + \sigma_n \tan \phi$$

where s is the shearing strength of the soil.

From the graph, we determine the intercept c, representing the soil's cohesiveness, to be 1 bar. The slope of the τ–σ function (representing $\tan \phi$) is seen to be about 0.5. Hence the angle of internal friction ϕ is about 26.5°.

We can now use these parameters to determine the normal and shear stresses on any plane in a sample subjected to any pair σ_1, σ_3 under the envelope. For example, if the lateral stress is 2.5 bar, the circle of failure, tangent to the envelope, is as drawn in Fig. 13.34 (the dotted circle), The deviator stress (equal to the circle's diameter) is seen to be 7.4 bar (i.e., an axial load of about 9.9 bar is needed to cause failure). We can determine the normal and shearing stresses on the failure plane graphically by reading the values of σ and τ at the point of the circle's tangency with the envelope line. These values turn out to be $\sigma_n = 4.5$ bar, $\tau = 3.25$ bar.

Note: The Mohr envelope for a confined system becomes horizontal if the soil is saturated. Additional cell and axial pressures cannot add to the effective stress, merely to pore-water pressure.

Sample Problems

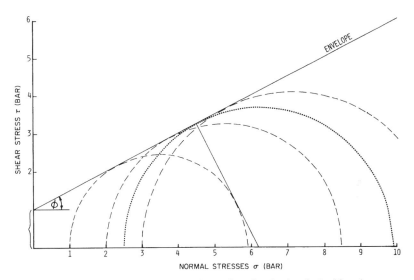

Fig. 13.34. Mohr circles and their envelope for the data in Problem 1.

2. Two briquets of soil crust were tested in modulus-of-rupture devices: (1) The first specimen had a width of 2 cm and a thickness of 0.5 cm, and was placed on 5-cm spaced supports and fractured by a force of 10^5 dyn. (2) The second specimen was 5 cm wide and 2 cm thick, and when placed on supports that were 10 cm apart required a force of 10^6 dyn to be fractured. Calculate which of the two specimens had the greater strength.

We calculate the modulus of rupture M by the following equation:

$$M = 3FL/2BD^2$$

where F is the force causing fracture, L is the length of the specimen between supports, B is breadth (width), and D is depth (thickness).

For sample (1):

$$M = \frac{3 \times 10^5 \text{ dyn} \times 5 \text{ cm}}{2 \times 2 \text{ cm} \times (0.5 \text{ cm})^2} = 1.5 \times 10^6 \text{ dyn/cm}^2 = 1.5 \text{ bar}$$

For sample (2):

$$M = \frac{3 \times 10^6 \text{ dyn} \times 10 \text{ cm}}{2 \times 5 \text{ cm} \times (2 \text{ cm})^2} = 7.5 \times 10^5 \text{ dyn/cm}^2 = 0.75 \text{ bar}$$

Note: The second sample, although it required a force ten times greater to be fractured, actually has only half the strength of the first sample. Incidentally, for proper testing of modulus of rupture, the ratio between the specimen's length (between supports) and its thickness should be at least 4.

> Our knowledge is fragmentary
> And so is our prophesying...
> I Corinthians 13:9

14 Soil Compaction and Consolidation

A. Introduction

When subjected to pressure, soil tends to compress; that is to say, it tends to increase its density. Hypothetically speaking, we can imagine several possible mechanisms by which soil compression, or densification, could take place. One rather unlikely mechanism is that the solid particles themselves might be compressed, or be crushed against each other. We can dismiss this possibility because the compressibility of most solid-phase components (soil minerals) is too low to provide appreciable compression, and their strength too great to allow crushing, in the range of pressures normally encountered or administered in agricultural and engineering practice. Another possible mechanism is the compression of the liquid phase, i.e., of soil water. Deep aquifers, under great overburden pressure, do indeed exhibit appreciable water compression. Not so, however, near the soil surface, where soil water is seldom confined (and hence is free to flow away when placed under pressure) and the pressures are seldom great enough to result in significant compression of water.

Still another mechanism, perhaps more likely to be of some importance, is the possible compression of confined air. In a very wet soil, air may be occluded (trapped) in isolated small pockets during the wetting phase, or it may effervesce out of solution to form bubbles when the temperature rises. Gases may also be released by biological or chemical reactions. Air is, of course, very much more compressible than either mineral solids or water. (The presence of compressible air bubbles has in fact been offered as an explanation for the "springiness" of wet soil so familiar to those who have had frequent occasion to walk in cultivated fields during or shortly after a rain.) However, the fractional volume of occluded air bubbles in a wet soil is generally small and, as the soil begins to drain, the air phase in the soil soon becomes continuous and open to the atmosphere so that air compression in an unsaturated soil is itself not likely to be significant except in special cases.

If the densities of none of the individual components of the soil changes materially, how then can the soil as a whole compress? What remains as the principal mechanism of soil compression is, obviously, the reduction of soil porosity through the partial expulsion of either or both of the permeating fluids, air and water, from the compressing soil body. Here we can consider two extreme cases and a whole range of intermediate conditions. The first case is that of a totally dry soil. Its compression under static pressure or by vibration causes the particles to reorient and to assume a closer packing arrangement, thereby reducing the fractional volume of air. In the opposite case of a water-saturated soil, any such decrease of porosity must necessarily take place at the expense of the fractional volume of water. The difference is, of course, that the viscosity of water is 50–100 times greater than that of air. Hence the expulsion of air is a rapid, almost instantaneous, phenomenon, whereas the expulsion of water is generally a very much slower process, particularly as progressive compression repeatedly closes the largest pores and requires subsequent flow to take place in narrower and more tortuous pores. Application of pressure to soil in any intermediate state between dryness and saturation, i.e., to a moist soil, will result first in expulsion of air and a gradual approach to saturation; only after essentially all of the air has been driven out and the soil in effect has become saturated will the further application of uncompensated compressive stress begin to result in removal of water.

To distinguish between the two processes or phases described, conventional wisdom requires that we designate them by different names. Traditionally, the term *compaction* has been applied to the compression of an unsaturated soil body resulting in reduction of the fractional air volume. The term *consolidation*, on the other hand, has long been used to signify the compression of a saturated soil by "squeezing out" water. (Note that all along we have been using the overall term compression to depict all processes of soil densification, including both compaction and consolidation.) We shall proceed to describe each of these processes in turn.

B. Two Opposing Views of Soil Compaction: Engineering and Agronomic

Soil compaction can be considered from the viewpoints of the civil engineer and of the agronomist. Traditionally, the civil engineer has considered the soil as a *construction material*, e.g., for building roadbeds, embankments, and landfills; as a *foundation* for dams and various buildings; or as a *tractable surface* for vehicles. Hence the engineer's interest lay in the possibility of manipulating the soil in ways designed to increase the strength and stability, and decrease the permeability, of soil layers. The process of soil compaction is an engineering technique to accomplish the desirable goal of forming the densest and tightest possible soil condition. To the agronomist interested in the soil as a medium for plant growth, on the other hand, soil compaction is a scourge, an undesirable consequence of mechanization, which must be avoided lest it result in the soil becoming unfit for crop production. We shall have to deal with each of these approaches on its own merits.

C. Soil Compactibility in Relation to Wetness

Having as their goal the attainment of high soil bulk density so as to minimize future subsidence, increase shearing strength, and reduce percolation, soil engineers have long sought methods for achieving the maximal possible degree of compaction with the least expenditure of energy. An empirical approach to this problem was developed nearly 50 years ago by R. R. Proctor, whose testing procedure, known as the *Proctor test*, was designed to determine the "optimal" soil wetness at which compaction of a given soil can be achieved most effectively by a given compactive effort. Although modified repeatedly and given different names, Proctor's basic procedure has enjoyed universal acceptance and has been adopted as the standard criterion for soil compaction in engineering practice (ASTM, 1958).

In principle, compactive stresses can be administered in several different ways. Perhaps the simplest is to confine a soil sample in a rigid-walled container and apply a *static load* (e.g., a weight) by means of a piston resting on the sample's surface. An alternative way is to apply a *dynamic load*, i.e., a time-variable load, as, for example, by means of an impacting or hammering device. Still another way, found to be most effective in the case of dry granular materials, is to vibrate the sample. Finally, one could attempt to compact soil by applying a space-variable and time-variable set of compressive and shearing stresses in combination. Such action, which in effect causes a churning (or "puddling") of the soil, is commonly called *kneading compaction*. It is this latter method which forms the basis of the Proctor test, which is, in essence, an imitation of the common engineering practice of compacting soil in the field by means of spiked rollers. The standard test is carried out in a cylindrical mold, 10 cm in diameter and having the capacity of 100 cm^3. The soil material is compacted in three layers, each layer being worked by a 5-cm diameter impact-driven tamper in a standardized manner.

For any given amount of compactive effort, the resulting bulk density is a function of soil moisture (wetness). This functional dependence, illustrated in Fig. 14.1, indicates that, starting from a dry condition, the attainable bulk density at first increases with increasing soil wetness, then reaches a peak called maximal density at a wetness value called optimum moisture, beyond which the density decreases. This behavior is readily explainable, at least qualitatively. Typically, a dry soil resists compaction because of its stiff matrix and high degree of particle-to-particle bonding, interlocking, and/or frictional resistance to deformation. As soil wetness increases, the moisture films weaken the interparticle bonds, cause swelling, and seem to reduce internal friction by "lubricating" the particles, thus making the soil more workable and compactible. As soil wetness nears saturation, however, the fractional volume of expellable air is reduced and the soil can no longer be compacted by a given compactive effort to the same degree as before. Henceforth, any further increase in soil moisture reduces, rather than increases, soil compactibility. Finally, at saturation, no amount of kneading can cause any increase in soil bulk density (unless water is expelled, as in consolidation, which, however, is not the subject of this section). At high wetness values, water is seen to prevent closer packing of the soil matrix. Rather, the water which hydrates the grains pushes them apart and causes swelling, thus reducing the attainable bulk density.

The function described, depicting the dependence of attainable bulk density upon soil moisture, does not constitute a single characteristic curve for a given soil but a family of curves, as shown in Fig. 14.2. For each level of compactive effort, there is a separate curve. With higher levels of compactive effort, the curve is shifted upward and leftward, indicating higher attainable bulk density at lower values of "optimal moisture."

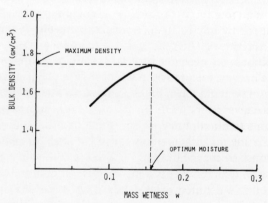

Fig. 14.1. Typical moisture-density curve for a medium-textured soil, indicating the maximum density obtainable with a particular compactive effort.

C. Soil Compactibility in Relation to Wetness

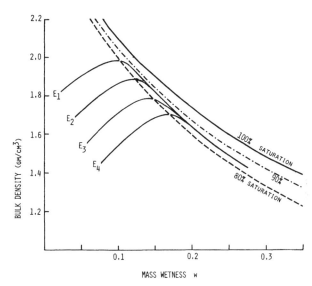

Fig. 14.2. Family of moisture-density curves for different compactive efforts ($E_1 > E_2 > E_3 > E_4$). Note that the 100% saturation line, representing zero air-filled porosity, was calculated assuming a particle density value of 2.65 gm/cm^3.

Experience shows that the curved line connecting the peaks of all the bulk density versus wetness curves corresponds approximately to the 80% degree-of-saturation line, and that the descending portions of these curves tend to converge on a curved line representing a degree of saturation of about 85–90%. To the extent that the Proctor test portrays or simulates compaction in the field, it can serve as a criterion by which to guide earth construction works. A typical job specification might require a contractor to compact an earth fill to, say, 90% of attainable maximal density as defined by the Proctor test (or by any of the derivative tests such the "AASHO" or "Modified AASHO" test [ASTM, 1958b]). The contractor, if he knows anything at all about soil mechanics, will then examine the results of the test to determine the optimal wetness value (or range of values) at which to carry out the compacting operation. In practice, an earth fill is laid by depositing and rolling successive layers while controlling soil moisture through measured applications (sprinkling) of water to the deposited material. On the other hand, if the soil material is too wet, it must be dried by drainage and, weather permitting, by evaporation.

In our discussion thus far we have made the implicit assumption that our "typical" soil is a fine-textured or medium-textured one (e.g., a clay, clay-loam, silt-loam, or loam) with some appreciable cohesiveness. If, however, the material to be compacted is coarse-textured (granular), then the most effective method of compaction is not by rolling or kneading but by vibration. Various types of ramming or jolting devices have been tried for this purpose.

D. Occurrence of Soil Compaction in Agricultural Fields

In the agronomic context, soils or soil layers are considered to be compacted when the total porosity, and particularly the air-filled porosity, are so low as to restrict aeration, as well as when the soil is so tight, and its pores so small, as to impede root penetration and drainage. Still another manifestation of soil compaction is the difficulty it creates in field management, particularly with respect to tillage.

Soils and soil layers may become compact naturally as a consequence of their textural composition, moisture regime, or the manner in which they were formed in place. *Surface crusts* can form over exposed soils under the beating and dispersing action of raindrops and the subsequent drying of a compacted layer of oriented particles. Naturally compact *subsurface* layers may consist of densely packed granular sediments, which may be partially cemented. Indurated layers, called *hardpans*, can be of variable texture and may, in extreme cases, exhibit rocklike properties (in which case they have been termed *fragipans*, or *ortstein*) and become almost totally impenetrable to roots, water, and air. Usually such hardpans are found at the junction of two distinctly different layers where penetration of water and/or dissolved or suspended materials is retarded by a clay layer, a water table, or bedrock (Lutz, 1952).

A *claypan* is a tight, restrictive subsoil layer of high clay content which tends to be plastic and relatively impermeable to water and air when wet. In humid climates, such layers may remain perpetually wet and give rise to perched water-table conditions above them, thus inducing anaerobic conditions within the root zone. Claypans may be depositional or may have developed in place. They occur at various depths in the profile. High clay content alone does not necessarily result in the formation of a claypan, as much depends on soil structure as well as on texture. Many good agricultural soils having clayey B horizons exhibit well-developed structure with large interped pores which permit the unobstructed passage of water and air. In claypans, however, structural development is poor and the clay may even be in a somewhat dispersed state. Claypans and hardpan conditions are difficult to rectify. Mechanical fragmentation by tillage alone may result in only temporary relief, as these layers tend to re-form spontaneously.

Another natural factor which can contribute to soil compaction is the tendency of clayey soil to shrink upon drying, particularly if it had been puddled in the wet state. Gill (1959) found that the bulk density of compressed samples increased from 1.54 gm/cm^3 at a mass wetness of 25% to 1.75 gm/cm^3 at a wetness of 20%. Well-structured soils, however, break into numerous small aggregates (crumbs) as they dry, so that even though individual aggregates shrink the layer as a whole becomes loose and porous.

Quite apart from the natural formation of compacted layers, soil compaction can, and all too often does, take place under the influence of man-induced mechanical forces applied to the soil surface. One such cause of soil compaction is trampling by livestock. In an example reported by Tanner and Mamaril (1959), grazing animals caused an increase in topsoil bulk density from 1.22 to 1.43 gm/cm^3, corresponding to a decrease of air-filled porosity from 17.3 to 7.2% and an increase in penetrometer resistance from 3.2 to 19.5 bar. However, by far the most common cause of soil compaction in modern agriculture is the effect of machinery, imposed on the soil by wheels, tracks, and soil-engaging tools.

E. Pressures Caused by Machinery

The magnitudes of pressures exerted on the soil surface by wheeled and tracked vehicles depend in a combined way on characteristics of the soil surface zone and of the wheels or tracks involved. The manner in which these pressures are distributed within the soil, and the deformations they cause, depend, in turn, on the pattern of surface pressure as well as on the mechanical or rheological characteristics of the soil in depth. This general topic was reviewed by Chancellor (1976). An invaluable store of data is available in the book by Gill and Vanden Berg (1967).

A general rule of thumb which applies to *pneumatic* (air-inflated) *tires* is that the pressure exerted upon the supporting surface is approximately equal to the inflation pressure. Since the total weight of a vehicle at rest should equal the sum of the products of the pressures exerted by the wheels and their

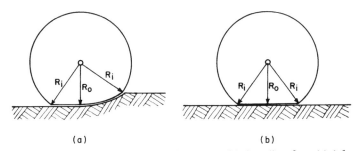

Fig. 14.3. Deformation of a pneumatic tire in contact with the soil surface: (a) deformation in the tire and soil; (b) deformation in the tire only. (From Gill and Vanden Berg, 1967.)

respective contact areas, one can reason that an increase in weight should be compensated by a flattening of the tires and a commensurate increase in contact area without change in inflation pressure (Fig. 14.3). In reality, however, there are considerable deviations from this principle, which disregards such factors as the stiffness of the tire walls, the presence of lugs or ribs, the shearing and slippage which are affected by drawbar pull, and the properties of the soil underfoot. Gill and Vanden Berg (1967) showed that the stiff walls of the tire carcass cause the pressures at the edges of the tire-to-soil contact area to be greater than those near the center (Fig. 14.4). More-

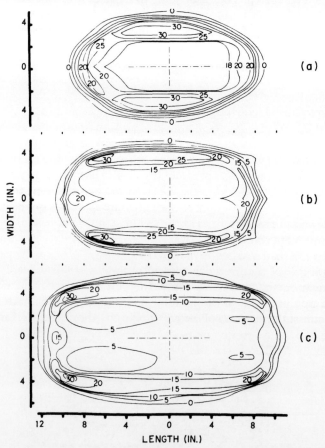

Fig. 14.4. Pressure distribution patterns under a smooth tire traveling from right to left on firm sand. Inflation pressures were (a) 14 lb/in.^{2}, (b) 10 lb/in.^{2}, and (c) 6 lb/in.^{2}. The centers of the patterns have pressures approximately equal to the inflation pressures, while the edge patterns show greater pressure due to tire carcass stiffness. Note that 14.7 lb/in.$^2 \approx$ 1 bar. (From Vanden Berg and Gill, 1962.)

E. Pressures Caused by Machinery

over, the longitudinal distribution of the pressures exerted during forward travel depends on the softness of the soil and the extent of sinkage. When inflation pressure is very high and the soil very soft, a pneumatic tire may behave like a rigid wheel and tire–soil contact pressures may all be below inflation pressure (Chancellor, 1976). Lugs on tires have the effect of concentrating a high pressure on a small fraction of the normal tire–soil contact area. This uneven distribution of pressures is, however, dissipated within the upper 15–25 cm of soil so that at greater depth there remains little difference between the pressures introduced by lugged and those by smooth tires. However, within the zone of influence, the kneading effect of the lugs may be a factor in soil puddling.

In the case of *rigid wheels*, the pressures transmitted to the soil depend very much on the softness or hardness of the soil itself. If the surface is firm, the contact area is reduced and the pressure is relatively high. If the soil is soft and easily deformable, the wheels sink into the ground so that the same load is distributed over a larger area. Hence soil pressures do not increase in proportion to the added load.

The ground pressure values for *crawler tracks* can be estimated by dividing the total weight of the vehicle by its contact area with the ground. There are, however, several factors which can cause deviations from this principle (Chancellor, 1976). During actual operation, tractors generally tilt backwards (Taylor and Vanden Berg, 1966), so that pressures on the rear side of the traveling track may be twice or thrice as great as the average pressure. This tilting and pressure shift, accompanied by shearing, tend to increase as drawbar pull by the crawler tractor is increased. Figure 14.5 illustrates this effect. A similar pattern of soil shearing and displacement has been observed under moving wheels, as shown in Fig. 14.6.

Determination of the magnitudes and distributions of the pressures exerted on the soil surface is the first step in any attempt to estimate the distribution of stresses within the soil body (Reaves and Cooper, 1960; Onafenko and Reece, 1967). Procedures for making such calculations were summarized by Chancellor (1976) on the basis of foundation theory (Tschebotarioff,

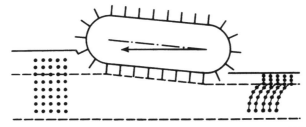

Fig. 14.5. Combined compaction and shearing deformation produced within the soil by a moving crawler tractor. (From Taylor and Vanden Berg, 1966.)

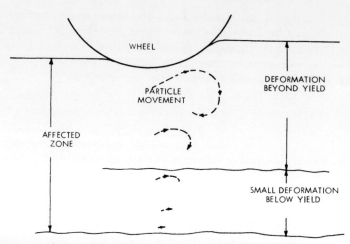

Fig. 14.6. Deformation and displacement patterns under a moving wheel. (From Yong and Osler, 1966.)

1973). The governing equations in this theory are known as the *Boussinesq* equations, which, though originally intended only to describe pressure distribution in uniform elastic materials, have long been used in soil mechanics [see, for example, the detailed description given by Plummer and Dore (1940)] and have been reported (Söhne, 1958) to be applicable to agricultural soils

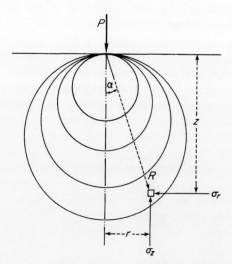

Fig. 14.7. Distribution of pressures under a concentrated vertical load applied to the soil surface, according to the Boussinesq theory. The circles represent lines of equal vertical stress σ_z.

E. Pressures Caused by Machinery

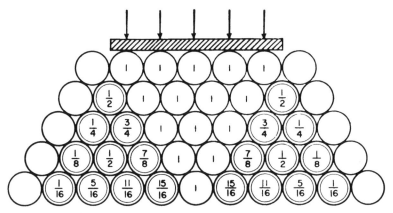

Fig. 14.8. Idealized vertical stress distribution in a granular material under partial-area load. (After Zelenin, 1950.)

also. The equations give the stresses at any point within a body of infinite extent due to a *concentrated load F* applied normal to the surface (Fig. 14.7). Disregarding the horizontal components of the stress, the equation for vertical pressure σ_z as a function of depth z and horizontal distance r (from the axis of the load) is

$$\sigma_z = 3Fz^3/2\pi(r^2 + z^2)^{5/2} \tag{14.1}$$

Note that directly under load ($r = 0$) the pressure decreases as the depth squared (i.e., $\sigma_z = 3F/2\pi z^2$).

To apply this theory to predict the downward propagation of pressures applied over a *finite area* of soil surface, we can divide the finite area into small units and assume each to be subject to a concentrated load, the effect of which spreads downward at an angle of 30° from the vertical. To estimate the total pressure acting at any point on any horizontal plane below the surface, one must add up the pressure contributions due to the forces acting on each surface element, as illustrated in Fig. 14.8 and 14.9. The theory predicts that, on any such plane at any depth, the pressure is maximal directly under the center of the loaded surface area and decreases toward the spreading edges. The magnitude of this maximal pressure felt at any depth within the soil depends not only on the magnitude of the pressure at the surface but also on the width of the surface area over which it is applied. Thus, increasing the surface area subjected to a given pressure will have the effect of increasing the maximal pressures experienced at all depths within the profile under the center of the loaded area. Consequently reducing the surface pressure by distributing the load over a larger surface area will decrease maximal sub-

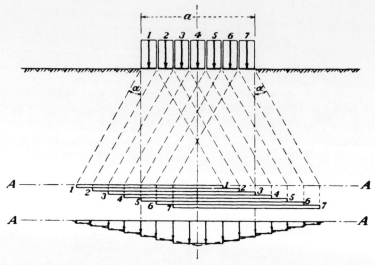

Fig. 14.9. Hypothetical distribution of vertical pressures within the soil, as under a crawler track, represented as a set of vertical blocks (1–7). Angle α is generally taken to be 30°. (From Tschebotarioff, 1973.)

surface pressures but to a degree less than proportional to the decrease in surface pressure. A pressure distribution pattern under a rear tractor tire, computed by Söhne (1958) on the basis of the theory described, is illustrated in Fig. 14.10.

Pressures are imparted to the soil not only by vehicles traveling on its surface, but also by tillage tools operating beneath the surface. As tools of various designs are thrust into and through the soil, several different effects may occur simultaneously, as the soil is cut, compressed, sheared, lifted, displaced, and mixed. Inevitably, some of the soil is pushed ahead by the moving tool against the resistance of the static soil body, and is thus compacted. Space does not permit an extended account of pressure distributions caused by different types of implements (such as moldboard plows, subsoilers, bulldozers, cultivators, rototillers, etc.), nor is the information available complete by any means. Reported pressures caused by tillage implements have been of the order of 1 bar for a bulldozer (Hettiaratchi *et al.*, 1966), 2–4 bar for plowshares and plowbottoms (Gill and McCreery, 1960; Mayanskas, 1959), and an estimated $5\frac{1}{2}$ bar for a subsoiler (Chancellor, 1976).

In the interest of simplification, we have thus far concentrated on the pressures applied by machinery to the soil. We must remember, however, that, in addition to pressures, machines quite inevitably impart shearing stresses as well. As shown by Taylor and Vanden Berg (1966), the very

F. Soil Compaction under Machinery-Induced Stresses

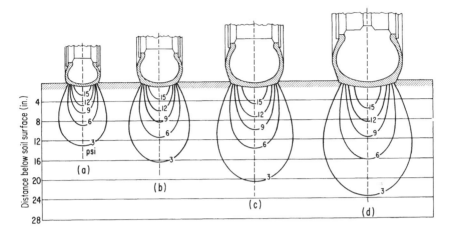

Fig. 14.10. Vertical pressures under tractor tires, computed on the basis of the Boussinesq theory, assuming a concentration factor $v = 5$ and soil with normal density and water content. (From Söhne, 1958.)

mechanism by which crawler tracks and drive wheels develop the traction needed to propel a vehicle forward over a soil base involves imparting shear stresses to the soil. An attempt to analyze the distributions of both normal and shearing stresses in the soil as affected by variously shaped and operated machines can indeed be a formidable task. Suffice it so say in the present context that the simultaneous application of pressures and shearing stresses can contribute greatly to soil compaction, as we have already mentioned in a preceding section in connection with the kneading compaction of soil samples in the laboratory.

F. Soil Compaction under Machinery-Induced Stresses

Having described the distribution of machinery-induced *stresses* within the soil, we must next consider how the soil *strains* under these stresses. The problem is complicated by the fact that the stresses acting on any element of soil in the field are seldom if ever *isotropic*, i.e., in a state of uniform ("hydrostatic") pressure equal in all directions without any shearing stresses. We recall from the preceding chapter that any difference between the principal stresses (that is to say, any occurrence of a *deviatoric stress*) necessarily gives rise to shearing stresses. Soil bodies straining under the combined influences of normal stresses (pressures) and shearing stresses are likely to undergo both compression and change of shape.

In the theories of elasticity and plasticity, the stress tensor is resolved into a spherical (isotropic) stress component which governs volume changes and a deviatoric stress component which causes shape changes (Hoffman and Sachs, 1953; Timoshensko and Goodier, 1951; Jaeger, 1964). Unfortunately, neither theory can provide a satisfactory description of soil compaction over a wide range of conditions encountered (Harris, 1971), owing to the fact that the behavior of soils differs greatly from that of the ideal materials upon which both theories are based. In attempting to apply the *theory of elasticity* to volume strain in soil, we encounter the difficulty that the soil's modulus of elasticity changes under the loading. Moreover, the theory is predicated upon the assumption of infinitesimal deformations (in practice, strains should not exceed 0.1%), yet in soil compaction strains can be anything but infinitesimal. Since the behavior of moist soils is more plastic than elastic, it is tempting to seek solace in the *theory of plasticity*. Unfortunately we find no complete satisfaction here either, since most of the working theories for typically plastic materials are concerned with predicting yield stress and the rate of change of shape rather than volume strain per se. Again, our unruly soil disobeys the neat assumptions of theory. Time can be an important factor in compaction. For example (Harris, 1971), the time required for a tractor wheel traveling at normal speed to rotate through the angle of effective soil contact (i.e., the duration of loading) is generally less than two-tenths of a second. Present theories of elasticity and plasticity are inadequate to describe the time dependence of the compaction process. The application of more complex rheological models to describe soil behavior is still at the rudimentary stage (McMurdie, 1963; Yong and Warkentin, 1975).

Progress in attempting to discover a rigorous and universal function between the state of compaction and the state of stress of a soil element has been slow owing to the inherently complicated space–time relations involved. In addition to the theoretical difficulties, there are serious experimental limitations. Our means of measurement are still too coarse. Insertion of measuring devices into the soil may affect soil reaction, particularly if an attempt is made to study natural soils in the field under dynamic (time variable, often short-lived) loadings. Imprecise spatial and/or temporal

F. Soil Compaction under Machinery-Induced Stresses

resolution of measurement often leads to gross averaging and disregard of nonuniform stress and strain patterns. The compaction of a soil body depends not only on the pattern of loading (e.g., static or dynamic, stable or vibratory) and the state of stress, but also—importantly—upon soil moisture, soil texture, and particle-size distribution (Bodman and Constantin, 1965; Gupta and Larson, 1979), the state and stability of soil structure, and initial (precompaction) soil conditions. Since the exact stress–strain–time relationship of soil as it varies in space is altogether too complex to define precisely at the present state of the art, an attempt to simplify the system by disregarding at least some of its complications may seem excusable for the moment. The specific question which poses itself at this stage is whether soil compaction as a volume strain can be related quantitatively to some measurable or calculable value of stress, be it the *mean normal stress* [i.e., the average of the three principal stresses, which in the notation of Chapter 13 would equal $\frac{1}{3}(\sigma_1 + \sigma_2 + \sigma_3)$], or the *major principal stress* (e.g., σ_1), or any other index of the stress state which might take account of the deviatoric stress. Since the possibility of developing usable rigorous relationships between stress and strain for soils by analytical means seems remote (Harris, 1971), investigators who have grappled with this question have offered various empirical or semi-empirical formulations for the stress–compaction function.

Perhaps the simplest approach to machinery-induced or vehicular compaction is to attempt to predict the sinkage of a soil surface under an applied load. We can presume that a loaded area will sink into the soil until the soil's resistive force (i.e., its bearing capacity) is in equilibrium with the applied external force. Ignoring for a moment the time dependence of the sinkage process and the distribution of stresses inside the soil mass, we can assume that compaction is directly related to the amount of sinkage. In his studies of soil compaction by agricultural implements, carried out nearly 70 years ago, Bernstein (cited by Bekker, 1956) proposed the following equation to relate ground-surface pressure P to sinkage z:

$$P = kz^n \qquad (14.2)$$

where k is a constant which has been called the *modulus of deformation* (its value depends on the size of the loaded area and on soil properties) and n is a constant which also depends on soil properties but was reported by Bernstein to have a value of 0.5 under what he considered average conditions.

Equation (14.2) suggests that the stress–compaction relationship is not linear, and that, exponent n being fractional, successive increments of pressure produce decreasing increments of compaction. This accords with the common observation that an increasing degree of compaction makes the soil progressively less amenable to subsequent pressures. The reason for this, we can conjecture, is that, with increasing density, the soil's cohesive strength increases as the particles come into closer contact and interlock against each other, thus also increasing frictional resistance to further deformation. Another factor involved is the change in matric potential of soil moisture resulting from the change in porosity and pore-size distribution. Soil compaction tends to increase matric potential (i.e., decrease matric suction, or tension) until, ultimately, all the air is expelled and the soil becomes saturated without an increase in mass wetness. We shall return to the subject of *porewater pressure* in our section on consolidation.

To take explicit account of the soil's cohesive and frictional properties, Bekker (1956, 1960) elaborated Bernstein's "modulus of deformation" parameter and related it to sinkage of vehicle wheels as follows:

$$k = k_c/b + k_\phi \tag{14.3}$$

where k_c is the *cohesive modulus of deformation*, k_ϕ is the *frictional modulus*, and b is the smaller dimension of the loaded area (i.e., the breadth). Accordingly,

$$P = (k_c/b + k_\phi)z^n \tag{14.4}$$

While these empirical relationships describe an aspect of soil compaction, they do not relate volume change directly to stress distribution within the soil.

A further modification of Bernstein's equation was offered by Cohron (1971) to describe soil compaction with depth (as represented by soil resistance to a penetrometer):

$$p_z = kz^n/b \tag{14.5}$$

in which p_z is soil resistance to penetration at any depth z (measured in terms of force per unit area of penetrometer tip); and k, n, and b are as defined previously. This equation was found to be a reasonable representation of experimental data in real field situations, regardless of whether the soil is purely frictional ($n = 1$), purely cohesive ($n = 0$) or exhibits intermediate behavior (Cohron, 1971).

A more fundamental approach to compaction was taken by Söhne (1953, 1958), who proposed the hypothesis that the degree of compaction is a function of the *major principal stress*, i.e., of the highest pressure applied in any direction to a body of soil. A negative semilogarithmic function was postulated for the relationship between soil porosity and the logarithm of the maximal pressure σ_{max} to which the soil has been subjected:

$$f = A - B \log \sigma_{max} \tag{14.6}$$

In this equation, A is a value which expresses the initial state of the soil and changes with soil wetness, and B is a characteristic constant for a given type of soil. This relationship has been used in attempts to describe compaction within the soil mass resulting from static or dynamic loads applied to the soil surface (Harris, 1960).

In a comprehensive series of tests reported by Vanden Berg (1958), changes in soil bulk density were related to the following indices of the stress state: *mean normal stress, maximum normal stress, deviatoric stress,* and *maximum shear stress.* Changes in soil bulk density could be related best to the mean normal stress, and the relationship was found to be exponential for all the soil classes tested, including a clay, a silty clay loam, and a sandy loam. In a subsequent report, Vanden Berg (1962) described using a modified triaxial apparatus to determine the principal stresses and strains of soil samples at various times during loading. The results suggested that, in addition to the mean normal stress, the maximal shearing stress contributed to increasing soil compaction. In still more recent work, Bailey and Vanden Berg (1968) attempted to describe compaction in terms of a three-dimensional diagram with coordinate axes of mean normal stress, maximum shearing stress, and the soil's specific volume (reciprocal of bulk density).

Knowledge of the distribution of pressures within the soil, and of the compaction response of the soil to such pressures, should in principle enable us to calculate the pattern of soil compaction throughout the affected profile when a portion of the surface is subjected to a load such as a crawler tractor track. A procedure for doing this was described by Chancellor (1976) on the basis of the Boussinesq equation.

Experimental evidence appears to corroborate the predicted patterns, at least for a soil whose pressure–porosity relationship is not sensitive to shearing deformation (Chancellor and Schmidt, 1962). For a soil sensitive to shearing deformation, the maximum porosity reduction seems to occur not at the surface, where pressure is maximal, but at some depth, where the effects of pressure and shearing stresses combine synergistically (Söhne, 1953, 1958).

G. Occurrence and Consequences of Soil Compaction

Instances of soil compaction are highly prevalent and becoming more so in modern agriculture. The trend to reduce labor on the farm by the use of larger and heavier machinery increases the likelihood and damage of untimely cultivation and the frequency of traffic over the soil. Expecially insidious is the common practice of plowing with the tractor wheel running over the bottom of the open furrow, where the soil is likely to be even more compactible than at the surface, and to greater depth, owing to higher moisture and lower organic matter content. Furthermore, compaction in depth is much more difficult to rectify and hence longer lasting than compaction at the surface.

Compaction is not limited to initial plowing (called *primary tillage*). As pointed out by Soane (1975), some 90% of the soil surface may be traversed by tractor wheels during the traditional preparation of seedbeds for close-growing crops such as cereal grains, followed by a further trampling of at least 25% during combine harvesting and as much as 60% where straw is baled and carted off. The compaction caused by all this traffic, particularly during seedbed operations, can increase bulk density to a depth of at least 30 cm and can remain throughout the life of the crop. Row crops such as cotton often sustain intensive wheel traffic, as repeated passes over the field are made for application of pesticides (Dumas *et al.*, 1973). Especially damaging is the practice of cultivating a clayey soil with heavy equipment when the soil is at a wet state and its strength is low. "Smearing" of the plow-layer bottom by plowshares creates a "plow sole" or "plow pan." The churning action by wheels, including the effects of wheel slip, can be more important than loading per se in bringing about degradation of soil structure and subsequent compaction (Davies *et al.*, 1973). Traffic associated with harvesting potatoes and sugar beets in wet conditions causes deep rutting, smearing, and compaction, which can also inhibit drainage (Swain, 1975).

Soil compaction reduces the volume and continuity of the larger pores. This effect is exemplified in Fig. 14.11 and in Table 14.1, derived from Bodman *et al.* (1958):

Table 14.1

PORE-SIZE DISTRIBUTION
IN A SANDY LOAM SOIL[a]

Pore diameter range (μm)	Bulk density (gm/cm^3)	
	1.63	1.79
>1000	3.8%	2.1%
100–1000	7.4%	4.2%
6–100	25.7%	12.5%
<6	26.1%	30.5%

[a] The numbers indicate percentage of total soil volume.

G. Occurrence and Consequences of Soil Compaction

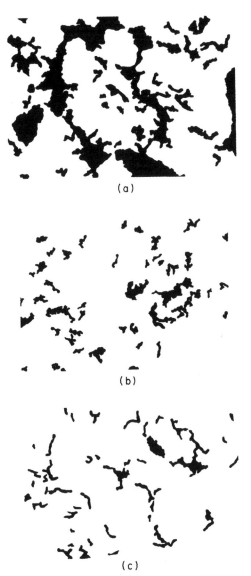

Fig. 14.11. The pore system of a soil. (a) Uncompressed soil. (b) Tilled soil compressed by subsequent interrow surface traffic. (c) Soil compressed during tillage (e.g., a "plow pan"). (After Trouse, 1971.)

As shown by Wiersum (1957), Taylor and Bruce (1968), and Cannell (1977), roots are unable to decrease in diameter to enter pores narrower than their root caps; thus, if they are to grow through compacted soil they must displace soil particles to widen the pores by exerting a pressure greater than the soil's mechanical strength. In addition to this mechanical constraint, soil compaction also impedes the movement of water and air through the soil by reducing the number of large pores. The resulting restriction of aeration and drainage thus exposes roots to several simultaneous stresses. Studies on the effect of mechanical impedance on root growth have been reported by, among others, Gill and Miller (1956), Barley (1962), Abdalla et al. (1969), Goss (1970), and Russell and Goss (1974). According to the latter, where barley roots must overcome externally imposed pressure of only 0.2 bar to enlarge pores, the rate of root elongation is reduced by 50%, and if the pressure in the soil is increased to just 0.5 bar, root extension is reduced by 80%. It is noteworthy, in this connection, that rigidly confined roots are capable of exerting pressures as great as 10 bar, a fact that was discovered as early as 1893 by Pfeffer [cited by Gill and Bolt (1955)] and since reconfirmed by several investigators (e.g., Stolzy and Barley, 1968; Taylor and Ratliff, 1969). This fact may explain the ability of roots to eventually penetrate compacted zones, as well as their oft-cited ability to penetrate, and enlarge, crevices in rocks, roads, and foundations. However, from the standpoint of promoting crop growth, we are generally interested in minimizing soil resistance to root elongation, rather than in mobilizing maximal root pressure.

The interactions among soil compaction, strength, moisture retention, aeration, and root growth have been studied by, among others, Taylor and Gardner (1963), Hopkins and Patrick (1969), Klute and Peters (1969), Greacen et al. (1969), and Kays et al. (1974). In many of the studies of root penetration in compact media, soil resistance has been characterized by means of penetrometers of various types. The usefulness and universality of penetrometer measurements is, however, still limited by the lack of a standard design and standardized procedures.

Where the soil is highly compact and rigid, root growth may be confined almost entirely to cracks and cleavage planes (Taylor et al., 1966). In seeking out such cracks, plant roots are not merely passive agents, as they promote differential shrinkage through preferred extraction of moisture from zones of greater penetrability.

As pointed out in a preceding section of this chapter, a wet soil has less resistance to compaction than a dry soil, provided the wet soil is not at or near saturation. Hence under the same pressure a wetter soil will compact to a lower porosity than a drier one. On the other hand, increasing soil wetness causes a decrease of air-filled porosity, thus the total porosity at which compaction ceases (as the air-filled pores are completely eliminated by soil densification) is higher for wetter than for drier soil. However, kneading (puddling) a saturated soil, although it will not result in significant compaction immediately, may result eventually in extreme compaction as the structurally degraded (puddled) soil shrinks during the subsequent drying process.

H. Control of Soil Compaction

Control of soil compaction is a continual requirement in modern agriculture. The performance of field operations necessarily involves *some* compaction. Hence a major task of soil management is, first, to minimize soil compaction to the extent possible, and, second, to alleviate or rectify that unavoidable measure of compaction caused by traffic and tillage once it occurs.

The most obvious approach to the prevention of soil compaction is the avoidance of all but truly essential pressure-inducing operations. This calls for reducing the number of operations involved in primary and secondary tillage (plowing and subsequent cultivations, respectively), using the most efficient implement at the most appropriate time so as to effect the desirable soil condition in a single pass rather than by a repeated sequence of passes. It has long been known that excessive soil manipulation, beyond what is required to prepare a seedbed and check weeds, leads to a loss of yield and deterioration of soil structure (Keen and Russell, 1973). Indeed overly intensive cultivation can cause disastrous results, such as accelerated erosion by water and wind.

In recent years, increasing awareness of these hazards has resulted in the development of radically new field management systems, variously called "zero tillage," "minimum tillage," or "conservation tillage" (Wittmus *et al.*, 1973, 1975; Lewis, 1973; McGregor *et al.*, 1975). Such systems reduce the number of operations, avoid unnecessary inversion of topsoil, and generally retain crop residues as a protective mulch over the surface. In former times, the repeated and frequent cultivation of orchards and of row-cropped fields was required for weed control. This requirement was reduced to a considerable extent by the introduction of sprayable herbicides. The potential benefits of reduced tillage go beyond the avoidance of compaction and preservation of soil structure, as they include savings in time, labor, and energy, the latter having become important enough in itself. However, excessive reliance on phytotoxic chemicals can pose serious environmental problems.

Since random traffic over the field by heavy machinery is a major cause of soil compaction, cultural systems have been proposed to confine vehicular traffic to permanent, narrow lanes and to reduce the fractional area trampled by wheels to less than 10% of the land surface if possible (Dumas et al., 1973). In row crops seedbed preparation, involving the formation of a highly pulverized top layer vulnerable to structural breakdown, should be confined to the narrow strips where planting of row crops takes place rather than be carried out over the entire surface as was the practice in former times. (In too many places, "former times" persist to this day.) The interrow zone can now be left in an open, cloddy condition which promotes water and air penetration and reduces erosion by water and wind. The field is thus divided into three (if possible, permanent) zones: (1) narrow planting strips, precision tilled; (2) narrow traffic lanes, permanently compacted; and (3) interrow water-management beds, maintained in a rough and cloddy condition and covered with a mulch of plant residues.

An extremely important factor is the timing of field operations in relation to soil moisture. Moist soils can be highly vulnerable to compaction. Operations which impose high pressures should, if possible, be carried out on dry soil, which is much less compactible. These include traffic in general, and—more specifically—such tillage methods as subsoiling, which in fact may be more effective in shattering dry soils than wet ones.

Much can be done by equipment designers to distribute loads more evenly over the tracked surface rather than concentrate them under the rear wheels and to provide independent power to implements (via the power take-off or a separate engine) as, for instance, in the case of rotary tillers. Powered tillage implements can reduce dependence on the tractor's draw-bar pull, which increases tractor wheel slippage and hence soil puddling. Tractors must generally be twice as heavy as the pulling forces they are expected to generate, so that reduction of draw-bar pull requirements can permit reduction of tractor weight. This and other means (e.g., decrease of tire inflation pressure) to reduce the stresses transmitted to the soil should constitute important design criteria for the manufacture and selection of tractors and implements. A basic discussion of tillage is given in Chapter 9 of Hillel (1980).

On the other side of the issue, it should be pointed out that not all instances of soil compaction are necessarily harmful. The practice of adding packing wheels to seeding machines has long been known to enhance germination, by ensuring better seed-to-soil contact and increasing the unsaturated hydraulic conductivity of the otherwise excessively loose and porous seedbed (Dasberg et al., 1966).

I. Soil Consolidation

Thus far in this chapter we have confined our attention to the process of soil compaction, which, we recall, has been defined as the compression of

I. Soil Consolidation

unsaturated soil due to reduction of its air-filled pore space without change in mass wetness. As already pointed out, continued compression of an unsaturated soil will ultimately result in the practically complete expulsion of the air and, if compression is continued still, in expulsion of water as well. This is analogous to pressing a moist sponge or wringing a moist piece of cloth until water is eventually squeezed out. So, the process by which a body of soil, either initially saturated or compacted to the point of saturation, is compressed in a manner that results in reduction of pore volume by expulsion of water, is called *consolidation*. Because of the slower nature of consolidation relative to compaction (water being 50–100 times more viscous than air at ordinarily encountered temperatures), consolidation is not an immediate response of soil to transient pressures such as those caused by episodic traffic or tillage. Rather, consolidation is manifested in the gradual settlement, or subsidence, of soil under long term loading such as that due to a permanent structure (e.g., a single building, a housing development, or even an entire city). Hence the practical importance of consolidation lies not so much in agriculture as in engineering, where the amount and uniformity of foundation settlement are of vital concern to those interested in the stability and safety of structures.

To arrive at a conceptual understanding of consolidation, let us consider a sample of water-saturated soil confined in a rigid-walled cylinder and subjected to compressive load applied to a piston in such a manner as to allow no possible outlet for water. If the confined water is assumed to be incompressible, then the applied pressure can cause no compression of the soil matrix and must be borne entirely by the water phase. If we now bore a narrow hole through the piston, some of the pressured water would tend to

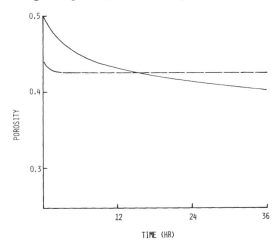

Fig. 14.12. Relative compressibilities and rates of consolidation for a sandy and a clayey soil (hypothetical). Solid line, clayey soil; dashed line, sandy soil.

squirt out, thus relieving the hydrostatic pressure inside the cell and gradually transmitting the load to the soil matrix, which would tend to compress at a rate commensurate with the volume of water extruded. The water contained in the voids of a saturated compressible layer of soil behaves under pressure in a manner similar to that of the water under the hypothetical piston just described. In this case, however, the water must find its way out through the exceedingly narrow and tortuous pores of the soil itself. Imagine all this taking place under an unevenly distributed load, resulting in different hydrostatic pressures at various points in the soil, and you will have an idea of the complex, space-variable, and time-variable nature of the consolidation process.

In general, granular soils such as dense sands are the least compressible soils, and such consolidation as does take place usually occurs relatively rapidly, thanks to these soils' high permeability. Clays and other fine-textured soils, with higher porosity, generally have a much greater total compressibility, but the rate of their consolidation is likely to be slow, sometimes extending over a period of months or even years. This difference is illustrated in Fig. 14.12.

A classical mechanical model, shown in Fig. 14.13, has been found to be useful in illustrating the principles involved in the consolidation process. The model consists of a perforated or porous piston pressed into a water-filled cylinder fitted with springs. When the piston is first loaded and before any downward motion has yet taken place, the pressure applied is transmitted to the water only. Soon, however, water begins to ooze out of the piston, and as the piston moves down, the springs begin to contract. As the process continues, more and more of the water escapes and the springs contract to an increasing degree. If the piston is placed under a constant load, the downward movement will finally stop and static equilibrium will prevail when the resistance of the compressed springs just equals the downward force acting on the piston. At this point and henceforth, the entire load will be borne by the springs, the excess water pressure having been relieved by the outflow. Thus at the beginning of the process the water is stressed but not the springs,

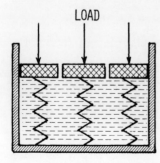

Fig. 14.13. Conceptual model illustrating the process of consolidation.

I. Soil Consolidation

and at the end the situation is reversed. At any intermediate point during the process, the load is carried partly by the water and partly by the springs. The springs are quite obviously analogous to the soil matrix, and the water in the cylinder to the water permeating soil pores.

The concept described in so many words can be summed up tersely in the form of an equation, known as *Terzaghi's effective stress equation* (Terzaghi, 1953):

$$\sigma_{tot} = p + \sigma_{eff}$$

or

$$\sigma_{eff} = \sigma_{tot} - p \tag{14.7}$$

Here σ_{tot} is the *total stress*, p is the hydrostatic pressure known as the *pore water pressure*, and σ_{eff} is the so-called *effective stress* borne by the solid matrix and therefore often termed the *intergranular stress*. The term p is a *neutral stress*, inasmuch as the liquid pressure in a saturated soil acts equally in all directions, hence it has no effect on compressing the matrix. When a total stress σ_{tot} is suddenly applied to a soil and then maintained for an indefinite period, p decreases and σ_{eff} increases as shown hypothetically in Fig. 14.14.

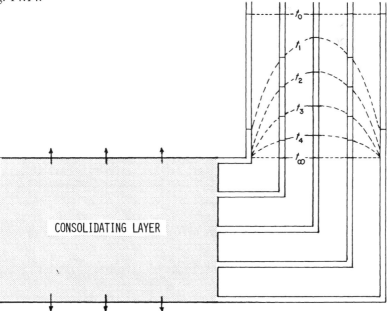

Fig. 14.14. A consolidating layer draining both upward and downward. The piezometers on the right indicate the isochrones at various times from the start (at t_0) to the eventual end (at t_∞) of the process.

The theory of consolidation, also due to Terzaghi, is based on several assumptions:

1. The soil is completely saturated.
2. The water and the soil particles are incompressible.
3. Water flow and hence consolidation occur in one dimension only (i.e., vertically).
4. Water movement in the consolidating layer obeys Darcy's law, and the hydraulic conductivity is constant.
5. Loading is instantaneous and the load constant, and the time lag of compression is caused entirely by the slow outflow of water from the compressing soil.
6. There is a linear relationship between the effective stress acting on the soil matrix and the volume change produced.

Since the soil volume decrement is due entirely to, and equal to, the change in fractional volume of voids V_v, which is, in turn, caused by the Darcian outflow of water, we can write an equation of continuity,

$$\partial V_v / \partial t = - K \, \partial^2 h / \partial z^2 \qquad (14.8)$$

wherein t is time, K hydraulic conductivity, h pressure head of soil water (actual pore pressure P divided by liquid density ρ and gravitational acceleration g: $h = P/\rho g$). The change in volume of voids can be related to the change in pore water pressure using the *modulus of volume change*, m_v, where

$$m_v = - \frac{\partial V_v / \partial t}{\partial P / \partial t} = - \frac{\partial V_v}{\partial P} \qquad (14.9)$$

Hence

$$m_v = \frac{\partial P}{\partial t} = \frac{K}{\rho g} \frac{\partial^2 P / \partial z^2}{\partial P / \partial t}$$

or

$$\frac{\partial P}{\partial t} = \frac{K}{\rho g m_v} \frac{\partial^2 P}{\partial z^2} = C_c \frac{\partial^2 P}{\partial z^2} \qquad (14.10)$$

in which C_c $(= K/\rho g m_v)$ is called the *coefficient of consolidation*.

Solution of Eq. (14.10) for different sets of boundary conditions can be found in standard texts on soil mechanics (e.g., Taylor, 1948; Lambe and Whitman, 1969). A typical example is the one-dimensional (vertical) consolidation of a saturated horizontal bed of clay contained between two layers of sand. The water is free to escape through both the upper and lower boundaries of the consolidating layer, and hence the layer is called an *open* layer.

I. Soil Consolidation

The total consolidation pressure σ_{tot} is assumed to be uniform everywhere within the clay layer. If we insert standpipes (called *piezometers*) to gauge the hydrostatic pressure distribution throughout the clay bed as it varies in time during the process of consolidation, we are likely to observe the pattern illustrated in Fig. 14.14. The water in each piezometer will rise to a height representing the excess neutral stress (hydrostatic pressure) at the point to which the bottom of the pipe is inserted. If the piezometers are equally spaced, then the lines connecting their water levels at regular intervals of time are a family of *isochrones*, indicating how the distribution of pore-water pressure with depth varies in time. The solution for the case described takes the form (Yong and Warkentin, 1975):

$$P = \sum_{n=1}^{n=\infty} \left(\frac{1}{H} \int_0^{2H} P_0 \sin \frac{n\pi z}{2H} \, dz \right) \left(\sin \frac{n\pi z}{2H} \right) e^{-n^2 \pi^2 c_c t / 4H^2} \quad (14.11)$$

where H is half the thickness of the clay stratum and n is any integer.

The differential equation of one-dimensional consolidation has been derived on the assumption that the total load causing settlement is applied to the soil instantaneously. In reality, the load may be applied gradually. Obviously the rate of consolidation, and possible also the total consolidation finally attained, will depend on the rate at which the load is applied. Other assumptions underlying the classical theory may also conflict with reality. If compression is considerable, the hydraulic conductivity (or permeability) cannot be regarded as constant in time and space and must be related functionally to the change in porosity. Darcy's law itself, strictly speaking, is valid for fluid flow relative to the soil matrix, rather than relative to absolute space, and if the matrix is shifting (i.e., compressing), this movement should be taken into account. Nevertheless, the theory presented has been found useful in obtaining estimates, at least, of expectable consolidation.

To characterize soils with respect to their consolidation behavior, lab-

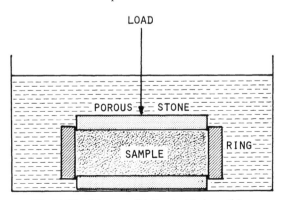

Fig. 14.15. The consolidation test (schematic).

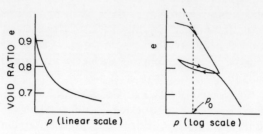

Fig. 14.16. Void ratio versus effective pressure curve for a compressible soil during consolidation. The semilogarithmic plot indicates the apparent preconsolidation pressure p_o and a pressure-release ("rebound")–reconsolidation loop.

oratory tests are generally carried out with small samples, preferably "undisturbed" cores, placed in an apparatus as shown in Fig. 14.15. Extraction of a sample from some depth in the field generally entails release of the original in situ, confining pressure, or overburden. The sample, however, retains what might be called a memory of its state of *preconsolidation* pressure. When reconsolidated, it will not compress significantly until its preconsolidation pressure is exceeded. Thereafter, its compression will tend to follow a straight line relation for void ratio versus the logarithm of pressure, at least until particle interlocking causes a significant increase of interparticle frictional resistance to further compression. Continued compression beyond this range is often called *secondary consolidation*. A technique for assessing a soil's preconsolidation pressure was developed by Casagrande and has been widely adopted (Casagrande and Fadum, 1940). The pattern of consolidation is illustrated in Fig. 14.16.

As long as the stress pattern does not exceed the yield strength of the soil, compression (volume strain) predominates over shear, and the latter can be ignored. If, however, the imposed shearing stresses exceed soil strength, shearing deformation may take place to an extent that can no longer be disregarded.

Sample Problems

1. Calculate and plot the functional dependence of bulk density upon mass wetness and upon volume wetness for a saturated soil, and for 90 and 80% degrees of saturation.

To derive the appropriate relationship, we can begin with the familiar equations relating bulk density ρ_b to porosity f [see Eqs. (2.16) and (2.17)]:

$$f = 1 - (\rho_b/\rho_s) \tag{14.12}$$

Sample Problems

or
$$\rho_b = \rho_s(1 - f) \tag{14.13}$$

By definition (Chapter 2) the degree of saturation s equals the volume of water V_w as a fraction of the volume of pores V_f:

$$s = V_w/V_f = \theta/f \tag{14.14}$$

where θ is volume wetness (V_w/V_t) and f is porosity (V_f/V_t), V_t being the total volume of a soil body.
Hence,

$$f = \theta/s \tag{14.15}$$

We incorporate the last expression in Eq. (14.13) to obtain a relationship between bulk density and volume wetness for any degree of saturation:

$$\rho_b = \rho_s[1 - (\theta/s)] \tag{14.16}$$

To obtain a relationship between bulk density and mass wetness w, we recall the well-known relation between volume wetness and mass wetness:

$$\theta = w\rho_b/\rho_w \tag{14.17}$$

and substitute this into Eq. (14.16):

$$\rho_b = \rho_s\left(1 - \frac{w}{s}\frac{\rho_b}{\rho_w}\right) \tag{14.18}$$

from which we can extract

$$\rho_b = \frac{\rho_w}{(\rho_w/\rho_s) + (w/s)} \tag{14.19}$$

Assuming that the density of solids $= 2.65$ gm/cm^3 and that of water 1 gm/cm^3, we can write

$$\rho_b = 1/[0.377 + (w/s)]\,(gm/cm^3) \tag{14.20}$$

For complete saturation use $s = 1.0$ and plot ρ_b against w. For the 90 and 80% degrees of saturation, plot ρ_b against w with s set equal to 0.9 and 0.8, respectively. The curves should resemble those of Fig. 14.2.

2. A Protor compaction test was performed on medium-textured soil with the following results:

Mass wetness w:	6%	9%	12%	15%	18%	21%
Bulk density ρ_b:	1.80	1.91	1.94	1.86	1.75	1.65 gm/cm^3

Plot the data and determine the optimal moisture and maximum density values. Calculate the volume wetness and degree of saturation at optimal moisture content.

An examination of the data, confirmed by the plotted graph, shows that the optimal moisture is at or about a mass wetness w value of 12%, and the maximal density ρ_b is 1.94 gm/cm^3.

Accordingly, the volume wetness is

$$\theta = w\frac{\rho_b}{\rho_w} = 0.12 \times \frac{1.94 \text{ gm/cm}^3}{1 \text{ gm/cm}^3} = 0.23 = 23\%$$

and the degree of saturation is

$$s = \theta/f$$

where f, the porosity, can be obtained from the soil's bulk and particle densities:

$$f = 1 - (\rho_b/\rho_s)$$

Hence

$$s = \frac{\theta}{1 - (\rho_b/\rho_s)} = \frac{0.23}{1 - 1.94/2.65} = 0.85 = 85\%$$

3. A concentrated load of 1000 kg [equivalent to a force of 10^4 N (newton), or 10^9 dyn] is applied to the soil surface. Estimate the vertical stress at depths of 5 and 10, directly under the load and at horizontal distances of 2 and 4 m.

We use the Boussinesq equation (14.1)

$$\sigma_z = \frac{3z^3 F}{2\pi(r^2 + z^2)^{5/2}}$$

where σ_z is the vertical stress, F the downward force acting on the surface, z the depth, and r the radial distance from the axis of the vertical force. For the sake of convenience, this equation can be rearranged as

$$\sigma_z = F\frac{3z^3}{2\pi\{z^2[(r^2/z^2) + 1]\}^{5/2}} = F\frac{3z^3}{2\pi z^5[(r/z)^2 + 1]^{5/2}}$$

$$= \frac{F}{z^2}\frac{3}{2\pi[1 + (r/z)^2]^{5/2}} = \frac{F}{z^2}A$$

wherein A is a composite function of r/z.

Sample Problems

For $z = 5$ m and $r = 0$, $A = 0.478$. Thus,

$$\sigma_z = \frac{0.478 \times 10^9 \text{ dyn}}{(500 \text{ cm})^2} = \frac{4.78 \times 10^8 \text{ dyn}}{2.5 \times 10^5 \text{ cm}^2}$$
$$= 1.91 \times 10^3 \text{ dyn/cm}^2 = 1.91 \times 10^{-3} \text{ bar}$$

For $z = 5$ m and $r = 2$ m, $A = 0.329$.

$$\sigma_z = \frac{0.329 \times 10^9 \text{ dyn}}{(500 \text{ cm})^2} = 1.32 \times 10^3 \text{ dyn/cm}^2 = 1.32 \times 10^{-3} \text{ bar}$$

For $z = 5$ m and $r = 4$ m, $A = 0.139$.

$$\sigma_z = \frac{0.139 \times 10^9 \text{ dyn}}{(500 \text{ cm})^2} = 0.56 \times 10^3 \text{ dyn/cm}^2 = 5.6 \times 10^{-4} \text{ bar}$$

For $z = 10$ m and $r = 0$, $A = 0.478$.

$$\sigma_z = \frac{0.478 \times 10^9 \text{ dyn}}{(1000 \text{ cm})^2} = 0.478 \times 10^3 \text{ dyn/cm}^2 = 4.78 \times 10^{-4} \text{ bar}$$

For $z = 10$ m and $r = 2$ m, $A = 0.433$.

$$\sigma_z = \frac{0.433 \times 10^9 \text{ dyn}}{(1000 \text{ cm})^2} = 0.433 \times 10^3 \text{ dyn/cm}^2 = 4.33 \times 10^{-4} \text{ bar}$$

For $z = 10$ m and $r = 4$ m, $A = 0.329$.

$$\sigma_z = \frac{0.329 \times 10^9 \text{ dyn}}{(1000 \text{ cm})^2} = 0.329 \times 10^3 \text{ dyn/cm}^2 = 3.29 \times 10^{-4} \text{ bar}$$

Note: In these calculations, we took no account of the soil's own overburden pressure. In reality, the vertical stresses caused by the load applied at the surface are incremental to the soil's overburden pressure, which tends to increase with depth.

Bibliography

Abdalla, A. M., Hetteriaratchi, D. R. P., and Reece, A. R. (1969). The mechanics of root growth in granular media. *J. Agr. Eng. Res.* **14,** 236–248.

Adam, K. M., Bloomsburg, G. L., and Corey, A. T., Diffusion of Trapped Gas from Porous Media, *Water Resour. Res.* **5,** 840–849.

Alexander, M. (1961, 1977). "Introduction to Soil Microbiology," Wiley, New York.

Allison, L. E. (1956). Soil and plant responses to VAMA and HPAN soil conditioners in the presence of high exchangeable sodium. *Soil Sci. Soc. Am. Proc.* **20,** 147–151.

American Society of Agronomy Monograph No. 9 (1965). "Methods of Soil Analysis, Physical and Mineralogical Properties," Part I.

Amerman, C. R., Hillel, D. I., and Peterson, A. E. (1970). A variable-intensity sprinkling infiltrometer. *Soil Sci. Soc. Am. Proc.* **34,** 830–832.

Anderson, M. S. (1926). Properties of soil colloidal material. U.S. Dept. Agr. Bull. 1452.

Arbhabhirama, A., and Kridakorn, C. (1968). Steady downward flow to a water table. Water Resour. Res. **4,** 116–121.

Aslyng, H. C. (1963). Soil physics terminology. *Int. Soc. Soil Sci. Bull.* **23,** 7.

ASTM (American Society for Testing Materials) (1956). *Symp. Vane Shear Testing Soils* Spec. Tech. Publ. 193. Am. Soc. Testing Mater., Philadelphia, Pennsylvania.

ASTM (American Society for Testing Materials) (1958a). Book of Standards, part II. pp. 217–224. Am. Soc. Testing Mater., Philadelphia, Pennsylvania.

ASTM (American Society for Testing Materials) (1958b). "Procedures for Soil Testing." Philadelphia, Pennsylvania.

Atterberg, A. (1911). Die Plastizitat der Tone. *Int. Mitt. Bodenk.* **1,** 10–43.

Atterberg, A. (1912). Die Konsistenz und die Bindigheit der Boden. *Int. Mitt. Bodenk.* **2,** 148–189.

Bailey, A. C., and Vanden Berg, G. E. (1968). Yielding by compaction and shear in unsaturated soils. *Trans. Am. Soc. Agr. Eng.* **11,** 307–311, 317.

Barber, E. S. (1965). Stress distribution. *In* "Methods of Soil Analysis," Part I. Monograph 9, Am. Soc. Agron., Madison, Wisconsin.

Barley, K. P. (1962). The effect of mechanical stress on the growth of roots. *J. Exp. Bot.* **13,** 95–110.

Bear, J. (1969). "Dynamics of Fluids in Porous Media," 453–457 Elsevier, Amsterdam.
Bear, J., Zaslavsky, D., and Irmay, S. (1968). "Physical Principles of Water Percolation and Seepage." UNESCO, Paris.
Bekker, M. G. (1956). "Theory of Land Locomotion." Univ. Michigan Press, Ann Arbor, Michigan.
Bekker, M. G. (1960). "Off-the-Road Locomotion: Research and Development in Terramechanics." Univ. Michigan Press, Ann Arbor, Michigan.
Bekker, M. G. (1961). Mechanical properties of soil and problems of compaction. *Trans Am. Soc. Agr. Eng.* **4**, 231–234.
Birkeland, P. W. (1974). "Pedology, Weathering, and Geomorphological Research." Oxford Univ. Press, London and New York.
Birkle, D. E., Letey, J., Stolzy, L. H., and Szuszkiewicz, T. E. (1964). Measurement of oxygen diffusion rates with the platinum microelectrode. *Hilgardia* **35**, 555–566.
Bishop, A. W., and Henkel, D. J. (1964). "The Measurement of Soil Properties in the Triaxial Test." Arnold, London.
Black, C. A. (ed.) (1965). "Methods of Soil Analysis," Part I. Am. Soc. Agron., Madison, Wisconsin.
Blake, G. R. (1965). Bulk density. *In* "Methods of Soil Analysis," pp. 374–390. Am. Soc. of Agron., Madison, Wisconsin.
Blake, G. R., and Page, J. B. (1948). Direct measurement of gaseous diffusion in soils. *Soil Sci. Soc. Am. Proc.* **13**, 37–42.
Bodman, G. B., and Constantin, G. K. (1965). Influence of particle size distribution in soil compaction. *Hilgardia* **36**, 567–591.
Bodman, C. B., Johnson, D. E., and Kruskal, W. H. (1958). Influence of VAMA and of depth of rotary hoeing upon infiltration of irrigation water. *Soil Sci. Soc. Am. Proc.* **22**, 463–468.
Boersma, L. (1965a). Field measurement of hydraulic conductivity below a water table. *In* "Methods of Soil Analysis," pp. 222–223. *Monograph 9*, Am. Soc. Agron., Madison, Wisconsin.
Boersma, L. (1965b). Field measurement of hydraulic conductivity above a water table. *In* "Methods of Soil Analysis," pp. 234–252. Monograph 9, Am. Soc. Agron., Madison, Wisconsin.
Bolt, G. H. (1956). Physico-chemical analysis of the compressibility of pure clays. *Geotechnique* **8**, 86–90.
Bolt, G. H. (1976). Soil physics terminology. *Bull. Int. Soc. Soil Sci.* **49**, 26–36.
Bolt, G. H., and Bruggenwert, M. G. M. (ed.) (1976). "Soil Chemistry." Elsevier, Amsterdam.
Bolt, G. H., and Frissel, M. J. (1960). Thermodynamics of soil moisture. *Neth. J. Agr. Sci.* **8**, 57–78.
Bomba, S. J. (1968). Hysteresis and time-scale invariance in a glass-bead medium. PhD Thesis, Univ. Wisconsin, Madison, Wisconsin.
Bouma, J., Hillel, D. I., Hole, F. D. and Amerman, C. R. (1971). Field measurement of unsaturated hydraulic conductivity by infiltration through artificial crusts. *Soil Sci. Soc. Am. Proc.* **32**, 362–364.
Bouwer, H. (1961). A double tube method for measuring hydraulic conductivity of soil in sites above a water table. *Soil Sci. Am. Proc.* **25**, 334–342.
Bouwer, H. (1962). Field determination of hydraulic conductivity above a water table with the double tube method. *Soil Sci. Soc. Am. Proc.* **26**, 330–335.
Bouyoucos, G. J. (1937). Evaporating the water with burning alcohol as a rapid means of determing moisture content of soils. *Soil Sci.* **44**, 377–383.
Bouyoucos, G. J., and Mick, A. H. (1940). An electrical resistance method for the continuous

measurement of soil moisture under field conditions. Michigan Agr. Exp. Sta. Tech. Bull. 172.
Bower, C. A., and Goertzen, J. O. (1959). Surface area of soils and clays by an equilibrium ethylene glycol method. *Soil Sci.* **87**, 289–292.
Bresler, E. (1978). Theoretical modeling of mixed-electrolyte solution flows for unsaturated soils. *Soil Sci.* **125**, 196–203.
Brooks, R. H., and Corey, A. T. (1966). Properties of porous media affecting fluid flow. *Proc. Am. Soc. Civ. Eng., J. Irrigation Drainage Div.* **IR2**, 61–88.
Brady, N. C. (1974). "The Nature and Properties of Soils." MacMillan, New York.
Bresler, E. (1972). Control of soil salinity. *In* "Optimizing the Soil Physical Environment Toward Greater Crop Yields" (D. Hillel, ed.) pp 102–132. Academic Press, New York.
Bresler, E. (1973). Simultaneous transport of solutes and water under transient unsaturated flow conditions. *Water Resour. Res.* **9**, 975–986.
Brooks, R. H., and Corey, A. T. (1964). Hydraulic properties of porous media. Hydrology Paper No. 3. Colorado State Univ., Ft. Collins, Colorado.
Brown, P. A. (1970). Measurement of water potential with thermocouple psychrometers: construction and application. USDA Forest Service Research Rep., INT-80.
Bruce, R. R., and Klute, A. (1956). The measurement of soil water diffusivity. *Soil Sci. Soc. Am. Proc.* **20**, 458–462.
Brunauer, S., Emmett, P. H., and Teller, E. (1938). Adsorption of gases in multimolecular layers. *J. Am. Chem. Soc.* **60**, 309–319.
Buckingham, E. (1904). Contributions to our knowledge of the aeration of soils. U. S. Bur. Soils Bull. 25.
Buckingham, E. (1907). Studies on the movement of soil moisture. U.S. Dept. of Agr. Bur. of Soils, Bull. 38.
Buras, N. (1974). Water management systems. *In* "Drainage for Agriculture" (J. van Schilfgaarde, ed.). Monograph 17, Am. Soc. Agron., Madison, Wisconsin.
Burdine, N. T. (1952). Relative permeability calculations from pore-size distribution data. *Trans. AIME* **198**, 35–42.
Burger, H. C. (1915). *Phys. Zs.* **20**, 73–76.
Burrows, W. C. (1963). Characterization of soil temperature distribution from various tillage-induced microreliefs. *Soil Sci. Soc. Am. Proc.* **27**, 350–353.
Buswell, A. M., and Rodebush, W. H. (1956). *Water. Sci. Am.* **202**, 1–10.
Cannell, R. Q. (1977). Soil aeration and compaction in relation to root growth and soil management. *Adv. Appl. Biol.* **2**, 1–86.
Carman, P. C. (1939) *J. Agr. Sci.* **29**, 262.
Carmen, P. C. (1956). "Flow of Gases Through Porous Media." Academic Press, New York.
Carslaw, J. S., and Jaeger, J. C. (1959). "Conduction of Heat in Solids." Oxford Univ. Press (Clarendon), London and New York.
Carson, J. E. (1961). Soil Temperature and Weather Conditions. Rep. No. 6470, Argonne National Laboratories, Argon.
Cary, J. W. (1963). Onsager's relations and the nonisothermal diffusion of water vapor. *J. Phys. Chem.* **67**, 126–129.
Cary, J. W. (1966). Soil moisture transport due to thermal gradients: Practical aspects. *Soil Sci. Soc. Am. Proc.* **30**, 428–433.
Cary, J. W., and Taylor, S. A. (1962a). The interaction of the simultaneous diffusions of heat and water vapor. *Soil Sci. Soc. Am. Proc.* **26**, 413–416.
Cary, J. W., and Taylor, S. A. (1962b). Thermally driven liquid and vapor phase transfer of water and energy in soil. *Soi. Sci. Soc. Am. Proc.* **26**, 417–420.

Casagrande, A. (1948). Classification and identification of soils. *Trans. Am. Soc. Civil Eng.* **113**, 901–930.

Casagrande, A., and Fadum, R. E. (1940). Notes on soil testing for engineering purposes. Harvard Univ. Grad. School of Eng. Soil Mech. Ser. No. 8, pp. 37–49.

Chancellor, W. J. (1976). Compaction of soil by agricultural equipment. Bull. 1881, Div. Agr. Sci., Univ. California, Richmond, California.

Chancellor, W. J., and Schmidt, R. H. (1962). Soil deformation beneath surface loads. *Trans. Am. Soc. Agr. Eng.* **5**, 204–246, 249.

Chapman, H. D. (1965). Cation exchange capacity. *In* "Methods of Soil Analysis" (C. A. Black, ed.). Monograph 9, Am. Soc. Agron., Madison Wisconsin.

Chepil, W. S. (1958). Soil conditions that influence wind erosion. U.S. Dept. Agriculture Tech. Bull. 1185.

Chepil, W. S. (1962). A compact rotary sieve and the importance of dry sieving in physical soil analysis. *Soil Sci. Soc. Am. Proc.* **26**, 4–6.

Childs, E. C. (1940). The use of soil moisture characteristics in soil studies. *Soil Sci.* **50**, 239–252.

Childs, E. C. (1969). "An Introduction to the Physical Basis of Soil Water Phenomena." Wiley (Interscience), New York.

Childs, E. C., and Collis-George, N. (1950). The permeability of porous materials. *Proc. R. Soc. London Ser. A* **201**, 392–405.

Chudnovskii, A. F. (1966). "Fundamentals of Agrophysics." Israel Program for Scientific Translations, Jerusalem.

Cohron, G. T. (1971). Forces causing soil compaction. *In* "Compaction of Agricultural Soils," (K. K. Barnes, ed.), pp. 106–122. Monograph, Am. Soc. Agr. Eng., St. Joseph, Michigan.

Colman, E. A., and Hendrix, T. M. (1949). Fiberglass electrical soil moisture instrument. *Soil Sci.* **67**, 425–438.

Corey, A. T. (1954). The interrelation between gas and oil relative permeabilities. Oil Producer's Monthly, Vol. XIX, No. 1. November.

Corey, A. T. and Brooks, R. H., Drainage characteristics of soils. *Soil Sci. Soc. Am. Proc.* **39**, No. 2, March–April 1975, pp. 251–255.

Crank, J. (1956). "The Mathematics of Diffusion." Oxford Univ. Press, London and New York.

Currie, J. A. (1961). Gaseous diffusion in porous media Part 3—Wet granular material. *Brit. J. Appl. Phys.* **12**, 275–281.

Currie, J. A. (1975). Soil respiration. *In* "Soil Physical Conditions and Crop Production." Tech. Bull. 29, Min. of Agr., Fisheries and Food, HMSO, London.

Dalton, F. N., and Rawlins, S. L. (1968). Design criteria for Peltier effect thermocouple psychrometers. *Soil Sci.* **105**, 12–17.

Dane, J. H. (1978). Calculation of hydraulic conductivity decreases in the presence of mixed Na-Cl$_2$ solutions. *Can. J. Soil Sci.* **58**, 145–152.

Dane, J. H., and Klute, A. (1977). Salt effects on the hydraulic properties of a swelling soil. *Soil Sci. Soc. Am. J.* **41**, 1043–1049.

Darcy, H. (1856). "Les Fontaines Publique de la Ville de Dijon." Dalmont, Paris.

Dasberg, S., and Bakker, J. W. (1970). Characterizing soil aeration under changing soil moisture conditions for bean growth. *Agr. J.* **62**, 689–692.

Dasberg, S., Hillel, D., and Arnon, I. (1966). Response of grain sorghum to seedbed compaction. *Agron. J.* **58**, 199–201.

Davidson, D. T. (1965). Penetrometer measurements. *In* "Methods of Soil Analysis," Part I. Monograph 9, Am. Soc. Agron., Madison, Wisconsin.

Davidson, J. M., Nielsen, D. R., Biggar, J. W., and Cassel, D. K. (1966). Soil water diffusivity and water content distribution during outflow experiments. Water in the unsaturated zone. Int. Assoc. Sci. Hydrology. *Proc. Wageningen Symp.* 214–223.

Davidson, J. M., Stone, L. R., Nielsen, D. R., and LaRue, M. E. (1969). Field measurement and use of soil water properties. *Water Resources Res.* **5,** 1312–1321.

Davies, D. B., Finney, J. B., and Richardson, S. J. (1973). Relative effects of tractor weight and wheel slip in causing soil compaction. *J. Soil Sci.* **24,** 399–408.

Day, O. R. (1965). Particle fractionation and particle size analysis. *In* "Methods of Soil Analysis," pp. 545–567. Monograph 9, Am. Soc. Agron, Madison, Wisconsin.

de Boer, J. H. (1953). "The Dynamical Charactor of Adsorption." Oxford Univ. Press, London and New York.

DeBoodt, M. (1972a). Improvement of soil structure by chemical means. *In* "Optimizing the Soil Physical Environment Toward Greater Crop Yields" (D. Hillel, ed.), pp. 43–55. Academic Press, New York.

DeBoodt, M. (ed.) (1972b). *Proc. Symp. Fundamentals Soil Conditioning, State Univ. of Ghent, Belgium.*

DeBoodt, M. and DeLeenheer, L. (1958). Proposition pour l'evaluation de la stabilite des aggregates sur le terrain. *Proc. Int. Symp. Soil Structure, Ghent, Belgium* pp. 234–241.

DeBoodt, M., DeLeenheer, L., and Kirkham, D. (1961). Soil aggregate stability indexes and crop yields. *Soil Sci.* **91,** 138–146.

de Groot, S. R. (1963). "Thermodynamics of Irreversible Processes." North-Holland Publ., Amsterdam.

de Jong, E. (1968). Applications of thermodynamics to soil moisture. *Proc. Hydrol. Symp., 6th* pp. 25–48. National Research Council of Canada.

DeLeenheer, L., and DeBoodt, M. (1954). Discussion on aggregate analysis of soils by wet sieving. *Trans. Int. Congr. Soil Sci., 5th, Leopoldville* **2,** 111–117.

Deresiewicz, H. (1958). Mechanics of granular matter. *Adv. Appl. Mech.* **5,** 233–306.

Deryaguin, B. V., and Melnikova, M. K. (1958). Mechanism of moisture equilibrium and migration in soils. Water and its conduction in soils. *Int. Symp. Highway Res. Board Spec. Rep.* 40, pp. 43–54.

de Vries, D. A. (1952). The thermal conductivity of soil. Med. Landbouwhogescho Wageningen.

de Vries, D. A. (1975). Heat transfer in soils. *In* "Heat and Mass Transfer in the Biosphere" (D. A. de Vries and N. H. Afgan, eds.), pp. 5–28. Scripta Book Co., Washington, D.C.

de Vries, D. A., and Afgan, N. H. (1975). "Heat and Mass Transfer in the Biosphere." Scripta, Washington, D.C.

de Vries, D. A., and Peck, A. J. (1958). On the cylindrical probe method of measuring thermal conductivity with special reference to soils. *Aust. J. Phys.* **11,** 255–271; 409–423.

Diamond, S. (1970). Pore size distribution in clays. *Clays Clay Mineral.* **18,** 7–24.

Dick, D. A. T. (1966). "Cell Water." Butterworth, London.

Dirksen, C., and Miller, R. D. (1966). Closed-system freezing of unsaturated soil. *Soil Sci. Soc. Am. Proc.* **30,** 168–173.

Dudal, R. (1968). Definitions of soil units for the soil map of the world. FAO, Rome.

Dumas, W. T., Trouse, A. C., Smith, L. A., Kummer, F. A., and Gill, W. R. (1973). Development and evaluation of tillage and other cultural practices in a controlled traffic system for cotton in the Southern Coastal Plain. *Trans. Am. Soc. Agr. Eng.* **16,** 872–875, 880.

Edlefsen, N. E., and Anderson, A. B. C. (1943). Thermodynamics of soil moisture. *Hilgardia* **15,** 31–298.

Edwards, R. S. (1956). A mechanical sieve designed for experimental work on tilths. *Empire J. Exp. Agr.* **24,** 317–322.

Eisenberg, D., and Kauzmann, W. (1969). "The Structure and Properties of Water." Oxford Univ. Press, London and New York.

Emerson, E. W. (1959). The structure of soil crumbs. *J. Soil Sci.* **10,** 235.

Emerson, W. W., and Grundy, G. M. F. (1954). The effect of rate of wetting on water uptake and cohesion of soil crumbs. *J. Agr. Sci.* **44**, 249–253.

Erickson, A. E., and van Doren, D. M. (1960). The relation of plant growth and yield to soil oxygen availability. *Trans. Int. Congr. Soil Sci., 7th, Madison, Wisconsin* **3**, 428–434.

Evans, D. D. (1965). Gas movement. *In* "Methods of Soil Analysis," Part I, pp. 319–330. Monograph 9, Am. Soc. Agron., Madison, Wisconsin.

Evans, I., and Sherratt, G. G. (1948). A simple and convenient instrument for measuring the shearing resistance of clay soils. *J. Sci. Inst.* **25**, 411–414.

Farrell, D. A. Greacen, E. L., and Gurr, C. G. (1966). Vapor transfer in soil due to air turbulence. *Soil Sci.* **102**, 305–313.

Ferguson H., and Gardner, W. H. 1962. Water content measurement in soil columns by gamma ray absorption. *Soil Sci. Soc. Am. Proc.* **26**, 11–14.

Ferraro, G. (1895). "Les Lois Psychologiques du Symbolisme." Publ., Paris.

Foth, H. (1978). "Fundamentals of Soil Science." Wiley, New York.

Fountaine, E. R., and Payne, P. C. J. (1951). The shear strength of top soils Natl. Inst. Agr. Engr. Tech. Memo. 42.

Frank, H. S., and Wen, W. (1957). Structural aspects of ion-solvent interaction in aqueous solutions: a suggested picture of water structure. *Dis. Faraday Soc.* **24**, 133–140.

Franzini, J. B. (1951). *Trans. Am. Geophys. Un.* **32**, 443.

Freeze, R. A., and Cherry, J. A. (1979). "Groundwater." Prentice Hall, Englewood Cliffs, N.J.

Fuchs, M., and Tanner, C. B. (1968). Calibration and field test of soil heat flux plates. *Soil Sci. Soc. Am. Proc.* **32**, 326–328.

Gairon, S., and Swartzendruber, D. (1973). Streaming potential effects in saturated water flow through a sand-kaolinite mixture. *In* "Physical Aspects of Soil Water and Salts in Ecosystems" (A. Hadas *et al.*, eds.). Springer-Verlag, Berlin and New York.

Gardner, W. (1920). The capillary potential and its relation to soil moisture constants. *Soil Sci.* **10**, 357–359.

Gardner, W. R. (1958). Some steady state solutions of the unsaturated moisture flow equation with application to evaporation from a water table. Soil Science, Vol. 85, No. 4.

Gardner, W. H. (1965). Water content. *In* "Methods of Soil Analysis," pp. 82–127. Monograph 9, Am. Soc. Agron., Madison, Wisconsin.

Gardner, W. H. (1972). Use of synthetic soil conditioners in the 1950's and some implications to their further development. *In Proc. Symp. Fundamentals Soil Conditioning, State Agr. Univ., Ghent, Belgium.* pp. 1046–1061.

Gardner, W. R. (1956). Representation of soil aggregate-size distribution by a logarithmic-normal distribution. *Soil Sci. Soc. Am. Proc.* **20**, 151–153.

Gardner, W. R. (1960). Soil water relations in arid and semi-arid conditions. *UNESCO* **15**, 37–61.

Gardner, W. R. (1968). Availability and measurement of soil water. *In* "Water Deficits and Plant Growth," Vol. 1, pp. 107–135. Academic Press, New York.

Gardner, W. R. (1970). Field measurement of soil water diffusivity. *Soil Sci. Soc. Am. Proc.* **34**, 832.

Gardner, W. R., and Brooks, R. H. (1956). A descriptive theory of leaching. *Soil Sci.* **83**, 295–304.

Gardner, W. R., and Mayhugh, M. S. (1958). Solutions and tests of the diffusion equation for the movement of water in soil. *Soil Sci. Soc. Am. Proc.* **22**, 197–201.

Gardner, W. R., Hillel, D., and Benyamini, Y. (1970). Post irrigation movement of soil water: I. Redistribution. *Water Resources Res.* **6**(3), 851–861; II. Simultaneous redistribution and evaporation. *Water Resources Res.* **6**(4), 1148–1153.

Gardner, W., Israelsen, O. W., Edlefsen, N. W., and Clyde H. (1922). The Capillary Potential Function and its Relation to Irrigation Practice. Physical Review second series, July–

December, p. 196.
Gee, G. W., Stiver, J. F., and Borchert, H. R. (1976). Radiation hazard from Americium-Beryllium neutron moisture probes. *Soil Sci. Soc. Am. J.* **40**, 492–494.
Giesel, W., Lorch, S., and Tenger, M. (1970). Water flow calculation by means of gamma absorption and tensiometer field measurement in the unsaturated soil profile. *In* "Isotope Hydrology, 1970," Int. Atomic Energy Agency, Vienna.
Gill, W. R. (1959). The effect of drying on the mechanical strength of Lloyd clay. *Soil Sci. Soc. Am. Proc.* **23**, 253–257.
Gill, W. R. (1968). The influence of compaction hardening of soil on penetration resistance. *Trans. Am. Soc. Agr. Eng.* **11**, 741–745.
Gill, W. R., and Bolt, G. H. (1955). Pfeffer's studies of the root growth pressures exerted by plants. *Agron. J.* **47**, 166–168.
Gill, W. R., and McCreery, W. F. (1960). Relation of size of cut to tillage tool efficiency. *Agri. Eng.* **41**, 372–374, 381.
Gill, W. R., and Miller, R. D. (1956). A method for study of the influences of mechanical impedance and aeration on the growth of seedling roots. *Soil Sci. Soc. Am. Proc.* **20**, 154–157.
Gill, W. R., and Vanden Berg, G. E. (1967). "Soil Dynamics in Tillage and Traction." Handbook 316, Agr. Res. Service, U.S. Dept. Agriculture, Washington, D.C.
Goss, M. J. (1970). Further studies on the effect of mechanical resistance on the growth of plant roots. Rept. Agr. Res. Coun. Letcombe Lab. ARCRL **20**, 43–45.
Grable, A. R. (1966). Soil aeration and plant growth. *Adv. Agron.* **18**, 57–106.
Grable, A. R., and Siemer, E. G. (1968). Effects of bulk density, aggregate size, and soil water suction on oxygen diffusion, redox potentials and elongation of corn roots. *Soil Sci. Soc. Am. Proc.* **32**, 180–186.
Greacen, E. L., Barley, K. P., and Farrell, D. A. (1969). The mechanics of root growth in soils with particular reference to the implications for root distribution. *In* "Root Growth" (W. J. Whittington, ed.), pp. 256–269. Butterworths, London.
Green, R. E., and Corey, J. C. (1971). Calculation of hydraulic conductivity: A further evaluation of predictive methods. *Soil Sci. Soc. Am. Proc.* **35**, 3–8.
Greenland, D. J. (1965). Interaction between clays and organic compounds in soils. Part I. Mechanisms of interaction between clays and defined organic compounds. *Soil Fert.* **28**, 415–425.
Greenland, D. J., Lindstrom, G. R., and Quirk, J. P. (1962). Organic materials which stabilize natural soil aggregates. *Soil Sci. Am. Proc.* **26**, 236–371.
Greenwood, D. J. (1971). Soil aeration and plant growth. *Rep. Prog. Appl. Chem.* **55**, 423–431.
Grim, R. E. (1953). "Clay Mineralogy." McGraw-Hill, New York.
Groenevelt, P. H., and Bolt, G. H. (1969). Non-equilibrium thermodynamics of the soil-water system. *J. Hydrol.* **7**, 358–388.
Grover, B. L. (1956). Simplified air permeameter for soil in place. *Soil Sci. Soc. Am. Proc.* **19**, 414–418.
Guggenheim, E. A. (1959). "Thermodynamics." North-Holland Publ., Amsterdam.
Gupta, S. C., and Larson, W. E. (1979). *Soil Sci. Soc. Am. J.* **43**, 758–764.
Gurr, C. G. (1962). Use of gamma rays in measuring water content and permeability in unsaturated columns of soil. *Soil Sci.* **94**, 224–449.
Gurr, C. G., Marshall, T. J., and Hutton, J. T. (1952). Movement of water in soil due to temperature gradients. *Soil Sci.* **74**, 333–345.
Hadas, A., and Fuchs, M. (1973). Prediction of the thermal regime of bare soils. *In* "Physical Aspects of Soil Water and Salts in Ecosystems" (A. Hadas, D. Swartzendruber, P. E. Rijtema, M. Fuchs, and B. Yaron, eds.). Springer-Verlag, Berlin and New York.

Hagan, R. M. (1952). Soil temperature and plant growth. In "Soil Physical Conditions and Plant Growth" (B. T. Shaw, ed.), pp. 367–462. Academic Press, New York.

Haines, W. B. (1930). Studies in the physical properties of soils. V. The hysteresis effect in capillary properties and the modes of moisture distribution associated therewith. *J. Agr. Sci.* **20,** 97–116.

Hanks, R. J., and Harkness, K. A. (1956). Soil penetrometer employs strain gages. *Agr. Eng.* **37,** 553–554.

Hanks, R. J., and Woodruff, N. P. (1958). Influence of wind on water vapor transfer through soil, gravel, and straw mulches. *Soil Sci.* **86,** 160–164.

Harris, R. F., Chester, G., and Allen, O. N. (1965). Dynamics of soil aggregation. *Adv. Agron.* **18,** 107–160.

Harris, W. L. (1960). Dynamic Stress Transducers and the Use of Continuum Mechanics in the Study of various Soil Stress-Strain Relationships. Ph.D. thesis, Michigan State Univ., East Lansing, Michigan.

Harris, W. L. (1971). The soil compaction process. In "Compaction of Agricultural Soils" (K. K. Barnes, ed.), pp. 9–44. Monograph, Am. Soc. Agr. Eng., St. Joseph, Michigan.

Helling, C. S. *et al.* (1964). Contribution of organic matter and clay to soil cation exchange capacity as affected by the pH of the saturated solution. *Soil Sci. Soc. Am. Proc.* **28,** 517–520.

Hettiaratchi, D. R. P., Whitney, B. D., and Reece, A. R. (1966). The calculation of passive pressure in two-dimensional soil failure. *J. Agr. Eng. Res.* **11,** 89–107.

Hillel, D. (1959). Studies of Loessial Crusts. Bull. 63, Israel Agr. Res. Inst., Beit Dagan.

Hillel, D. (1960). Crust formation in loessial soils. *Trans. Int. Congr. Soil 7th Sci.*, **I,** 339–339.

Hillel, D. (1968). Soil Water Evaporation and Means of Minimizing It. Hebrew Univ. Faculty Agriculture Res. Rept., Rehovot, Israel.

Hillel, D. (1971). "Soil and Water: Physical Principles and Processes." Academic Press, New York.

Hillel, D. (1972). Soil moisture and seed germination. In "Water Deficits and Plant Growth" (T. T. Kozlowski, ed.). pp. 65–89. Academic Press, New York.

Hillel, D. (1977a). Computer Simulation of Soil Water Dynamics. Int. Dev. Res. Center, Ottawa, Canada.

Hillel, D. (1977b). Soil Management. In "Yearbook of Science and Technology." pp. 372–374. McGraw-Hill, New York.

Hillel, D. (1980). "Applications of Soil Physics." Academic Press, New York.

Hillel, D., and Gardner, W. R. (1970). Measurement of unsaturated conductivity diffusivity by infiltration through an impeding layer. *Soil Sci.* **109,** 149.

Hillel, D., and Mottes, J. (1966). Effect of plate impedance, wetting method and aging on soil moisture retention. *Soil Sci.* **102,** 135–140.

Hillel, D., Krentos, V. D., and Stylianou, Y. (1972). Procedure and test of an internal drainage method for measuring soil hydraulic characteristics *in situ*. *Soil Sci.* **114,** 395–400.

Hillel, D. (1974a). Methods of laboratory and field investigation of physical properties of soils. *Trans. Int. Soil Sci. Congr., 10th, Moscow* **I,** 301–308.

Hillel, D., and Benyamini, Y. (1974b). Experimental comparison of infiltration and drainage methods for determining unsaturated hydraulic conductivity of a soil profile in situ. *Proc. FAO/IAEA Symp. Isotopes and Radiation Techniques in Studies of Soil Physics, Vienna*, pp. 271–275.

Hillel, D. and Berliner, P. (1974c). Water-proofing surface zone soil aggregates for water conservation. *Soil Sci.* **118,** 131–135.

Hillel, D., Talpaz, H., and van Keulen, H. (1975). A macroscopic model of water uptake by a non-uniform root system and of water and salt movement in the soil profile. *Soil Sci.* **121**, 242–255.

Hoffman, O., and Sachs, G. (1953). "Introduction to the Theory of Plasticity." McGraw-Hill, New York.

Holmes, J. W. (1956). Calibration and field use of the neutron scattering method of measuring soil water content. *Aust. J. Appl. Sci.* **7**, 45–58.

Hopkins, R. M., and Patrick, W. H. (1969). Combined effects of oxygen content and soil compaction on root penetration. *Soil Sci.* **108**, 408–413.

Hubbert, M. K. (1956). Darcy's law and the field equations of the flow of underground fluids. *Am. Inst. Min. Met. Petl. Eng. Trans.* **207**, 222–239.

Jackson, R. D. (1960). The Importance of Convection as a Heat Transfer Mechanism in Two-Phase Porous Materials. Unpublished Ph.D. dissertation, Colorado State Univ., Fort Collins, Colorado.

Jackson, R. D. (1964). Water vapor diffusion in relatively dry soil: I. Theoretical considerations and sorption experiments. *Soil Sci. Soc. Am. Proc.* **28**, 172–176.

Jackson, R. A. (1972). On the calculation of hydraulic conductivity. *Soil Sci. Soc. Am. Proc.* **36**, 380–383.

Jackson, R. D., and Taylor, S. A. (1965). Heat transfer. *In* "Methods of Soil Analysis," pp. 349–360. Monograph No. 9, Am. Soc. Agron., Madison, Wisconsin.

Jackson, R. D., Reginato, R. J., Kimball, B. A., and Nakayama, F. S. (1974). Diurnal soil-water evaporation: comparison of measured and calculated soil-water fluxes. *Soil Sci. Soc. Am. Proc.* **38**, 861–866.

Jacob, C. E. (1946). *Trans. Am. Geophys. Un.* **27**, 198.

Jaeger, J. C. (1964). "Elasticity, Fracture, and Flow." Wiley, New York.

Jamil, A. (1976). Effect of Clay Swelling on the Hydraulic Parameters of Porous Sandstones. Ph.D. thesis submitted to Colorado State Univ., Fort Collins, Colorado.

Janert, H. (1934). The application of heat of wetting measurements to soil research problems. *J. Agr. Sci.* **24**, 136–145.

Jenny, H. F. (1941). "Factors of Soil Formation." McGraw-Hill, New York.

Johnson, H. P., Frevert, R. K., and Evans, D. D. (1952). Simplified procedure for the measurement and computation of soil permeability below the water table. *Agr. Eng.* **33**, 283–289.

Jury, W. A. (1973). Simultaneous Transport of Heat and Moisture Through a Medium Sand. Ph.D. Thesis, Univ. of Wisconsin, Madison, Wisconsin.

Katchalsky, A., and Curran, P. F. (1965). "Nonequilibrium Thermodynamics in Biophysics." Harvard Univ. Press, Cambridge, Massachusetts.

Kavanau, J. L. (1965). Water. *In* "Structure and Function in Biological Membranes," pp. 170–248. Holden-Day, San Francisco, California.

Kays, S. J., Nicklow, C. W., and Simons, D. H. (1974). Ethylene in relation to the response of roots to mechanical impedance. *Plant Soil* **40**, 565–571.

Keen, B. A. (1931) "The Physical Properties of Soil." Longmans, Green, New York.

Keen, B. A., and Russell, E. W. (1937). Are cultivation standards wastefully high? *J. R. Agr. Soc.* **98**, 53–60.

Kemper, W. D. (1965). Aggregate stability. *In* "Methods of Soil Analysis." Am. Soc. Agron., Madison, Wisconsin.

Kemper, W. D., and Amemiya, M. (1957). Alfalfa growth as affected by aeration and soil moisture stress under flood irrigation. *Soil Sci. Soc. Am. Proc.* **21**, 657–660.

Kemper, W. D., and Chepil, W. S. (1965). Size distribution of aggregates. *In* "Methods of Soil Analysis." Am. Soc. Agron., Madison, Wisconsin.

Kemper, W. D., and Evans, N. A. (1963). Movement of water as affected by free energy and pressure gradients. II. Restriction of solutes by membranes. *Soil Sci. Soc. Am. Proc.* **27,** 485–490.

Kemper, W. D., and Maasland, D. E. L. (1964). Reduction in salt content of solution on passing through thin films adjacent to charged surfaces. *Soil Sci. Soc. Am. Proc.* **28,** 318–323.

Kersten, M. S. (1949). Thermal Properties of Soils. Bull. 28, Univ. Minnesota Inst. Technol., St. Paul, Minnesota.

Kezdi, A. (1974). "Handbook of soil Mechanics," Vol. I, Soil Physics. Elsevier, Amsterdam.

Kimball, B. A., and Lemon, E. R. (1971). Air turbulence effects upon soil gas eschange. *Soil Sci. Soc. Am. Proc.* **35,** 16–21.

Kimball, B. A. and Lemon, E. R. (1972). Theory of air movement due to pressure fluctuations. *Agr. Meteorol.* **9,** 163–181.

Kirkham, D. (1946). Field method for determination of air permeability of soil in its undisturbed state. *Soil Sci. Soc. Am. Proc.* **11,** 92–99.

Kirkham, D., DeBoodt, M. F., and DeLeenheer, L. (1959). Modulus of rupture determination on undisturbed soil core samples. *Soil Sci.* **87,** 141–144.

Klotz, I. M. (1974). Water. *In* "Horizons in Biochemistry" (M. Kasha and B. Pullman, eds.). pp. 253–550. Academic Press, New York.

Klute, A. (1965a). Laboratory measurement of hydraulic conductivity of saturated soil. *In* "Methods of Soil Analysis," pp. 210–221. Monograph 9. Am. Soc. Agron., Madison, Wisconsin.

Klute, A. (1965b). Laboratory measurement of hydraulic conductivity of unsaturated soil. *In* "Methods of Soil Analysis," pp. 253–261. Monograph No. 9, Am. Soc. Agron., Madison, Wisconsin.

Klute, A., and Peters, D. B. (1969). Water uptake and root growth. *In* "Root Growth" (W. J. Whittington, ed.), pp. 105–133. Buttersworths, London.

Kohnke, H. (1968). "Soil Physics." McGraw-Hill, New York.

Kondner, R. L. (1958). A non-dimensional approach to the vibratory cutting, compaction and penetration of soils. Tech. Rep. 8 by Johns Hopkins Univ. to U.S. Army Corps of Eng., Waterways Exp. Sta., Vicksburg, Mississippi.

Kramer, P. J. (1956). Physical and physiological aspects of water absorption. *In* "Handbuch der Pflanzenphysiologie, Vol. III, Pflanze and Wasser, pp. 124–159. Springer-Verlag, Berlin and New York.

Kravtchenko, J., and Sirieys, P. M. (eds.) (1966). "Rheology and Soil Mechanics." Springer-Verlag, Berlin and New York.

Kristenson, K. J., and Lemon, E. R. (1964) Soil aeration and plant root relations: III. Physical aspects of oxygen diffusion in the liquid phase of the soil. *Agron. J.* **56,** 295–301.

Kunze, R. J., and Kirkham, D. (1962). Simplified accounting for membrane impedance in capillary conductivity determinations. *Soil Sci. Soc. Am. Proc.* **26,** 421–426.

Kunze, R. J., Uehara, G., and Graham, K. (1968). Factors important in the calculation of hydraulic conductivity. *Soil Sci. Soc. Am. Proc.* **32,** 760–765.

Lagerwerff, J. V., Nakayama, F. S., and Frere, M. H. (1969). Hydraulic conductivity related to porosity and swelling of soil. *Soil Sci. Soc. Am. Proc.* **33,** 3–11.

Laliberte, G. E. (1969). A mathematical function for describing capillary pressure-desaturation data. *Bull. Int. Ass. Sci. Hydrol.* **14**(2), 131–149.

Lambe, T. W., and Whitman, R. V. (1969). "Soil Mechanics." Wiley, New York.

Langmuir, I. (1918). The adsorption of gases on plane surfaces of glass, mica, and platinum. *J. Am. Chem. Soc.* **40,** 1361–1402.

Langmuir, I. (1938). Distribution of cations between two charged plates. *Science* **88,** 430–433.

Lemon, E. R. (1962). Soil aeration and plant root relations. I. Theory. *Agron. J.* **54,** 167–170.

Lemon, E. R. and Erickson, A. E. (1952). The measurement of oxygen diffusion in the soil with a platinum microelectrode. *Soil Sci. Soc. Am. Proc.* **16**, 160–163.
Leopold, L. B. (1974). "Water: A Primer." Freeman, San Francisco, California.
Letey, J. (1968). Movement of water through soil as influenced by osmotic pressure and temperature gradients. *Hilgardia* **39**, 405–418.
Letey, J. and Stolzy, L. H. (1964). Measurement of oxygen diffusion rates with the platinum microelectrode: I. Theory and equipment. *Hilgardia* **55**, 545–554.
Letey, J., Stolzy, L. H., and Kemper, W. D. (1967). Soil aeration. *In* "Irrigation of Agricultural Lands," pp. 943–948. Amer. Soc. Agron., Madison, Wisconsin.
Lettau, H. H. (1962). A theoretical model of thermal diffusion in nonhomogeneous conductors. *Gerlands. Beitr. Geophys.* **71**, 257–271.
Levy, R., and Hillel, D. (1968). Thermodynamic equilibrium constants of Na-Ca exchange in some Israeli soils. *Soil Sci.* **106**, 393–398.
Lewis, W. M. (1973). No-till systems. *In* "Conservation Tillage," pp. 182–187. Soil Conserv. Soc. Am., Ames, Iowa.
Low, P. F. (1961). Physical chemistry of clay-water interactions. *Adv. Agron.* **13**, 269–327.
Low, P. F. (1965). The effect of osmotic pressure on the diffusion rate of water. *Soil Sci.* **80**, 95–100.
Luthin, J. N. (ed.) (1957). "Drainage of Agricultural Lands." Monograph 7. Am. Soc. Agron., Madison, Wisconsin.
Lutz, J. F. (1952). Mechanical impedance and plant growth. *In* "Soil Physical Conditions and Plant Growth" (B. T. Shaw, ed.), pp. 43–71. Academic Press, New York.
Lutz, J. F. and Leamer, R. W. (1939). Pore-size distribution as related to the permeability of soils. *Soil Sci. Soc. Am. Proc.* **4**, 28–31.
Maasland, M., and Kirkham, D. (1955). Theory and measurement of anisotropic air permeability in soil. *Soil Sci. Soc. Am. Proc.* **19**, 395–400.
Marshall, C. E. (1964). "The Physical Chemistry and Mineralogy of Soils." Wiley, New York.
Marshall, T. J. (1958). A relation between permeability and size distribution of pores. *J. Soil Sci.* **9**, 1–8.
Marshall, T. J. (1959). The diffusion of gases through porous media. *J. Soil Sci.* **10**, 79–82.
Mayanskas, I. S. (1959). Investigation of the pressure distribution on the surface of a plow share in work. *J. Agr. Eng. Res.* **4**, 186–190.
Mazurak, A. P. (1950). Effect of gaseous phase on water-stable synthetic aggregates. *Soil Sci.* **69**, 135–148.
McCalla, T. M. (1944). Water-drop method of determining stability of soil structure. *Soil Sci.* **58**, 117–121.
McClelland, J. H. (1956). Instrument for measuring soil condition. *Agr. Eng.* **37**, 480–481.
McGregor, K. C., Greer, J. D., and Gurley, G. E. (1975). Erosion control with no-till cropping practices. *Trans. Am. Soc. Agr. Eng.* **18**, 918–920.
McIntyre, D. S. (1970). The platinum microelectrode method for soil aeration measurement. *In* "Advances in Agronomy," Vol. 22, pp. 235–283. Academic Press, New York.
McIntyre, D. S., and Philip, J. R. (1964). A field method for measurement of gas diffusion into soils. *Aust. J. Soil Res.* **2**, 133–145.
McMurdie, J. L. (1963). Some characteristics of the soil deformation process. *Soil Sci. Soc. Am. Proc.* **27**, 251–254.
McNeal, B. L. (1974). Soil salts and their effects on water movement. *In* "Drainage for Agriculture" (J. van Schilfgaarde, ed.). Monograph 17, Am. Soc. Agron., Madison, Wisconsin.
McNeal, B. L., and Coleman, N. T. (1966). Effect of solution composition on soil hydraulic conductivity. *Soil Sci. Soc. Am. Proc.* **30**, 308–312.
McNeal, B. L., Layfield, D. A., Norvell, W. A., and Rhoades, J. D. (1968). Factors influencing

hydraulic conductivity of soils in the presence of mixed-salt solutions. *Soil Sci. Soc. Am. Proc.* **32,** 187–190.
Meiuner, H., and Sheriff, D. W. (1976). "Water and Plants." Halsted Press, New York and Wiley, New York.
Miller, E. E., and Elrick, D. E. (1958). Dynamic determination of capillary conductivity extended for non-negligible membrane impedance. *Soil Sci. Soc. Am. Proc.* **22,** 483–486.
Miller, E. E., and Klute, A. (1967). Dynamics of soil water. Part I—Mechanical forces. In "Irrigation of Agricultural Lands," pp. 209–244. Monograph 11, Am. Soc. Agron., Madison, Wisconsin.
Miller, E. E., and Miller, R. D. (1955a). Theory of capillary flow: I. Practical implications. *Soil Sci. Soc. Am. Proc.* **19,** 267–271.
Miller, E. E., and Miller, R. D. (1955b). Theory of capillary flow: II. Experimental information. *Soil Sci. Soc. Am. Proc.* **19,** 271–275.
Miller, E. E., and Miller, R. D. (1956). Physical theory for capillary flow phenomena. *J. Appl. Phys.* **27,** 324–332.
Miller, R. J., and Low, P. F. (1963). Threshold gradient for water flow in clay systems. *Soil Sci. Soc. Am. Proc.* **27,** 605–609.
Millington, R. J. (1959). Gas diffusion in porous media. *Science* **130,** 100–102.
Millington, R. J., and Quirk, J. P. (1959). Permeability of porous media. *Nature (London)* **183,** 387–388.
Monteith, J. L. (1973). "Principles of Environmental Physics." American Elsevier, New York.
Moore, R. E. (1939). Water conduction from shallow water tables. *Hilgardia* **12,** 383–426.
Morgan, J., and Warren, B. E. (1938). X-ray analysis of the structure of water. *J. Chem. Phys.* **6,** 666–673.
Morin, J., Goldberg, D., and Seginer, I. (1967). A rainfall simulator with a rotating disk. *ASAE Trans.* **10,** 74–77.
Mortland, M. M., and Kemper, W. D. (1965). Specific surface. In "Methods of Soil Analysis," pp. 532–544. Monograph 9, Am. Soc. Agron., Madison, Wisconsin.
Morton, C. T., and Buchele, W. F. (1960). Emergence energy of plant seedlings. *Agr. Eng.* **41,** 428–431.
Mualem, Y. (1976). A new model for predicting the hydraulic conductivity of unsaturated porous media. *Water Resour. Res.* **12,** 513–522.
Muskat, M. (1946). "The Flow of Homogeneous Fluids Through Porous Media." Edwards, Ann Arbor, Michigan.
Narten, A. H., and Levy, H. A. (1969). Observed diffraction pattern and proposed model of liquid water. *Science* **165,** 447–454.
Nemethy, G., and Scheraga, H. A. (1962). Structure of water and hydraulic bonding in proteins. I. A model for the thermodynamic properties of liquid water. *J. Chem. Phys.* **36,** 3382–3400.
Nerpin, S., Pashkina, S., and Bondarenko, N. (1966). The evaporation from bare soil and the way of its reduction. *Symp. Water Unsaturated Zone, Wageningen.*
Nichols, M. L., and Randolph, J. W. (1925). A method of studying soil stresses. *Agr. Eng.* **6,** 134–135.
Nielsen, D. R., and J. W. Biggar, (1961). Miscible displacement in soils: I. Experimental information. *Soil Sci. Soc. Am. Proc.* **25,** 1–5.
Nielsen, D. R., and Biggar, J. W. (1962). Miscible displacement: III. Theoretical consideration. *Soil Sci. Soc. Am. Proc.* **26,** 216–221.
Nielsen, D. R., Jackson, R. D., Cary, J. W., and Evans, D. D. (eds.) (1972). "Soil Water." Am. Soc. Agron., Madison, Wisconsin.
Nutting, P. G. (1943). Some standard thermal dehydration curves of minerals. U.S.G.S. Professional paper 197-E.

Ogata, G., and Richards, L. A. (1957). Water content changes following irrigation of bare field soil that is protected from evaporation. *Soil Sci. Soc. Am. Proc.* **21**, 355–356.

Olsen, H. W. (1965). Deviations from Darcy's law in saturated clays. *Soil Sci. Soc. Am. Proc.* **29**, 135–140.

Onafenko, O., and Reece, A. R. (1967). Soil stresses and deformations beneath rigid wheels. *J. Terramech.* **4**, 59–80.

Overbeek, J. T. G. (1952). *In* "Colloid Science" (H. R. Kruyt, ed.), Chapters 4–6. Elsevier, Amsterdam.

Page, J. B. (1948). Advantages of the pressure pycnometer for measuring the pore space in soils. *Soil Sci. Soc. Am. Proc.* **12**, 81–84.

Papendick, R. I. and Runkles, J. R. (1965). Transient-state oxygen diffusion in soil. I. The case when rate of oxygen consumption is constant. *Soil Sci.* **100**, 251–261.

Parker, J. J., and Taylor, H. M. (1965). Soil strength and seedling emergence relations. *Agron. J.* **57**, 289–291.

Peck, A. J. (1969). Entrapment, stability, and persistence of air bubbles in soil water. *Aust. J. Soil Res.* **7**, 79–90.

Penman, H. L. (1940). Gas and vapor movements in the soil: I. The diffusion of vapors through porous solids. *J. Agr. Sci.* **30**, 437–461.

Philip, J. R. (1955). Numerical solution of equations of the diffusion type with diffusivity concentration dependent. *Trans. Faraday Soc.* **51**, 885–892.

Philip, J. R. (1960). Absolute thermodynamic functions in soil-water studies. *Soil Sci.* **89**, 111.

Philip, J. R. (1964). Similarity hypothesis for capillary hysteresis in porous materials. *J. Geophys. Res.* **69**, 1553–1562.

Philip, J. R. (1969). Hydrostatics and hydrodynamics in swelling soils. *Water Resources Res.* **5**, 1070–1077.

Philip, J. R. (1975). Water movement in soil. *In* "Heat and Mass Transfer in the Biosphere." Scripta, Washington, D.C.

Philip, J. R., and deVries, D. A. (1957). Moisture movement in porous materials under temperature gradients. *Trans. Am. Geophys. Un.* **38**, 222–228.

Plummer, F. L., and Dore, S. M. (1940). "Soil Mechanics and Foundations." Pitman, New York.

Poulovassilis, A. (1962). Hysteresis of pore water, an application of the concept of independent domains. *Soil Sci.* **93**, 405–412.

Prigogine, I. (1961). "Introduction to Thermodynamics of Irreversible Processes." Wiley, New York.

Pringle, J. (1975). The assessment and significance of aggregate stability in soil. *In* "Soil Physical Conditions and Crop Production," pp. 249–260. Tech. Bull. 29, Min. Agr., Fisheries and Food, London.

Proctor, R. R. (1948). Laboratory compaction methods, penetration resistance measurements, and indicated saturation penetration resistance. *Proc. Int. Conf. Soil Mech. Found. 2nd* **5**, 242–245.

Puri, A. N., and Puri, B. R. (1939). Physical characteristics of soils: II. Expressing mechanical analysis and state of aggregation of soils by single values. *Soil Sci.* **47**, 77–86.

Quastel, J. H. (1954). Soil conditioners. *Res. Plant Physiol.* **5**, 75–92.

Quirk, J. P., and Schofield, R. K. (1955). The effect of electrolyte concentration on soil permeability. *J. Soil Sci.* **6**, 163–178.

Radhakrishnan, S. (1958). "Indian Philosophy." Macmillan, New York.

Raney, W. A. (1950). Field measurement of oxygen diffusion through soil. *Soil Sci. Soc. Am. Proc.* **14**, 61–65.

Rawitz, E., Margolin, M., and Hillel, D. (1972). An improved variable intensity sprinkling

infiltrometer. *Soil Sci. Am. Proc.* **36**, 533–535.
Rawlins, S. L., and Raats, P. A. C. (1975). Prospects for high frequency irrigation. *Science* **188**, 604–610.
Reaves, C. A., and Cooper, A. W. (1960). Stress distribution in soils under tractor loads. *Agr. Eng.* **41**, 20–21, 31.
Reeve, R. C. (1965a). Air-to-water permeability ratio. *In* "Methods of Soil Analysis." Am. Soc. Agron., Madison, Wisconsin.
Reeve, R. C. (1965b). Modulus of rupture. *In* "Methods of Soil Analysis," Pt. I. Monograph 9, Am. Soc. Agron., Madison, Wisconsin.
Reeve, R. C., and Bower, C. A. (1960). Use of high salt water as a flocculant and source of divalent cations for reclaiming sodic soils. *Soil Sci.* **90**, 139–144.
Reid, C. E. (1960). "Principles of Chemical Thermodynamics." Van Nostrand-Reinhold, Princeton, New Jersey.
Reitemeyer, R. F., and Richards, L. A. (1944). Reliability of the pressure-membrane method for extraction of soil solution. *Soil Sci.* **57**, 119–135.
Retzer, J. L., and Russell, M. B. (1941). Differences in the aggregation of a Prairie and a Gray-Brown Podzolic soil. *Soil Sci.* **52**, 47–58.
Rhoades, J. D. (1974). Drainage for salinity control. *In* "Drainage for Agriculture" (J. van Schilfgaarde, ed.). Monograph 17, Am. Soc. Agron., Madison, Wisconsin.
Richards, L. A. (1928). The usefulness of capillary potential to soil-moisture and plant investigators, *J. Agri. Res.* **37**, 719–742.
Richards, L. A. (1931). Capillary conduction of liquids in porous mediums. *Physics* **1**, 318–333.
Richards, L. A. (1953). Modulus of rupture of soils as an index of crusting of soil. *Soil Sci. Soc. Am. Proc.* **17**, 321–323.
Richards, L. A. (ed.) (1954). "Diagnosis and Improvement of Saline and Alkali Soils." U.S. Dept. Agr. Handbook 60.
Richards, L. A. (1965). Physical condition of water in soil. *In* "Methods of Soil Analysis," pp. 128–152. Monograph 9, Am. Soc. Agron., Madison, Wisconsin.
Richards, L. A., and Weaver, L. R. (1944). Moisture retention by some irrigated soils as related to soil moisture tension. *J. Agri. Res.* **69**, 215–235.
Richards, S. J. (1965). Soil suction measurements with tensiometers. *In* "Methods of Soil Analysis," pp. 153–163. Monograph 9, Am. Soc. Agron., Madison, Wisconsin.
Richards, S. J., and Weeks, L. V. (1953). Capillary conductivity values from moisture yield and tension measurements on soil columns. *Soil Sci. Soc. Am. Proc.* **17**, 206–209.
Rijtema, P. E. (1959). Calculation of capillary conductivity from pressure plate outflow data with non-negligible membrane impedance. *Neth. J. Agr. Sci.* **7**, 209–215.
Robinson, F. E. (1964). A diffusion chamber for studying soil atmosphere. *Soil Sci.* **83**, 465–469.
Rose, C. W., and Stern, W. R. (1967). The drainage component of the water balance equation. *Aust. J. Soil Res.* **3**, 95–100.
Rose, C. W., Stern, W. R., and Drummond, J. E. (1965). Determination of hydraulic conductivity as a function of depth and water content for soil *in situ*. *Aust. J. Soil Res.* **3**, 1–9.
Rubin, J. (1966). Theory of rainfall uptake by soils initially drier than their field capacity and its applications. *Water Resources Res.* **2**, 739–749.
Russel, E. J. (1912). "Soil Conditions and Plant Growth." Longman, London.
Russell, E. W. (1973). "Soil Conditions and Plant Growth," 10th ed. Longman, London.
Russell, M. B. (1941). Pore-size distribution as a measure of soil structure. *Soil Sci. Soc. Am. Proc.* **6**, 108–112.
Russell, M. B. (1949). Methods of measuring soil structure and aeration. *Soil Sci.* **68**, 25–35.
Russell, M. B. (1952). Soil aeration and plant growth. *In* "Soil Physical Conditions and Plant Growth" (B. T. Shaw, ed.), pp. 253–301. Academic Press, New York.

Russell, M. B., and Feng, C. L. (1947). Characterization of the stability of soil aggregates. *Soil Sci.* **63**, 299–304.

Russell, R. S., and Goss, M. J. (1974). Physical aspects of soil fertility—the response of roots to mechanical impedance. *Neth. J. Agr. Sci.* **22**, 305–318.

Russell, E. J., and Russell, E. W. (1950). "Soil Conditions and Plant Growth." Longman, London.

Russo, D. and Bresler, E. (1977). Analysis of the saturated-unsaturated hydraulic conductivity in a mixed sodium-calcium soil system. *Soil Sci. Soc. Am. J.* **41**, 706–710.

Salberg, J. R. (1965). Shear strength. *In* "Methods of Soil Analysis," Pt. I. Monograph 9, Am. Soc. Agron., Madison, Wisconsin.

Schamp, N. (1971). Soil conditioning by means of organic polymers. *Pedologic* **21**, 100.

Scheidegger, A. E. (1957). "The Physics of Flow through Porous Media." Macmillan, New York.

Schofield, C. S. (1940). Salt balance in irrigated areas. *J. Agr. Res.* **61**, 17–39.

Schofield, R. K. (1935). The pF of the water in soil. *Trans. Int. Cong. Soil Sci., 3rd.* **2**, 37–48.

Scotter, D. R. and Raats, P. A. C. (1968). Dispersion in porous mediums due to oscillating flow. *Water Resources Res.* **4**, 1201–1206.

Seed, H. B., Woodward, R. J., and Lundgren, R. (1964). Fundamental aspects of the Atterberg limits. *Proc. Am. Soc. Civil Eng. J. Soil Mech. Found. Div.* **90**, No. SM6.

Sellers, W. D. (1965). "Physical Climatology." Univ. of Chicago Press, Chicago, Illinois.

Shainberg, I. (1973). Ion exchange properties in irrigated soils. *In* "Arid Zone Irrigation" (B. Yaron, *et al.*, eds.). Springer-Verlag, Berlin and New York.

Shaw, R. H., and Buchele, W. F. (1957). The effect of the shape of the soil surface profile on soil temperature and moisture. *Iowa State College J. Sci.* **32**, 95–104.

Slatyer, R. O. (1967). "Plant-Water Relationships." Academic Press, New York.

Slichter, C. S. (1899). U.S. Geol. Sur. Ann. Rep. 19-II, pp. 295–384.

Smiles, D. E., and Rosenthal, M. J. (1968). The movement of water in swelling materials. *Aust. J. Soil Res.* **6**, 237–248.

Smith, A. (1932). Seasonal subsoil temperature variations. *J. Agr. Res.* **44**, 421–428.

Smith, G. D., Newhall, F., Robinson, L. H., and Swanson, D. (1964). "Soil Temperature Regimes, Their Characteristics and Predictability." U.S. Dept. Agr. Soil Conservation Service SCS-TP-144, Washington, D.C.

Soane, B. D. (1970). The effect of traffic and implements on soil compaction. *Agr. Eng.* **25**, 115–128.

Soane, B. D. (1975). Studies on some soil physical properties in relation to cultivations and traffic. *In* "Soil Physical Conditions and Crop Production," pp. 249–260. Tech. Bull. 29, Min. Agr., Fisheries and Food, London.

Söhne, W. H. (1953). Pressure distribution in the soil and soil deformation under tractor tires. *Grundlagen Landtech.* **5**, 49–63.

Söhne, W. H. (1958). Fundamentals of pressure distribution and soil compaction under tractor tires. *Agr. Eng.* **39**, 276–281, 290.

Söhne, W. H. (1966) Characterization of tillage tools. Särtryck ur Grundföbättring **1**, 31–48.

Soil (1957). Yearbook of Agriculture, U.S. Dept. of Agriculture.

Soil Survey Manual (1951). U.S. Dept. of Agriculture Handbook No. 18.

Sor, K., and Kemper, W. D. (1959). Estimation of surface area of soils and clays from the amount of adsorption and retention of ethylene glycol. *Soil Sci. Soc. Am. Proc.* **23**, 105–110.

Sourisseau, J. H. (1935). Determination and study of physico-mechanical properties of soil. *Organ. Raps. II Congr. Int. Genie Rural, Madrid* pp. 159–194.

Sowers, G. F. (1965). Consistency. *In* "Methods of Soil Analysis," pp. 391–399. Monograph No. 9., Am. Soc. Agron., Madison, Wisconsin.

Stanhill, G. (1965). Observations on the reduction of soil temperature. *Agr. Met.* **2,** 197–203.
Steinhardt, R., and Hillel, D. (1966). A rainfall simulator for laboratory and field use. *Soil Sci. Soc. Am. Proc.* **30,** 680–682.
Stotzky, G. (1965). Microbial respiration. *In* "Methods of Soil Analysis," pp. 1550–1569. Monograph 9, Am. Soc. Agron., Madison, Wisconsin.
Stolzy, L. H., and Barley, K. P. (1968). Mechanical resistance encountered by roots entering compact soil. *Soil Sci.* **105,** 297–301.
Stolzy, L. H., and Letey, J. (1964). Characterizing soil oxygen conditions with a platinum microelectrode. *Agr. J.* **16,** 249–279.
Su, C. and Brooks, R. H. (1975). Soil hydraulic properties from infiltration tests. Watershed Managemend Proceedings, Irrigation and Drainage Division, ASCE. Logan, Utah. August 11–13, pp. 516–542.
Su, C. and Brooks, R. H. (1979). Measurement of water retention data for soils. Submitted for publication in ASCE.
Suklje, L. (1969). "Rheological Aspects of Soil Mechanics." Wiley (Interscience), New York.
Sutcliffe, J. (1968). "Plants and Water." Arnold, London.
Swain, R. W. (1975). Subsoiling. *In* Soil Physical Conditions and Crop Production, Tech. Bull. 29, 00. 189–204. Min. Agr. Fish Food. HMSO, London.
Swarzendruber, D. (1962). Non Darcy behavior in liquid saturated porous media. *J. Geophys. Res.* **67,** 5205–5213.
Talsma, T. (1960). Comparison of field methods of measuring hydraulic conductivity. *Trans. Congr. Irrigat. Drainage* **4,** C145–C156.
Tanner, C. B., and Elrick, D. E. (1958). Volumetric porous (pressure) plate apparatus for moisture hysteresis measurements. *Soil Sci. Soc. Am. Proc.* **22,** 575–576.
Tanner, C. B., and Mamaril, C. P. (1959). Pasture soil compaction by animal traffic. *Agron. J.* **51,** 329–331.
Taylor, D. W. (1948). "Fundamentals of Soil Mechanics." Wiley, New York.
Taylor, H. M., and Bruce, R. R. (1968). Effect of soil strength on root growth and crop yield in the southern United States. *Trans. Int. Congr. Soil Sci., 9th* **1,** 803–811.
Taylor, H. M., and Gardner, H. R. (1963). Penetration of cotton seedling tap roots as influenced by bulk density, moisture content, and strength of soil. *Soil Sci.* **96,** 153–156.
Taylor, H. M., and Ratliff, L. F. (1969). Root elongation rates of cotton and peanuts as a function of soil strength and soil water content. *Soil Sci.* **108,** 113–119.
Taylor, H. M., Robertson, G. M., and Parker, J. J. (1966). Soil strength-root penetration relations for medium- to coarse-textured soil materials. *Soil Sci.* **102,** 18–22.
Taylor, J. H., and VandenBerg, G. E. (1966). Role of displacement in a simple traction system. *Trans. Am. Soc. Agr. Eng.* **9,** 10–13.
Taylor, S. A., and Cary, J. W. (1960). Analysis of the simultaneous flow of water and heat or electricity with the thermodynamics of irreversible processes. *Int. Congr. Soil Sci. Trans., 7th* **1,** 80–90.
Taylor, S. A., and Cary, J. W. (1964). Linear equations for the simultaneous flow of matter and energy in a continuous soil system. *Soil Sci. Soc. Am. Proc.* **28,** 167–172.
Taylor, S. A., and Jackson, R. D. (1965). Soil Temperature. *In* "Methods of Soil Analysis," pp. 331–344. Monograph 9, Am. Soc. Agron., Madison, Wisconsin.
Terry, C. W., and Wilson, H. M. (1953). The soil penetrometer in soil compaction studies. *Agr. Eng.* **34,** 831–834.
Terzaghi, K. (1953). "Theoretical Soil Mechanics." Wiley, New York.
Terzaghi, K., and Peck, R. B. (1948). "Soil Mechanics in Engineering Practice." New York.
Timoshensko, S., and Goodier, J. N. (1951). "Theory of Elasticity." McGraw-Hill, New York.
Tiulin, A. F. (1928). Questions on soil structure: II. Aggregate analysis as a method for determining soil structure. *Perm. Agr. Exp. Sta. Div. Agr. Chem. Rep.* **2,** 77–122.

Topp, G. C. (1969). Soil water hysteresis measured in a sandy loam and compared with the hysteresis domain model. *Soil Sci. Soc. Am. Proc.* **33,** 645–651.

Topp, G. C. and Miller, E. E. (1966). Hysteresis moisture characteristics and hydraulic conductivities for glass-bead media. *Soil Sci. Soc. Am. Proc.* **30,** 156–162.

Trouse, A. C. (1971). Soil Conditions as they affect plant establishment, development, and yield. *In* "Compaction of Agriculture Soils." Am. Soc. Agr. Eng., St. Joseph, Michigan.

Tschebotarioff, G. P. (1973). "Foundations, Retaining and Earth Structures." McGraw-Hill, New York.

van Bavel, C. H. M. (1949). Mean weight diameter of soil aggregates as a statistical index of aggregation. *Soil Sci. Soc. Am. Proc.* **14,** 20–23.

van Bavel, C. H. M. (1952). Gaseous diffusion and porosity in porous media. *Soil Sci.* **73,** 91–104.

van Bavel, C. H. M. (1963). Neutron scattering measurement of soil moisture: Development and current status. *Proc. Int. Symp. Humidity Moisture, Washington, D.C.* pp. 171–184.

van Bavel, C. H. M. (1972). Soil temperature and crop growth. *In* "Optimizing the Soil Physical Environment Toward Greater Crop Yields" (D. Hillel, ed.), pp. 23–33. Academic Press, New York.

van Bavel, C. H. M., and Hillel, D. (1975). A simulation study of soil heat and moisture dynamics as affected by a dry mulch. *Proc. Summer Simulat. Conf., San Francisco, California.* Simulation Councils, La Jolla, California.

van Bavel, C. H. M., and Hillel, D. (1976). Calculating potential and actual evaporation from a bare soil surface by simulation of concurrent flow of water and heat. *Agr. Meteorol.* **17,** 453–476.

van Bavel, C. H. M., Stirk, G. B., and Brust, K. J. (1968a). Hydraulic properties of a clay loam soil and the field measurement of water uptake by roots: I. Interpretation of water content and pressure profiles. *Soil Sci. Soc. Am. Proc.* **32,** 310–317.

van Bavel, C. H. M., Brust, K. J., and Stirk, G. B. (1968b). Hydraulic properties of a clay loam soil and the field measurement of water uptake by roots: II. The water balance of the root zone. *Soil Sci. Soc. Am. Proc.* **23,** 317–321.

Vanden Berg, G. E. (1958). Application of Continuum Mechanics to Compaction in tillable soils. Ph.D. thesis, Michigan State Univ., East Lansing, Michigan.

Vanden Berg, G. E. (1962). Requirements for a soil mechanic *Trans. Am. Soc. Agr. Eng.* **4,** 234–238.

Vanden Berg, G. E., and Gill, W. R. (1962). Pressure distribution between a smooth tire and soil. *Trans. Am. Soc. Agr. Eng.* **5,** 105–107.

van der Molen, W. H., (1956). Desalinization of saline soils as a column process. *Soil Sci.* **81,** 19–27.

van Duin, R. H. A. (1956). "On the Influence of Tillage on Conduction of Heat Diffusion of Air, and Infiltration of Water in Soil," p. 62. Versl. Landbouwk, Onderz.

van Genuchten, R. (1978). Calculating the unsaturated hydraulic conductivity with a new closed-form analytical model. Publication of the Water Resour. Prog. Dept. Civ. Eng., Princeton Univ., Princeton, New Jersey 08540. September, 1978.

Van Olphen, H. (1963). "An Introduction to Clay Colloid Chemistry." Wiley (Interscience), New York.

van Rooyen, M., and Winterkorn, H. F. (1959). Structural and textural influences on thermal conductivity of soils. *Highway Res. Bd. Proc.* **38,** 576–621.

van Schilfgaarde, J. (ed.) (1974). "Drainage for Agriculture." Monograph 17, Am. Soc. Agron., Madison, Wisconsin.

van Wijk, W. R., and de Vries, D. A. (1963). Periodic temperature variations in homogeneous soil. *In* "Physics of Plant Environment" (W. R. van Wijk, ed.). North-Holland Publ., Amsterdam.

Vilain, M. (1963). L'aeration du sol. Mise au point bibliographique. *Ann. Agron.* **14,** 967–998.
Visser, W. C. (1966). Progress in the knowledge about the effect of soil moisture content on plant production. Inst. Land Water Management, Wageningen, Netherlands, Tech. Bull. 45.
Vomocil, J. A. (1954). In situ measurement of soil bulk density. *Agr. Eng.* **35,** 651–654.
Vomocil, J. A. (1965). Porosity. *In* "Methods of Soil Analysis," pp. 299–314. Am. Soc. Agron., Madison, Wisconsin.
Vomocil, J. A., and Flocker, W. J. (1961). Effect of soil compaction on storage and movement of soil air and water. *Trans. Am. Soc. Agr. Eng.* **4,** 242–246.
Vomocil, J. A., Waldron, L. J., and Chancellor, W. J. (1961). Soil tensile strength by centrifugation. *Soil Sci. Soc. Am. Proc.* **25,** 176–180.
Wadleigh, C. H., and Ayers, A. D. (1945). Growth and biochemical composition of bean plants as conditioned by soil moisture tension and salt concentration. *Plant Physiol.* **20,** 106–132.
Waggoner, P. E., Miller, P. M., and DeRoo, H. C. (1960). "Plastic Mulching—Principles and Benefits." Bull. No. 634, Connecticut Agr. Exp. Sta., New Haven, Connecticut.
Watson, K. K. (1966). An instantaneous profile method for determining the hydraulic conductivity of unsaturated porous materials. *Water Resources Res.* **2,** 709–715.
Wesseling, J. (1962). Some solutions of the steady-state diffusion of carbon dioxide through soils. *Neth. J. Agr. Sci.* **10,** 109–117.
Wesseling, J. (1974). Crop growth and wet soils. *In* "Drainage for Agriculture" (J. van Schilfgaarde, ed.), pp. 7–38. Monograph 17, Am. Soc. Agron., Madison, Wisconsin.
White, N. F., Duke, H. R., Sunada, D. K., and Corey, A. T. (1970). Physics of desaturation in porous materials. *J. IR Div.*, ASCE Proc. **IR-2,** 165–191.
White, N. F., Duke, H. R., Sunada, D. K., and Corey, A. T. (1972). Boundary effects in the desaturation of porous media, *Soil Sci.* **113,** 7–12.
Whittig, L. D. (1965). X-ray diffraction techniques for mineral identification and mineralogical composition. *In* Methods of Soil Analysis, Part I, Chap. 49, pp. 671–697. ASA, Madison, Wisconsin.
Wierenga, P. J., and de Wit, C. T. (1970). Simulation of heat transfer in soils. *Soil Sci. Soc. Am. Proc.* **32,** 326–328.
Wiersum, L. K. (1957). The relationship of the size and structural rigidity of pores to their penetration by roots. *Pl. Soil* **9,** 75–85.
Willey, C. R., and Tanner, C. B. (1963). Membrane-covered electrode for measurement of oxygen concentration in soil. *Soil Sci. Soc. Am. Proc.* **27,** 511–515.
Winger, R. J. (1960). In-place permeability tests and their use in subsurface drainage. Off. of Drainage and Ground Water Eng., Bur. of Reclamation, Denver, Colorado.
Winterkorn, H. F. (1936). Surface chemical factors influencing the engineering properties of soil. *Proc. Nat. Res. Council Highway Res. Board Ann. Meeting 16th, Washington, D.C.* pp. 293–301.
Wise, M. E. (1952). Dense random packing of unequal spheres. *Philips Res. Rep.* **7,** 321–343.
Wittmus, H. D., Triplett, G. B., Jr., and Greb, B. W. (1973). Concepts of conservation tillage using surface mulches. *In* "Conservation Tillage," pp. 5–12. Soil Conserv. Soc. Am., Ames, Iowa.
Wittmus, H., Olson, L., and Delbert, L. (1975). Energy requirements for conventional versus minimum tillage. J. Soil Water Conserv. **30,** 72–75.
Wong, J. Y. (1967). Behavior of soil beneath rigid wheels. *J. Agr. Eng. Res.* **12,** 257–269.
Woodside, W. (1958) Probe for thermal conductivity measurement of dry and moist materials. *Am. Soc. Heating and Air-Conditioning Eng. J. Sect., Heating, Piping, and Air Conditioning,* 163–170.
Yamaguchi, M., Howard, F. D., Hughes, D. L., and Flocker, W. J. (1962). *Soil Sci. Soc. Am. Proc.* **26,** 512–513.

Bibliography

Yavorsky, B., and Detlaf, A. (1972). "Handbook of Physics." Mir, Moscow.

Yoder, R. E. (1936). A direct method of aggregate analysis and a study of the physical nature of erosion losses. *J. Am. Soc. Agron.* **28**, 337–351.

Youngs, E. G. (1964). An infiltration method of measuring the hydraulic conductivity of unsaturated porous materials. *Soil Sci.* **109**, 307–311.

Yong, R. N., and Osler, J. C. (1966). On the analysis of soil deformation under a moving rigid wheel. *Proc. Int. Conf. Soc. Terrain-Vehicle Sys.* p. 341.

Yong, R. N., and Warkentin, B. P. (1975). "Soil Properties and Behaviour." Elsevier, Amsterdam.

Youker, R. E., and McGuinness, J. L. (1956). A short method of obtaining mean weight-diameter values of aggregate analyses of soils. *Soil Sci.* **83**, 291–294.

Zelenin, A. N. (1950). "Basic Physics of the Theory of Soil Cutting." U.S.S.R. Academy of Sciences, Moscow.

Index

A

Adhesion, 8
Adsorbed water, 84, 182, 197
Adsorption, 8, 65–66, 73–74, 244
Adsorption isotherm, 244
Aeration, soil, 4, 7, 13, 263–284
Aggregates (clods)
   shapes, 104–105
   size distribution, 105
   stability, 98, 108–111
Aggregation, 93–119
Air entrapment, 102, 111, 152, 154, 179, 180, 212
Air-entry value (of suction), 148
Air-filled porosity, 13, 18, 266–268, 273, 280, 309
Air permeability, 281, 282
Albedo (reflectivity), 289, 310–313
Alkalinity, soil, 250–251
Aluminosilicate minerals, 71–73
Anaerobic conditions, 266, 277
Anaerobiosis, 277
Angle of internal friction, 338
Anion exclusion, 237, 244
Anisotropy, *see* Isotropy and anisotropy
Atterberg limits, 348–351

B

Balance of energy (bare soil), 290–291
Balance of water, (soil profile), 162–163, 220
Bernoulli equation, 172
BET equation, adsorption, 66
Bingham liquid, 182, 327
Black body, 288
Boltzmann transformation, 207–209
Bonding of aggregates, 100–101
Boundary conditions, flow processes, 190, 205
Boussinesq equation (pressure distribution), 364
Bouyoucos hydrometer, 64
Breakthrough curves, 240–242
Bulk density, 11, 17, 18, 95–96
Bulk modulus, 323
Bulk specific gravity, 124

C

Capillarity, 8, 45–47, 152
Capillary conductivity, 201
Capillary model, capillary hypothesis, 211
Capillary potential, 142
Capillary rise, 46, 52, 252
Capillary water, 135
Cation exchange, 73, 77
Cementation of aggregates, 98, 100, 101, 113, 249
Characteristic curve of soil moisture, 148–152
   measurement of, 161–162
Chemical potential, 139
Childs and Collis–George equation, 210
Clausius–Clapeyron equation, 38

407

Clay domains, 100, 101
Clay minerals, 71–76
Clay micelles, 62
Clay pan, 248
Clay, properties and behavior, 71–92
Clay skins, 248
Clogging, 87, 179, 248
Cohesiveness, soil ("cohesion"), 338
Colloids, 75, 76
Compactibility of soil in relation to wetness, 357–360
   moisture–density curves, 358–359
Compaction, 97, 99, 150–151, 355–376
   in agricultural fields, 360–361
   control of, 375–376
   engineering and agronic views of, 356–357
   under machinery-induced stresses, 367–371
   occurrence and consequences of, 371–375
Composite column, flow in, 176
Compressibility, coefficient of, 323
Compression of soil, 355–356
Conduction of heat, 289
Conductivity, hydraulic, *see* Hydraulic conductivity
Conductivity, thermal, *see* Thermal conductivity
Conservation, soil and water, 7
Consistency, soil, 347–351
Consolidation, 376–382
   conceptual model of, 378
   testing of, 381–382
   theory of (Terzaghi's), 380–381
Consumptive use of water by crops, 255
Contact angle (liquid on solid), 43–45, 149, 154
Continuity equation, 188, 202–203, 237, 242–243, 271, 292
Coordination number, 95
Coulomb forces, 88
Coulomb's law, 338
Cross-coefficients, Onsager, 302
Crust, surface, 112, 216, 346
   effect on infiltration, 216–217
Cracked soil, 87, 112, 257

D

Darcy's law, 170, 203, 245
   limitations of (deviations from), 181–183, 246

Density, bulk, *see* Bulk density
Density of solids (particles), 10, 17, 18
Deviator stress, 334
deVries model of thermal conductivity, 298
Dew point, 40, 159
Differential water capacity, *see* Specific water capacity
Diffusion, 158, 205, 236–238, 263, 272–277
Diffusion coefficient for gases in soil, 273
Diffusivity, hydraulic, 204–207, 220
Dilatancy, 339
Direct shear test, 338, 340
Disperse systems, 8
Dispersing agent, 62
Dispersion (hydrodynamic) coefficient, 239–240
Dispersion of clay, 247, 256
Divergence of flux, 202
Double-layer, electrical, 85
Double-probe gamma-ray apparatus, 133
Double-tube method (hydraulic conductivity), 188
Drainage, 7, 217–221, 233, 254
Drop method (aggregate stability), 111
Dry sieving analysis, 107

E

Earthworms, 101, 102
Einstein–Stokes equation, 34
Elasticity, theory of, 322–325
Electrical resistance blocks, 126–128
Electrostatic double layer, 77–81, 90, 143
Eluviation and illuviation, 15
Emissivity, 288, 291
Energy balance for a bare soil, 290–291
Energy conservation principle, 292
Energy exchange, 7, 307
Energy state of soil water, 134–136
Energy transfer, modes of, 288–290
Entrapped air, 102, 111, 154, 179, 180
Entropy, 139
Entry of water into soil, *see* Infiltration
Erosion, 3, 16, 109
Evaporation, 223, 233, 293
Evapotranspiration, 217, 250
Exchangeable ions, 73, 81, 82, 143
Exchangeable sodium percentage (ESP), 112, 247
Exchange capacity (cations), 81
Exchange complex, 82, 86, 244, 249

F

Failure of soil bodies, 334–337
Fick's laws, 172, 205, 206, 236, 238, 272, 292
Field capacity, 267
Flocculation–dispersion of clay, 8, 82, 83, 87–90
Flocculation limit, 348
Flocs, 98
Flow of fluids
   laminar, 167, 181
   turbulent, 168
Flow of soil air (convection), 269–272
Flow of water
   in saturated soil, 166–194
   in unsaturated soil, 195–232
Fluid mechanics, 171, 325–327
Fluidity, 48, 180–181
Flux and flux density, 170, 177
Fourier's law, 172, 291
Free energy, Gibbs, 139
Freezing point depression, 33
Friction, 8

G

Gamma-ray attenuation, 132–134
Gamma-ray scanner, 132, 213
Gapon equation, 83
Gaseous exchange, soil–atmosphere, 263–284
Geometric mean diameter (of aggregates), 107
Germination, 7, 309
Gibbs free energy, 139
Gouy–Chapman theory, 79, 246
Grain size distribution, *see also* Particle size distribution, 60–62, 184
Granular soils, structure of, 95–97
Gravitational potential, 140–141
Gravitational water, 135
Greenhouse effect, 310
Groundwater, 250
Groundwater drainage, *see also* Drainage, 250

H

Haines jumps, 154
Head (gravitational, pressure, and total hydraulic), 172–174
Health hazard, radiation devices, 131–132, 134
Heat capacity (soil), 292, 294
Heat conduction equations, 205, 291, 292

Heat (soil), 7, 287–317
Heat fluxes (sensible, latent, and soil), 291
Heat flux plates, 307
Heat of wetting, 59
Henry's law, 51
Heterogeneity, 162, 213, 214, 219, 254
Homogeneity and isotropy, 185–186, 214
Hooke's law, 167, 322, 327
Horizons (soil profile), 15
Humus, 55, 76–77, 81, 98
Hydration, 29, 31, 82, 84–87, 275, 276
Hydraulic conductivity, 155, 170, 178–180
   related to pore geometry, 183–185
   related to solutes, 247–249
   related to suction and wetness, 198–201
   of saturated soil, 178–180
   theoretical calculation of, 209–212, 223–225
   of unsaturated soil, 196–197
Hydraulic gradient, 158, 169, 170
Hydraulic head, 170, 172–174
Hydraulic radius, 184
Hydraulic resistance, 176, 214, 216
Hydraulic resistivity, 176
Hydrodynamic dispersion, 235, 238–240
Hydrogen bonds, 24–25, 100
Hydrophobic (water-repellent) soil, 44, 115–117
Hysteresis, 127, 152–155, 201, 203, 206, 212

I

Ice, structure of, 25–26
Illuviation, 15
Imbibition, 247
Impedance (*see* Hydraulic resistance)
Infiltrability, 215
Infiltration of water into soil, 215–217, 272
Infiltrometers, 215, 217
Ink-bottle effect, 153–154
Instantaneous profile method, 213, 219, 227–232
Intensive vs. extensive properties, 139
Interfacial phenomena, 8
Internal drainage, *see also* Redistribution, 217–221
Intrinsic permeability, 181
Ion exchange, *see also* Cation exchange, 8
Irreversible thermodynamics, *see* Thermodynamics of irreversible processes
Irrigation, 7, 158, 162, 233, 250, 256, 272, 309

Isomorphous replacement, 73
Isotropy and anisotropy, 185–186, 189

K

Kaolinite, 67, 72
Kelvin equation, 38
Klinkenberg effect, 181
Kozeny–Carman equation, 184, 248
Krilium, 113

L

Lamellae, 73
Laminar flow in narrow tubes, 166–168
Langmuir equation (adsorption), 66
Laplace equation, 190
Latent heat, 289, 298
Layers and layered soils, 132, 134, 190, 192–193, 198, 206, 215, 216–217
Leaching, 242, 250, 251, 254–257
Leaching fraction, attainable, 256
Leaching requirement, 255, 256, 260, 261
Light and heavy soils, 59
Liquid limit, 348
Liquid ratio, 125
Liquids, rheology of, 325–327

M

Major principal stress, 369
Marshall's equation, 185
Matric potential, 143, 148
Matric suction, 143, 148
Mean normal stress, 369
Mean weight diameter (of aggregates), 107
Measurement of hydraulic conductivity,
 of saturated soil, 187–188
 of unsaturated soil, 212–221
Measurement of soil strength, 338–346
Measurement of soil water
 content (wetness), 125–134
 potential, 155–162
 soil-moisture characteristic, 161–162
Measurement of thermal conductivity, 298–299
Mechanical analysis, 56, 62–64
Mechanical composition, *see also* Texture, 56
Membranes, 32, 33, 34, 148, 157, 158, 245
Menisci of water, 45–47
Microaggregates and macroaggregates, 101
Micropores and macropores, 104, 151, 198, 267, 312

Microrelief, 310
Migration of clay, 248
Miscible displacement, 240–242, 257
Models of porous media, 184
Modulus of deformation, 369
Modulus of rupture, 112, 343, 344
Modulus of shearing, 323
Mohr circle of stresses, 331–334
 envelope of, 337, 342
Mohs scale of hardness, 58
Monodisperse and polydisperse systems, 95
Montmorillonite (smectite), 67, 72
Mulch and mulching, 309, 310

N

Navier–Stokes law, 171
Net radiation, 289, 291
Neutron moisture meter, 128–132, 217
 calibration of, 131
 safety of, 131
Neutron scattering and thermalization, 128
Newtonian liquid, 182, 326, 327
Nonisothermal conditions, 222, 300–303
Numerical solution of flow equations, 195, 253, 306

O

Ohm's law (analogous to Darcy's law), 171, 176
Onsager's reciprocal relations, 302
Organic matter, 77, 83, 102, 294
Organic soils, 77
Organosilicones, 117
Osmometer, 157
Osmosis, 32
Osmotic efficiency factor, 246
Osmotic (solute) potential, 144
Osmotic pressure, 31–34, 85, 245
Oven-dry state, 84
Overburden potential and pressure, 142
Outflow method (hydraulic conductivity), 213
Oxygen diffusion rate (ODR), 283

P

Packing of spheres, 95–97
Pan (soil layer), 16, 248
Particle size distribution, *see also* Grain size distribution, 60–62
Pedogenesis, 15
Pedology, 16

Peds, 94, 98
Peltier effect, 160
Penetrometers, 344–346, 370
Perched water table, 215
Periodic temperature wave in soil, 303–308
  damping depth, 305
  phase shift, 305
Permeability, 104, 110–111, 180–181, 247, 281
  relation to pore geometry, 180, 183–185
Permeameters, 187, 282
pF, 147
pH, 27–29, 77, 83, 86, 89
Phenomenological cross-coefficients, 302
Philip and deVries theory (heat and water transport), 301
Piezometers, 174
Piezometric head, 141
Planck's law, 288
Plasticity, 322–325
Plasticity index, 351
Plastic limit, 349
Platinum electrodes (O_2 diffusion), 283
Pneumatic potential, 144
Poiseuille's law, 168, 169, 197, 210, 211, 277
Poisson's ratio, 323
Pores, intraaggregate and interaggregate, 149, 151, 197, 211, 246, 247, 267, 274–275
  micro- and macro-, 104, 151, 198, 267, 312
Pore size distribution, 12, 102, 142, 150, 162, 180, 184, 185, 209, 277
Pore water pressure, 341
Porosity, 11, 12, 14, 17, 18, 96
Potential energy of soil water, 134–135
Potential of soil water, 136–138
Pressure intrusion method, 104
Pressure plate or membrane apparatus, 142, 162
Pressure potential, 141–144
Pressures caused by field machinery, 361–367
Primary minerals, 58, 71
Proctor test, 359
Profile (of soil), *see* Soil profile
Properties, intensive and extensive, 139
Properties, static and dynamic, 56
Psychrometry, 158–160
Puddling, 347

R

Radiation, 288
Radiation exchange in the field, 289
Radiation, net, 289, 291
Radiation thermometer, 307
Raindrop impact, 215
Reclamation of saline soils, 251, 256
Redistribution of soil water, 219
Redox potential, 283
Reflection coefficient, osmotic, 246
Reflectivity, *see* Albedo
Relative humidity, 146, 159, 160
Residence time, average, 235
Residual (immobile) water, 212
Resistance blocks, electrical, 126–128
Resistance, hydraulic, 176, 214, 216
Respiration in soil, 4, 263, 275, 277–280
Respirometer, 278–279
Retention curve of soil moisture, *see* Soil-moisture characteristic
Reverse osmosis, 245
Reynolds number, 168, 181
Rheological models, 327–328
Rheology, 318
Ridging, 310
Root growth and development, 7, 114–115, 309
Root systems, 254
Root zone, 195, 221, 234, 252, 263
Runoff, surface, 250

S

Shrinkage, 351
Salinity, 242, 250–251
Salinization, 3, 250, 254
Salt balance, 251–254
Salt-sieving effect, 245
Salt tolerance, 255
Saturated conductivity, *see* Hydraulic conductivity, of saturated soil
Saturated soil, flow of water in
  equations of, 188–190
  vs. unsaturated flow, 195–198
Scanning curves, 153, 155
Secondary minerals in soils, 58, 71
Sedimentation analysis, 63, 111
Seepage, 179
Selectivity coefficient, 83
Sensible heat, 291
Shearing (tangential) stress, 48
Shrinkage limit, 350
Simultaneous flow (water, heat, solutes), 300–303
Sodium adsorption ratio (SAR), 248
Soil, defined, 6

Soil air, composition of, 268–269
Soil as a three-phase system, 8–9
Soil class, 59
Soil conditioners, 113–115
Soil constituents, volume and mass relationships, 9–14
Soil heterogeneity, 8
Soil formation, 6, 16
Soil matrix, 8, 59, 97, 125
Soil mechanics, 7
Soil moisture, *see* Soil water
Soil-moisture characteristic (retention or release curve), 148–152, 155, 161–162, 205, 209
Soil physics, defined, 4, 7
Soil–plant–water system, 4
Soil productivity, 7–8
Soil profile, 14–17, 19
Soil solution, 9, 38, 83, 127, 145, 152, 179, 234, 238, 250, 255
Soil structure, 16, 29, 77, 82, 93–119, 154, 178, 179
Soil texture, 56, *see also* Texture of soil
Soil water
  content and potential, 123–165
  energy state, 134–136
  "forms" of, 135–136
  measurement of content, 125–134
  measurement of potential, 155–162
  movement in saturated soil, 166–194
  movement in unsaturated soil, 195–232
  potential, 136–137
Soil-water movement
  Darcy's law, 170, 203
  infiltration, 215–217
  with thermal gradients, 300–303
Soil-water potential, 136–138
  components, 137
  definition, 136
  gravitational potential, 140–141
  matric potential, 142
  osmotic (solute) potential, 144
  pneumatic potential, 144
  pressure potential, 141–144
  quantitative expression, 145–148
  submergence potential, 141–142
Soil wetness, 12–13, 14, 18, 124–125, 145
  degree of saturation, 13, 14, 18
  mass wetness, 12
  volume wetness, 13

Soils, basic physical properties, 6–20
  adsorption phenomena, 65–66
  behavior of clay, 71–91
  mechanical analysis, 62–64
  ratios of constituents, 9–14
  specific surface, 65–68
  structure, 93–119
  texture, 56–60
  three phases, 8–9
Solar radiation, 289
Solonchak, 250
Solonetz, 251
Solute potential, *see* Osmotic potential
Solutes, effect on water movement, 245–249
Solutes, movement in soil, 233–261
Sources and sinks, 243, 274, 292, 293
Specific heat
  of soil, 292, 294
  of water, 294
Specific surface, 65–68, 86, 181
Specific volume, 11
Specific water capacity, 152, 205, 206
Stability of aggregates, 108–111
Steady (stationary) state flow, 188, 212, 299
Stefan–Boltzmann law, 288
Stern layer, 78, 83
Stokes' law, 62–64
Strain, 319–320
Streaming potential, 246
Stress, 321
Stress distribution in soil, 328–331
Stress–strain relations, 318–337
Suction, 141
Surface tension, 8, 40–41, 143
Swelling and shrinking, 8, 13, 83, 84–87, 125, 247, 249, 351

T

Tactoids of clay, 87, 88
Temperature amplitude at soil surface, 307, 309, 312
Temperature gradients, 196, 291
Temperature profile of soil, 307–308
Temperature wave in soil, 304–308
Tensile strength, 343
Tensiometers, 142, 156–158, 217
Tension, 141
Tension plate assembly, 162
Terzaghi's effective stress equation, 379

Index

Terzaghi's theory of consolidation, 380–381
Texture of soils, 55–68, 179
  textural fractions (sand, silt, and clay), 56–59
  textural triangle, 59–60
Theories and models (general discussion), 4–5
Thermal conductivity, 291, 295–300
Thermal diffusivity, 293, 300
Thermal properties of soil constituents, 296
Thermal regime of soil profiles, 303–308, 310–313
  modification of, 309–313
Thermodynamics
  of irreversible processes, 302–303
  and the potential concept, 138–140
  laws of, 138–139
  of soil water, 140
Thixotropy, 337
Threshold gradient, 182
Threshold salt concentration, 248
Tillage, 7, 99, 281, 312, 347
Tilth, 94
Tortuosity, 177, 180, 183, 197, 221, 236
Tortuosity factor, 177, 236
Traffic, overland, 108, 281, 346
  trafficability, 346
Transient (unsteady) state, 188, 212, 299
Triaxial shear test, 338, 341
Turbulent flow, 168

U

Uniformity index, 62
Unsaturated conductivity
  relation to suction and wetness, 198–201
Unsaturated soil, flow equations for, 202–204
Unstable flow, 217

V

van der Waals forces, 36, 88, 100
Vane shear test, 342, 343
Van't Hoff equation, 86
Vapor enhancement factor, 298
Vapor movement (diffusion), 196, 207, 221–223

Vapor pressure of water, 33, 37–40, 222
Vertisols, 87
Vibration, compaction by, 97
Viscosity, 167, 326
  kinematic, 48, 326
  of water, 48–50, 197
Viscous forces, 181
Void ratio, 12, 14, 18
Volumetric heat capacity of soils, 293–295, 300

W

Waste heat, disposal of, 312
Water balance of soil profile, 220
Water capacity, differential or specific, 152
Water harvesting, 115
Water-logged soils, 255, 256, 263, 280
Water, models of, 27
Water, physical properties of, 21–52
  adsorption on solid surfaces, 35–37
  capillarity, 45–47
  change of state, 25–27
  contact angle on solid surfaces, 43–45
  curvature of surfaces, 42–43
  density and compressibility, 47–48
  ionization and pH, 27–29
  molecular structure, 22–23
  osmotic pressure, 31–34
  solubility of gases, 34–35
  solvent properties, 29–31
  surface tension, 40–41
  vapor pressure, 37–40
  viscosity, 48–50
Water repellency, 44, 115
Water table, 174, 254
  perched, 215
Weathering, 6, 15
Wetness, 12–13, 14, 18, 124–125, 145
Wet sieving, 109
Wetting front, 196
Wien's law, 288

Y

Young's modulus, 323

Printed and bound by CPI Group (UK) Ltd, Croydon, CR0 4YY
11/06/2025
01899189-0004